Blessing and Curse in Syro-Palestinian Inscriptions of the Iron Age

American University Studies

Series VII
Theology and Religion
Vol. 120

PETER LANG
New York • San Francisco • Bern • Baltimore
Frankfurt am Main • Berlin • Wien • Paris

Timothy G. Crawford

Blessing and Curse in Syro-Palestinian Inscriptions of the Iron Age

PETER LANG
New York • San Francisco • Bern • Baltimore
Frankfurt am Main • Berlin • Wien • Paris

Library of Congress Cataloging-in-Publication Data

Crawford, Timothy G.
 Blessing and curse in Syro-Palestinian inscriptions of the Iron Age / Timothy G. Crawford.
   p. cm. — (American university studies. Series VII, Theology and religion ; vol. 120)
 Includes bibliographical references.
 1. Semites—Religion. 2. Blessing and cursing—History.
 3. Inscriptions, Semitic. 4. Palestine—Religion. 5. Syria—Religion.
 I. Title. II. Series.
 BL1600.C73   1992   299'.2—dc20   91-19808
 ISBN 0-8204-1662-2   CIP
 ISSN 0740-0446

Die Deutsche Bibliothek-CIP-Einheitsaufnahme

Crawford, Timothy G.:
Blessing and curse in Syro-Palestinian inscriptions of the Iron Age / Timothy G. Crawford.—New York; Berlin; Bern; Frankfurt/M.; Paris; Wien; Lang, 1992
 (American university studies : Ser. 7, Theology and religion; Vol. 120)
 ISBN 0-8204-1662-2
NE: American university studies / 07

The paper in this book meets the guidelines for permanence and durability of the Committee on Production Guidelines for Book Longevity of the Council on Library Resources.

© Peter Lang Publishing, Inc., New York 1992

All rights reserved.
Reprint or reproduction, even partially, in all forms such as microfilm, xerography, microfiche, microcard, offset strictly prohibited.

Printed in the United States of America.

To Janet

ACKNOWLEDGMENTS

There have been many people who have helped me and influenced me in the process of researching and writing this study. It is thanks to Dr. Joel F. Drinkard, Jr., my dissertation supervisor, that I came upon this topic at all. In the Fall of 1986, I began my formal study of the inscriptions under his direction and it was in conversation with him a year later that I finally came upon the topic that has held my interest for these last two-and-a-half years. Without his original guidance and continued interest and input, this study could not have been completed. As I have prepared the manuscript for publication he has been a great further encouragement.

It is thanks to the excellent library of The Southern Baptist Theological Seminary, Louisville, Kentucky, that I have had the materials necessary for this study. The Seminary's Albright Collection gives it one of the best collections of materials up to the beginning of the 1970s. The Seminary's acquisitions personnel continue to make it a significant repository of resources for the study of the ancient Near East and its languages. Virtually all of the materials I needed, which were not held by the library, were obtained for me by our excellent interlibrary loan staff. This tradition of helpfulness has been continued by the library of Bluefield College where I now teach.

Thanks is due to Bluefield College, Bluefield, Virginia, as well for significant financial support in the preparation of this manuscript.

I am grateful to Steven E. Smith of Apple Computers for his patient technical assistance and encouragement. It is thanks to him that I was able to produce this dissertation on a Macintosh Computer and print it on an Apple LaserWriter. The Latin alphabet portions of the final manuscript are printed in Times font; the Hebrew portions in the Hebraica font available from Linguist's Software, Inc., P.O. Box 580, Edmonds, WA 98020-0580, tel. (206) 775-1130.

Most of all, I am grateful to Janet my wife. She has made the crucial difference to me during this work. She has patiently supported me. She has encouraged me at critical points. Her faith in me has given me strength.

TABLE OF CONTENTS

	Page
TABLE OF ABBREVIATIONS	xiii
LIST OF TABLES	xvii

Chapter
1. INTRODUCTION 1
   - Statement of Purpose 2
   - Methodology 2
   - Notes to Chapter 1 5
2. SEMANTIC SURVEY AND HISTORY OF INTERPRETATION ... 7
   - Semantic Survey 7
     - Akkadian 7
     - Ugaritic 9
     - Aramaic 10
     - Phoenician and Punic 10
     - Inscriptional Hebrew 11
     - Edomite 12
     - Biblical Hebrew 12
     - Summary 15
   - History of Interpretation 16
     - Johannes Pedersen 17
     - Sigmund Mowinckel 18
     - Johannes Hempel 19
     - Claus Westermann 20
     - Gerhard Wehmeier 20
     - Josef Scharbert 21
     - Herbert Brichto 22
     - Anthony Thistleton 23
     - Christopher Mitchell 24
     - Summary 25
   - Notes to Chapter 2 26
3. BLESSING 35
   - שלמ and ברכ 35
     - With Yahweh 35
       - Arad 16 35
       - Arad 21 37
       - Arad 40 38
       - Kuntillet 'Ajrud 4 38
       - Ketef Hinnom 1 and 2 40
     - With Other DN 43

Ḥorvat 'Uza	43
With No DN	44
Samaria C1101	44
Summary	45
ברכ Alone	46
With Yahweh	46
Kuntillet 'Ajrud 5	46
Kuntillet 'Ajrud 3	46
Kuntillet 'Ajrud 2	47
Khirbet el-Qôm	48
With Other DN	53
Karatepe A.i.1	53
Kuntillet 'Ajrud 1	54
Ivory Box from Ur	55
With No DN	56
Samaria C1220	56
Summary	57
ברכ Without Context	58
Arad 28	58
Lachish 31	58
שלמ Alone	58
With Yahweh	58
Arad 18	58
Lachish 2	60
Lachish 7	60
Lachish 9	61
Lachish 3	61
Lachish 5	61
Lachish 6	61
Without DN	62
Wadi Muraba'ât	62
Summary	62
Substantive Blessing	63
With Yahweh	63
Kuntillet 'Ajrud 6	63
With Other DN	64
Yeḥimilk	65
Abiba'al	66
Eliba'al	66
Shipiṭba'al	66
Kilamuwa 2	67
Without DN	68

Gezer Calendar	68
Tell Siran Bottle	70
Summary	73
Conclusions	73
Notes to Chapter 3	74

4. CURSE 97
   ארר 97
      With Yahweh 97
         Lachish 5 97
      Without DN 98
         Siloam Tomb 98
         Khirbet Beit Lei 4 99
         Khirbet Beit Lei 7 100
         Khirbet Beit Lei (unnumbered) 101
      Summary 101
   Other Words for Curse 101
      אלה 101
         Panammu line 2 102
         Arslan Tash 1 103
      קבב 106
         Deir 'Alla II.17 107
         Deir 'Alla IX.a.3 108
         Deir 'Alla X.a.3 108
      Summary 109
   Substantive Curse 109
      With DN 109
         Kilamuwa 1.13–16 109
         Sefire III 112
         Panammu lines 21–23 113
         Nerab 2 114
      Without DN 116
         Aḥiram 116
         Nimrud Hebrew 121
         Deir 'Alla I 122
         Deir 'Alla II 129
      Summary 132
   Conclusions 133
   Notes to Chapter 4 134

5. BLESSING AND CURSE 157
   ברכ and ארר 157
      En Gedi 157
      Summary 159

ברכ and שלמ with Substantive Curse	159
Karatepe A	159
Karatepe B	163
Karatepe C	164
Summary	165
Substantive Blessing and Curse	166
Tell Fekherye lines 7-12	166
Tell Fekherye lines 13-23	169
Amman Citadel	174
Zakir B 16–C 2	177
Sefire I	181
Sefire II	195
Hadad	200
Nerab 1	204
Summary	207
Conclusions	208
Notes to Chapter 5	208
6. CONCLUSIONS	231
Notes to Chapter 6	234
APPENDIX	237
BIBLIOGRAPHY	241

TABLE OF ABBREVIATIONS

ABL	Robert F. Harper, *Assyrian and Babylonian Letters Belonging to the Kouyunjik Collections of the British Museum*. *ABL* followed by a number designates a text in this collection, volume not given.
AD	G. R. Driver, *Aramaic Documents of the Fifth Century B.C.*
ADAJ	Annual of the Department of Antiquities of Jordan
EpigAnat	Epigraphica Anatolia
AfO	Archiv für Orientforschung
AHw	Wolfram von Soden, *Akkadische Handwörterbuch*
AIR	Patrick D. Miller, Jr.; Paul D. Hanson; and S. Dean McBride; *Ancient Israelite Religion*
AJSL	American Journal of Semitic Languages and Literatures
ANEP	James B. Pritchard, ed. *The Ancient Near East in Pictures Relating to the Old Testament*, 2nd. ed.
ANET	James B. Pritchard, ed. *Ancient Near Eastern Texts Relating to the Old Testament*, 3rd ed.
AP	A. Cowley, *Aramaic Papyri*. *AP* followed by a number designates the text of that number in the collection.
ASAE	Annales du service des antiquitiés de l'Egpte (Cairo)
ATDA	Jacob Hoftijzer and G. van der Kooij, eds., *Aramaic Texts from Deir 'Alla*
AUSS	Andrews University Seminary Studies
BA	Biblical Archaeologist
BAR	Biblical Archaeology Review
BASOR	Bulletin of the American Schools of Oriental Research
BDB	Francis Brown; S.R. Driver; and Charles A. Briggs, *A Hebrew and English Lexicon of the Old Testament*
BeO	Bibbia e orienta
BH	Biblical Hebrew
BMAP	Emil Kraeling, ed., *Brooklyn Museum Aramaic Papyri*. *BMAP* followed by a number designates a text from this collection.
BMB	Bulletin du Musée de Beyrouth
BN	Biblische Notizen
BO	Bibliotheca Orientalis
BTDA	Jo Ann Hackett, *Balaam Text from Deir 'Alla*
BZAW	Beiheft zur Zeitschrit für die Alttestamentliche Wissenschaft
CBQ	Catholic Biblical Quarterly
CML	John C. L. Gibson, ed. *Canaanite Myths and Legends*
CRAIBL	Comptes rendus des séances de l'académie des inscriptions et belles-lettres
CT	*Cuneiform Texts from Babylonian Tablets in the British Museum*
CTA	A. Herdner, ed., *Corpus des tablettes en cunéiformes alphabétiques découvertes à Ras Shamra-Ugarit*. *CTA* followed by a number desig-

	nates a text in this collation.
DJD	P. Benoit; J. T. Milik; R. de Vaux; eds. Discoveries in the Judean Desert
DOTT	D. Winton Thomas, ed., *Documents from Old Testament Times*
DISO	C.F. Jean and J. Hoftijzer, eds., *Dictionnaire des inscriptions sémitiques de l'ouest*
EphCar	Ephemerides Carmeliticae
HebrSt	Hebrew Studies
HTR	Harvard Theological Review
HUCA	Hebrew Union College Annual
IEJ	Israel Exploration Journal
IJT	The Indian Journal of Theology
JQR	Jewish Quarterly Review
JANES	The Journal of the Ancient Near Eastern Society of Columbia University
JAOS	Journal of the American Oriental Society
JBL	Journal of Biblical Literature
JNES	Journal of Near Eastern Studies
JPOS	Journal of the Palestine Oriental Society
JRAS	Journal of the Royal Asiatic Society
JS	Journal for Semitics
JSS	Journal of Semitic Studies
JSOT	Journal for the Study of the Old Testament
JTS	Journal of Theological Studies
KAI	H. Donner and W. Röllig, eds., *Kanaanäische und aramäische Inschriften*. *KAI* followed by a number refers to the reference number of the inscription.
KB	Ludwig Koehler and Walter Baumgartner, *Lexicon in Veteris Testamenti Libros*
MDB	Monde de la Bible
MPAIBL	Mémoires présentés à l'académie des inscriptions et belles-lettres
MUSJ	Mélanges de l'Université Saint Joseph
NSI	G. A. Cooke, *A Textbook of North-Semitic Inscriptions*
OLP	Orientalia Lovaniensia Periodica
OLZ	Orientalische Literaturzeitung
PEFQS	Palestine Exploration Fund Quarterly Statement
PEQ	Palestine Exploration Quarterly
PKTA	E. Ebling, *Parfümrezepte und kultische Texte aus Assur*
RA	Revue d'assyriologie et d'archaéologie orientale
RB	Revue biblique
RÉS	Répertoire d'épigraphie sémitique. *RÉS* followed by a number designates a text in included in this series, no volume numbers.
RHR	Revue de l'histoire des religions
RivBibl	Rivista Biblica
RS	Revue sémitique d'epigraphie et d'histoire ancienne
RSF	Rivista di studi fenici

SBL	Society of Biblical Literature
SEL	Studi epigraphici e linguistici
TA	Tel Aviv
TDNT	G. Kittel and G. Friedrich, eds., *Theological Dictionary of the New Testament*
TDOT	G. Johannes Botterweck and Helmer Ringgren, eds., *Theological Dictionary of the Old Testament*
TSSI	John C. L. Gibson, *Textbook of Syrian Semitic Inscription*
UF	Ugarit-Forschungen. Internationales Jahrbuch für die Altertumskunde Syrien-Palästinas
UL	Cyrus Gordon, *Ugaritic Literature*
UT	Cyrus Gordon, *Ugaritic Textbook*. *UT* followed by a number designates the text of that number.
VA	Vorderasiatische Abteilung, Thontafelsammlung. *VA* followed by a number refers to a text in this collection.
VAT	*Vorderasiatische Schriftdenkmäler der Königlichen Museen zu Berlin.* *VAT* followed by a number designates a text in the collection.
VetChr	Vetera Christianorum
VT	Vetus Testamentum
WO	Die Welt des Orients
WZKM	Wiener Zeitschrift für die Kunde des Morgenlandes
ZA	Zeitschrift für Assyriologie und vorderasiatische Archäologie (earlier editions, Zeitschrift für Assyriologie und verwandte Gebiete)
ZAW	Zeitschrift für die Alttestamentliche Wissenschaft
ZDMG	Zeitschrift der Deutschen Morgandländischen Gesellschaft
ZDPV	Zeitschrift des Deutschen Palästinavereins

LIST OF TABLES

	Page
Table 1, Comparison of the Plaques with the Massoretic Text	41
Table 2, Graffito Inscriptions Comparison	51
Table 3, Karatepe Royal Blessing (A.iii.2-6) Structure	161

Chapter 1

Introduction

The sizeable and increasing number of biblical period inscriptions makes it surprizing that a study of the religious elements in these inscriptions has not been made previously. Part of the reason may be that until recently an insufficient number of inscriptions having religious significance was known. In the past thirty years, the extant corpus of Hebrew, Ammonite, Aramaic, Edomite, Moabite, and Phoenician materials has grown considerably and inscriptions bearing texts of religious import make up a significant part of that total. These new discoveries have also made possible improved interpretations of inscriptions found previously. While the inscriptions have been individually compared and contrasted to the biblical materials, with widely varying degrees of thoroughness, they have not been studied as a group to discern their relationship to each other nor adequately examined as the window they are (if clouded) on the religion of the ancient world in which biblical Yahwism was born.[1]

Relevancy of this study is further demonstrated by the fact that since 1979, well-over 150 scholarly articles, books, and dissertations, as well as popular magazine and newspaper articles, have been written about inscriptions from Syria-Palestine. Recent important contributions to the study of these inscriptions include John Gibson's three volume *Textbook of Syrian Semitic Inscriptions*, 1971–1985; Joseph Naveh's *Early History of the Alphabet*, 1982; Randall Garr's *Dialect Geography of Syria-Palestine, 1000–586 B.C.E.*, 1985; and Joel Drinkard's unpublished Ph.D. dissertation "Vowel Letters in Pre-Exilic Palestinian Inscriptions," 1980. Studies of religious aspects of the inscriptions, with the exception of the mentioned articles (see note 1), have been generally limited to treatments of personal names, e.g., Robert Lawton's dissertation,[2] Jeffrey Tigay's much more extensive monograph *You Shall Have No Other Gods*,[3] and Jeaneane D. Fowler's *Theophoric Personal Names in Ancient Hebrew*.[4] On-going work includes research being done by Pierce Matheney, Jr., "The Relation Between Iron Age Artifacts and Inscriptions and the Religion of Israel During the Kingdoms," under an ASOR grant and Graham Davies's work on a concordance of Hebrew epigraphic materials.[5]

Many gaps remain in our knowledge of biblical Yahwism. According to the biblical writers it seems to be the religion of a self-professed minority.[6] Kings who received the writers' approbation as propagators of right religion were portrayed as reformers cutting across the societal norm, e.g., destroying the ubiquitous asherah and high places and repairing the Temple of Yahweh fallen into disrepair. Obscuring ancient religious beliefs for the modern reader is the fact that the biblical writers criticized practices well-known to the people of their times while moderns are left to wonder, for example what or who an "asherah" was. We are left with tantalizing bits and pieces of religious information.[7] Thus, there is still ample room for this kind of research.

Statement of Purpose

The purpose of this work is to examine blessing and curse in Syro-Palestinian epigraphic materials contemporaneous with the Israelite monarchy (northern and southern kingdoms) in comparison with each other and the Old Testament. Having examined the inscriptions and sorted through their various blessings and curses, conclusions will be drawn concerning patterns of meaning of blessing and curse within and between the various cultures.

This study will show that "blessings" are of four types in these inscriptions: 1) a blessing may be a general invocation of a deity (or deities) to bless (ברכ) a person or people in some non-specified way, or 2) it may be a call for a deity (or deities) to bestow some perceived good on a person or people without the use of the word ברכ, 3) it may be a mixture of these first two: a call for a deity (or deities) to bless (ברכ) in some specific way, for example with long life, or 4) the invoked deity may be implied rather than mentioned by name. The types of curses are similar, though the first and the third are scarcer: a general word for curse being found much less often than a specific curse. In cases where blessing or curse are called for but these specific words (ברכ or ארר) are not used, categories of "substantive blessing," "substantive curse," and "substantive blessing and curse" will be included. Comparisons of inscriptions containing "substantive blessings" and "substantive curse" with other inscriptions and the Bible will illustrate that these elements were understood as blessing and curse. Passages in which a state of blessedness or cursedness are described as extant conditions (for example, where economic prosperity already exists and is described as the gift of the gods) rather than asked of the gods, will generally not be included for this study. Personal names occurring in the inscriptions, seals, or bullae will not be included in this study, though some may be interpreted as general calls for blessing.[8]

The best established dates of the inscriptions will be used, as will the readings of the *editio princeps* in each case, unless the scholarly community has generally come to accept a later reading in place of the *editio princeps*.

Methodology

The target inscriptions for this study are the alphabetic inscriptions of the Iron II period, the period of the United and Divided Monarchies of Israel and Judah. Following Aharoni's Iron Age chronology, Iron I is dated 1200–1000 B.C. (the period of the Judges); Iron II is dated 1000–586 B.C. (the period of the Monarchy); and Iron III is dated from about 586–330 B.C. (the Persian Period).[9] As is noted in the next paragraph, most of the biblical materials themselves do not originate from this period. Because of their contemporaneity, the inscriptions, whether Hebrew, Phoenician, Aramaic or other, may provide stronger parallels to each other than to the biblical texts and vice versa. Biblical passages will be suggested for comparison, however, wherever they seem to provide legitimate parallels to the inscriptions. The target inscriptions are limited to those of this period in order to concentrate the comparison on the religion of monarchic Israel and its contemporaries.

While the inscriptional sources can be dated with a fair degree of certainty by stratigraphic, ceramic, and paleographic analyses, dealing with the dating of the biblical sources is vastly more complicated. Kyle McCarter has stated this problem well,

> When we attempt to reconstruct the religion of Israel before the destruction of Solomon's Temple, we find ourselves in a predicament. Our primary source—and almost our only source—is the Bible. It is the consensus of modern scholarship that the Hebrew Bible as we know it came into existence in the centuries following the Babylonian destruction of Jerusalem amid the crises provoked by that disaster. The present form of its various parts is the work of writers living at the time of the exile and later, and as such it is subject to many limitations as a source for the reconstruction of the religion of the earlier period.
>
> This predicament is not ameliorated by the probability that the Bible, though a product of the exile and later periods in its present form, contains substantial preexilic material, as most scholars continue to believe.[10]

Since this is the case, and since there have been no large scale discoveries of Iron Age documents comparable to Qumran or Ugarit,

> we must be satisfied with filling out the biblical references to hostile religious parties on the basis of information drawn, first, from analogic comparison with what data we possess about other religious traditions and, second, from study of the small corpus of surviving First Temple inscriptions.[11]

The target inscriptions of this study will also be limited, almost completely, to epigraphic materials found in Syria-Palestine: the southern-most site being Kuntillet 'Ajrud and the northern-most being Karatepe.[12] The histories of the nations which composed Syria-Palestine were so intertwined and communication between them so pervasive, that it would be artificial to exclude inscriptions from nearby territories simply because they were outside of Israel's political influence.

All of the published alphabetic inscriptions—in Hebrew, Ammonite, Moabite, Phoenician, Aramaic, Edomite, and uncertain languages, e.g., Arslan Tash and Deir 'Alla—from this period and within the prescribed geographical boundaries will be examined for the presence of blessing and/or curse and those determined to include blessing and/or curse will be utilized for this study.

Chapter 2 consists of two parts. The first part is a semantic survey of Akkadian, Ugaritic, Aramaic, Phoenician, Punic, Inscriptional Hebrew, Edomite, and Biblical Hebrew words for blessing and curse. Establishing the usages of such words will be of foundational value for the rest of the study. The second part of the chapter is a short history of interpretation of biblical blessing and curse seen through summaries of the ideas of nine scholars.

Chapter 3 treats inscriptions containing a blessing without accompanying curse. The first division presents texts which use ברכ and שלמ. The second section will contain texts which utilize ברכ but not שלמ. The third division consists of forms of ברכ which are without decipherable context. The fourth part will present inscriptions containing שלמ but not ברכ. The last major division will be comprised of texts which have a substantive blessing, not having ברכ or שלמ. In each section, those inscriptions having the general word or words or substantive blessing combined with Yahweh will be treated first, followed by those which combine the general word or words or substantive blessing with another DN, and finally those with the general word or words or substantive blessing but without a DN.

Chapter 4 includes those inscriptions containing a curse without a blessing. The first division will be comprised of inscriptions containing ארר. The second division will deal with inscriptions using a word for curse other than ארר. The last major section of the chapter will treat inscriptions containing a substantive curse. In each of these divisions, the general word or substantive curse with Yahweh will be treated first, followed by the general word or substantive curse with another DN, and finally the general word or substantive curse with no DN.

Chapter 5 presents texts in which blessing and curse appear together. The first division consists of inscriptions having ברכ and ארר together. Inscriptions having ברכ and שלמ with a substantive curse will be the next section. The last section treats inscriptions containing substantive blessing and curse. In each section, the general words for substantive blessing and curse with Yahweh will be examined first, followed by the general words or substantive blessing and curse with another DN, and last, the general words or substantive blessing and curse without a DN.

Chapter 6 consists of conclusions drawn from the study of the inscriptions. Also, some suggestions for further study in this area will be presented.

The appendix of the dissertation will be composed of charts which will give basic information on each of the inscriptions or each group thereof (some

Introduction: Chapter 1

locations have yielded only a single inscription while others have been the source of several). Where available the following information will be included: the date of the particular inscription, its language, its date of discovery, the place of original publication, and the location where photographs or drawings of each may be found.

---

[1] Several excellent brief studies have been made of individual inscriptions including those of P. Kyle McCarter, Jr. largely concerning the Kuntillet 'Ajrud inscriptions ("Aspects of the Religions of the Israelite Monarchy: Biblical and Epigraphic Data," in *Ancient Israelite Religion*, ed. Patrick D. Miller, Jr., Paul D. Hanson, S. Dean McBride [Philadelphia: Fortress Press, 1987]), Jo Ann Hackett concerning the Deir 'Alla inscriptions ("Religious Traditions in Israelite Transjordan," also in *Ancient Israelite Religion* [*AIR*] and Jonas C. Greenfield ("Aspects of Aramean Religion," also in *AIR*). Brian Peckham's article "Phoenicia and the Religion of Israel" (also in *AIR*) is an excellent, if brief, broader comparison and contrast between Phoenician and biblical religions.

[2] Robert B. Lawton, Jr., "Israelite Personal Names on Hebrew Inscriptions Antedating 500 B. C. E.," Ph.D. dissertation, Harvard University, 1977. Lawton collates the names but provides very little analysis and no synthesis.

[3] Jeffrey H. Tigay, *You Shall Have No Other Gods: Israelite Religion in the Light of Hebrew Inscriptions*, Harvard Semitic Studies, no. 31 (Atlanta: Scholars Press, 1986). The second part of the title would be better named "Israelite Religion in the Light of Epigraphic Onomastics," since Tigay devotes only fifteen pages (pp. 21–36) to other inscriptions of religious value. Also, his limitation "Israelite" is questionable (e.g., he excludes some of the inscriptions from Kuntillet 'Ajrud because their script is described as Phoenician): with the contact between nations and movement of peoples in the ancient world that is becoming ever better known, defining an "Israelite" is not so easy. Israel, and even mountainous Judah, shared their territories, whether by choice or not, with numerous other peoples as the Bible reports. He seems content with summary treatments of non-onomastic inscriptions, basing his understanding of "asherah" in the "Yahweh and His Asherah" inscriptions (from Kuntillet 'Ajrud and Khirbet el-Qôm) on their non-conformity with BH grammar (pp. 26–27). His treatment of names is more thorough, though Olyan emphasizes the gap that can exist between popular piety and the frequency of PNs (*Asherah and the Cult of Yahweh in Israel*, SBL Monograph Series, no. 34 [Atlanta: Scholars Press, 1988], p. 36).

This is not the place for an examination of "asherah" nor even for a treatment of the possible "relationship" between Yahweh and Asherah though it occurs in inscriptions treated below with blessing and curse. For such treatments, the reader is directed for preliminary information to J. A. Emerton, "New Light on Israelite Religion: The Implications of the Inscriptions from Kuntillet 'Ajrud," *ZAW* 94 (1982), 2–20; and for further information to Saul Olyan, cited just above; and Walter A. Maier, III, *'Ašerah: Extra Biblical Evidence*, Harvard Semitic Monographs, no. 37 (Atlanta: Scholars Press, 1986); as well as the articles cited for the particular inscriptions below.

[4] The full title is *Theophoric Personal Names in Ancient Hebrew, A Comparative Study*, JSOT Supplement Series, no. 49 (Sheffield: 1988).

[5] Matheney's grant was announced in the *American Schools of Oriental Research Newsletter* 39/1 (1987), 8. Professor Ernest Nicholson told me of Graham Davies's project in a conversation during the Spring of 1989. I understand that Davies is providing a catalogue of the Hebrew inscriptions that will be available from Cambridge University Press in late 1991 or early 1992.

[6] Compare Bernhard Lang, *Monotheism and the Prophetic Minority, An Essay in Biblical History and Sociology*, The Social World in Biblical Antiquity, no. 1 (Sheffield: Almond Press, 1983), especially pages 57–59.

[7] McCarter notes ("Aspects of the Religion of the Israelite Monarchy," p. 137) that, "Of the competing parties in preexilic Yahwism only one was vindicated by history, and its thought is preserved in the Bible."

⁸This group has been excluded because of space considerations as well as the availability of other onomastic studies like those mentioned above.

⁹Yohanan Aharoni, *The Archaeology of the Land of Israel*, trans. Anson F. Rainey, (Philadelphia: Westminster, 1982), p. xix.

¹⁰McCarter, "Aspects of the Religion of the Israelite Monarchy," p. 137.

¹¹Ibid., p. 138.

¹²A very few inscriptions from the time frame described but from just beyond the territorial limits, i.e., Karatepe, the Nimrud Hebrew, and Arslan Tash inscriptions will be included because they were found just beyond the geographical limit and provide valuable parallels or because they may well have originated within the territorial limits.

Chapter 2

Semantic Survey and History of Interpretation

In order to understand the inscriptions better, the semantic field of blessing and curse in the NW semitic languages, Akkadian and the biblical literature will be introduced and a brief history of interpretation of blessing and curse in the Bible will be made. The history of interpretation section will consist of a summary of the thoughts of nine scholars about blessing and curse: Pedersen, Mowinckel, Hempel, Westermann, Wehmeier, Scharbert, Brichto, Thistleton, and Mitchell.

Semantic Survey

The semantic study will start with Akkadian *birku/burku*, *karābu* (noun and verb) for blessing and *arāru*, *arru*, and *qalālu* for curse. Ugaritic *brk* (noun and verb) and *mrr* for blessing will follow. Specific words for curse are extremely rare in Ugaritic but some suggestions will be noted. Next will be Aramaic ברכ for blessing and אלה for curse. Phoenician and Punic words ברכ for blessing and קלה, קבב, and אלה for curse will be fourth. Inscriptional Hebrew instances of ברכ and ארר will be treated fifth. Edomite ברכ will be treated sixth. The Biblical Hebrew roots ברכ for blessing and ארר, קלל, קללה, אלה, and קבב for curse will be treated last.[1]

Akkadian
The Akkadian words for blessing and curse will be presented below and examples will be given to demonstrate their meanings. This background will be valuable in later showing the parallels with the other inscriptional materials and the Bible.

birku or *burku*. The noun *birku* or *burku* has three meanings. First, meaning "knee" it is used in medical terminology and in literature apparently in the same sense as feet *ul āniḫa šēpāki lāsima birkāki* "tireless are your feet, swiftly running your knees." Second, it may refer to the lap of a human being *šarru bēlī marmārešu ina burkiešu lintuḫu* "may the king, my lord, be able to lift even his grandchildren to his lap"; to the lap of gods *mārka mārmārika šarrūtu ina burki ša DN uppaš* "your son and your grandson will exercise kingship on the lap of Ninurta";[2] or to the laps of Assyrian images where fines were deposited 10 *MA.NA kaspa* 1 *MA.NA ḫurāṣa ina burki ša Ištar āšibat Arba'ili išakkan* "(whoever comes in the future and makes a claim against PN) deposits ten minas of silver (or) one mina of gold in the lap of (the image of the) Ištar, who dwells in Arbela." Third, it may also be used as a euphemism for male and female sexual organs kāmītu ša pī ilī kāsītu ša birki ištarāti "she (the witch) who gags the mouth even of gods, stops the womb even of goddesses."[3]

karābu. The noun karābu has two meanings: 1) "prayer" and 2) "blessing." With the meaning "prayer" it is used, for example, in the following tašemme Šamaš sulâ suppâ u karabi (var. -bu) "you, Šamaš, listen to demands, supplication and prayers"; in the DN ^{d}Išmekarabu and the PN Sinkárabīš [me]. It is used with the meaning "blessing" for example in the following Šamaš u Ištar [ana] ahija karaba ra[bâ] hidûta banī[ta] liddinuniššu "may Šamaš and Adad [sic] give my brother great blessings and pure joy."[4]

The verb karābu has several shades of meaning. 1) It was used of gods pronouncing formulas of blessing: DN ana epēšika annîm līkrubu "may...give his blessing to this undertaking of yours."[5] 2) It was used to pronounce formulas of praise, adoration, homage, or greeting which were directed to gods ihdû ikrubu Marduk ma šarru "they (the gods) rejoiced, did homage (declaring): Marduk is king" or to kings mahar Aššur u ilāni mātišunu ikrubu šarrūtī the Manneans "did homage to me as king in the presence of (the images) of Aššur and their own gods"; or greetings to private individuals ([šīb]ūtum ikarabušu [milik] harrāni imalliku Gilgāmeš "the aldermen bid him godspeed and give Gilgāmeš advice concerning the journey."[6] 3) It was used to invoke blessings upon other persons (for a specific purpose) before the images of the gods and to pray to the gods šūt iktarraba ikribīšina tamtahhar "you always receive graciously the prayers of those who assiduously pray to you." With this force it was used also in private letters ikrīb aktanarrabu mahar bēlija u bēltija ana abija kâta lībašū "May the blessings I constantly invoke upon you, my father, remain in effect before my lord and my lady."[7] When used to invoke gods it could be accompanied by cultic acts niq'am mahar ilika iqīma kurubām "make a libation before your god and invoke blessings upon me." It is also found with this meaning in private letters bēlī annam īpulannima mahar DN u DN$_2$ ana bēlija akrūb "my lord consented and so I invoked blessings on my lord before" DN and DN$_2$.[8] 4) Karābu was used for the gesture of adoration or greeting 3-šú ana Esagila ikarrab dalāti ipetti "three times he raises his hand towards Esagila, (then) opens the door."[9] 5) It could be used to dedicate an offering by pronouncing the formula ṣalam šarrūtija epēša...ina mahar Aššur...ana šazzuzi lu akrūb "I vowed to make a stela representing me as king (and) to place (it) in front of (the image of) Aššur (at the threshold of my city)."[10]

Though the Akkadian forms do correspond in meaning to BH ברך, they are probably not directly related as metathesis of the first and third radicals is extremely rare. The Akkadian is probably related to the Old South Arabic krb "to consecrate, to sacrifice."[11]

arāru. arāru "curse" can be a noun or a verb. "The great gods, having assembled, elevated the position of Marduk and did obeisance, while they pronounced

Chapter 2: Semantic Survey and History of Interpretation

upon themselves an imprecation, swore by water and oil, touching (?) their throats" *ipḫurunimma ilū rabûti šīmat Marduk ullû šunu uškinnu uzakkiruma ana ramanišunu araru ina mē u šamni itmû ulappitu napšāte*.[12] It can also be used for "blaspheme."[13]

The verb has several different meanings. It can mean "to curse, treat with disrespect, insult, disavow, or disown." It is used, with a DN as subject to curse people on 1) royal inscriptions *ilū rabûtum ša šamê u ereṣetim...šuāti zērašu māssu ṣabšu nišīšu u ummānšu erretam maruštam liruru errētim anniātim Enlil ina pīšu ša la uttakkaru lirūršuma arḫis likšudašu* "may the great gods of heaven and nether world curse him, his descendants, his land, his soldiers, his people, and his army with a baleful curse, may Enlil with his unalterable utterance curse him with these curses so that they speedily affect him"; 2) legal texts *arrat la pašāri lirurušu* "may they [the gods depicted on the *kudurru* [14]] curse him with a curse which cannot be dispelled"; 3) other occasions *ilāni...arrat la napšuri marušti lirarušuma* "may the gods curse with an evil curse which cannot be dispelled him (who destroys the tablet)." It is used in contracts PN *ina ušuzzi ša aḫḫēšu eqla iknukma arrata irūrma...iddin* "PN deeded the field [to PN₂] in a sealed document with the appropriate curse formula in the presence of his brothers." There are also secondary and tertiary meanings of *arāru*: "to fear, become agitated" or "to rot, discharge a putrid liquid, or defecate (of birds)."[15] Note that KB gives the meanings "bind," "enchant," and "ban" for *arāru*.[16]

In addition, there are the further forms: *arru*, a verbal adjective meaning "accursed,"[17] *ariru* a noun meaning "a priest who pronounces a curse," and the nouns *arratu* and *erretu* "curse."[18]

qalālu. *qalālu* can mean "to become weak, thin, light," "to lose importance, become discredited," and in the form *qullulu* "to make an inferior-quality product, to reduce diminish," "to discredit, to be discredited." Examples of meanings of interest for this study are "to lose importance, become discredited" as in *šarru itti kabtūtišu <<qal>> iqallil* "the king will become discredited through the connivance of his noblemen"; "to discredit" as in *šeṭūssa la teleqqe u qáqásà ana awatim la tuqálāl* "in view of the circumstances do not hold her in contempt or treat her with disrespect."[19]

Ugaritic [20]

brk. As a noun, *brk* occurs several times in the Ugaritic materials. It has three meanings: 1) "knee," 2) "pool," or 3) "blessing." 1) The gods "lowered their heads onto their knees," in apparent dismay, when faced with the demand of Yammu and Naharu.[21] 2) In *UT* 67:I:16 *brky* is a pool where animals drink.[22] 3) It can also mean "blessing" *brkm* in the emphatic phrase [*ymn*] *brkm ybrk* "richly he blessed his [servant].[23] It is also possible for *brk* to be a euphemism for "curse."[24]

The verb *brk* "to bless" occurs several times in the Ugaritic materials. Four times in *UT* 128 (*ARTU*, KRT II:ii-iii:17) the verb occurs: El is called upon by the other gods to bless Keret at his wedding banquet, a blessing that is specifically for the fertility of the bride (ii:14); El does so (ii:18, 19); and so do the other gods as they leave (iii:17). Also, in 2 Aqht I:24 and 35 (*ARTU*, Aqhat I.i.24, 35), a third party calls upon El to bless (*ltbrknn*) son-less Dani'ilu with a son because he has been a faithful worshipper; El does so (*ybrk*).²⁵

mrr . Another verb for "bless," *mrr*, also occurs in Ugaritic. Though its etymological antecedents are disputed, it definitely has the meaning "bless" in Ugaritic. Perhaps its meaning is more specialized and refers to a specific kind of blessing, strength to avenge or vigor to produce children.²⁶ In *UT* 128:II:14-16, a form of *mrr* is used inparallel to *brk*, *ltbrk* (15) [*krt*] *t`ltmr n'mn* (16) [*ǵlm*] *il* "Surely you will have to bless Kirtu, the nobleman, surely you will have to fortify the gracious lad of Ilu."²⁷ Gordon translates *ltmr* as "protect."²⁸

Though there is no specific word for curse in the extent literature,²⁹ numerous examples of substantive curses occur.³⁰

Aramaic

ברך occurs as both a noun and a verb in epigraphic Aramaic. Six of the eight Hermopolis papyri, dating around 500 B.C., use an extended greeting/blessing formula. The sentence ברכתכי לפתח זי יחוני אפיכ בשלמ ("I bless you by Ptaḥ who has caused me to see your face in peace") occurs in i.2; ii.2; iii.2; iv.2; v.1-2 (ברכנכנ "we"); and vi.1-2. A fifth century letter from Egypt makes use of an extended greeting or greeting/blessing combination very similar to Arad 16: the sender wishes the recipient health (שלמ) and in the next line "now be you blessed [by the god Yahu]."³¹

Three times on grave inscriptions the formula *bryk* PN is found: once on the 482 B.C. grave inscription from Saqqara (*KAI* 267) and twice on the Carpentras stele (*KAI* 269): line 1 *brykh* PN and line 3 *kdm* DN (Osiris) *brykh*.

Forms of ברך occur six times in the Aramaic portions of the Bible. Four times it has the meaning "praise" (Daniel 2:19, 20; 3:28; and 4:31; in this last passage Nebuchadrezzar says he "blessed,""praised," and "honored" "the Most High...the One who lives forever"). In Daniel 6:11, the form occurs twice, as the verb "to kneel" and the plural noun "knees": Daniel "kneeled upon his knees."³²

The word אלה occurs for "curse" in the Aramaic materials solely in the Panammu inscription where it seems to describe difficult situations caused by a broken treaty.³³

The root קבב is found as many as four times in the Deir 'Alla plaster inscriptions. Though the contexts are unclear or broken, these occurrences seem to speak of the power to utter curses.³⁴

Phoenician and Punic

ברכ. In Phoenician and Punic only the verb form of ברכ is attested, though in *KAI* 147 it might be a noun. The verb occurs in the Qal and should be

Chapter 2: Semantic Survey and History of Interpretation

translated "to bless" with the sense of granting "happiness, vitality, success." It is often a concluding formula in inscriptions. Two interesting formulas occur: 1) A (DN) brk/ybrk/tbrk B (PN) "A (DN) blesses B" (e.g., *KAI* 10:8-9) and 2) A (PN$_1$) brk B (PN$_2$) *l* C (DN) "A commends B to C (deity) for a blessing" (e.g., the Saqqara letter *KAI* 50:2).

Several instances of the verb ברכ are found in the middle to late eighth century Azitiwada inscription from Karatepe, calling for Ba'al KRNTRYŠ to bless Azitiwada with "life, health, and strength" (A.i.2 and C.iii.16-17). Elsewhere in the inscription, Azitiwada is called "the blessed of Ba'al" הברכ בעל (A.i.1).[35] ברכ is found in a seventh century inscription on an ivory box from Ur (*KAI* 29) in which Ashtart is asked to bless the subject's days. The sixth century B.C. papyrus from Saqqara (*KAI* 50) combines greeting with blessing (or extended greeting ברכ and שלמ) in a striking parallel to Hebrew and Edomite greetings. In the late fifth century inscription of Yehaumilk, "the Lady of Byblos" is called upon to bless the king with a long rule and favor with the gods.[36]

קלת. The adjective קלת is found in two second century B.C. Neo-Punic texts found in southern Algeria (*KAI* 162:5; 163:3). The meaning in both places is understood as "a cursed brood."[37]

קבב. The root קב occurs once in a Punic votive inscription (*CIS* I 4945) where תנת פנ בעל ("Tanit face of Ba'al") is called upon to curse (קבת) anyone who disturbs the dedicated gift.[38]

אלה. The noun אלה is found in the form אלת, in the first of the two Arslan Tash incantation texts. It is found four or perhaps five times and seems best translated "oath or pact."[39] In BH each oath (אלה) included a curse should the oath be broken.[40] In this inscription, that interpretation may not apply.[41]

Inscriptional Hebrew

ברכ. The verb ברכ occurs sixteen times in the Hebrew inscriptions of the target period. These occurrences range from letter greetings or probable greetings (Arad 16, 21, 40; Kuntillet 'Ajrud 4, 3; Samaria C1101, C1220) to dedications, benedictions, or praise (Kuntillet 'Ajrud 1, 5; Ketef Hinnom 1, 2; Khirbet el-Qôm; En Gedi). Some of these categories are uncertain and there are two instances where the word ברכ is found in isolation (Arad 28; Lachish 31). In eleven of these cases ברכ occurs with Yahweh either in formula or in the same context. In two of these occurrences there is no DN preserved. In one of these Hebrew inscriptions ברכ occurs with the DN Ba'al (Kuntillet 'Ajrud 1). Twice it occurs with the combination "Yahweh and his Asherah" (Kuntillet 'Ajrud 3 and Khirbet el-Qôm).

ארר. Over half-dozen times the root ארר is found in target period Hebrew inscriptions.[42] All of these instances are found with the name Yahweh in context or in a context which is clearly Yahwistic. Aside from the questionable

Chapter 2: Semantic Survey and History of Interpretation

curse in Lachish 5, all of the uses of ארר were intended to protect graves or their inscriptions.

Edomite

There is a single occurrence of the root ברכ in the known Edomite inscriptions.[43] The form הברכתכ occurs on an ostracon from Ḥorvat ʿUza and is part of a letter greeting similar to that found in the Arad letters treated below. No general words for curse or specific curses have been found in the inscriptions.

Biblical Hebrew

ברך. There are two Biblical Hebrew roots spelled ברך. The first root consists of a verb found twice in the Qal and once in the Hiphʿil meaning "kneel." The noun form from this root means "knee" and occurs 24 times.[44] An additional noun ברכה "pool, water resevoir, basin" also occurs several times; its relation to the two roots mentioned here is disputed.[45]

The second root has the basic meaning "bless." It occurs in the Qal 72 times,[46] usually as the passive participle "blessed, praised." Three times it is found in the Niphʿal "to be blessed, bless oneself." In the Piʿel it appears 235 times "to bless, greet, praise."[47] Puʿal forms appear fourteen times "to be blessed." Hithpaʿel forms occur seven times "to bless oneself, bless one another." The noun ברכה "blessing, praise" appears 71 times.[48] In addition, personal names are found which incorporate ברך, including six different people named ברוך.[49]

The following three ברוך formulas are used in the Old Testament of one person blessing another: 1) *baruk* + second person pronoun/PN+ *ʾasher* "blessed are you who..." (e.g., 1 Sam. 25:33; 2) *baruk* + (second person pronoun) + *lYhwh* "blessed be thou by Yahweh" (e.g., 1 Sam. 15:13; Ruth 3:10; Ps. 115:15; and Judg. 17:2, actually *baruk beni lYhwh*, *lYhwh* should be interpreted as a request to Yahweh to bless someone); 3) *baruk* + (second person pronoun) *lYhwh* *ʾasher/ki* "blessed are you by Yahweh who/because..." the "who/because" refers to the one blessed, they are blessed because of a deed specified in the context (e.g., 1 Sam. 23:21; 2 Sam. 2:5; Ruth 2:20). According to Mitchell, there are also four ברוך formulas with Yahweh as subject :

1) ברוך יהוה אשר\כי; the relative clause describes the beneficent act of God for which he is praised; 2) ברוך יהוה, with a clause describing God's beneficent actions following in apposition to יהוה or asyndetically; 3) ברוך followed by a title of God as the subject; the title gives the reason for the praise; and 4) ברוך יהוה with no reason following; a clause modifying ברוך such as "forever and ever" may follow.[50]

The root ברך occurs most frequently in the Piʿel stem. When used in the Piʿel, as sometimes in the Qal passive participle, the reference to God as the

Chapter 2: Semantic Survey and History of Interpretation 13

source of blessing may be oblique or simply understood, but in virtually every case, that implication is quite strong. The following formulas appear: 1) PN₁ (superior) *brk* PN₂ (inferior) "A blessed B" (e.g., Gen. 27 [thirteen times]; 28:1, 6; 32:1; 49:28; Num. 22:6ff; Deut. 10:8; Josh 22:6; 2 Sam. 6:8; 19:40; 1 Kgs. 8:14, 55). 2) PN₁ *brk* PN₂ "A blessed B" between equals (e.g., Gen. 12:3; 27:29; Num. 24:9). This formula may also be a greeting 1 Sam. 13:10; 2 Kgs. 4:29; 10:15; the greeting is not specified but may be along the lines of Ruth 2:4. 3) PN₁ (inferior) *brk* PN₂ (superior) is relatively rare (e.g., Gen. 47:7, 10; Deut. 24:13; 2 Sam. 14:22; Ps. 72:15) the person blessing is actually asking for God to give success, health, etc. to the superior. 4) PN *brk* Yhwh "A blessed Yahweh" or better "A praised Yahweh" is used only in Gen. 24:48 of the pre-Deuteronomic materials, elsewhere Deut. 8:10; Josh. 22:33; Judg. 5:2; 1 Chr. 29:10; Pss. 66:8; 68:27(26) et al.[51] ברך is also found in the Puʻal stem where it simply functions as the passive of the Piʻel (Num. 22:6; Deut. 33:13; Judg. 5:24; 2 Sam. 7:29; Job 1:21; Pss. 37:22; 113:2). In a few cases, ברך is used as an euphemism for curse (e.g., "Naboth cursed God and the king" ומלך ברך נבות אלהים 1 Kgs. 21:13; also Job 1:5, 9-11; 2:9).[52]

The root also occurs in the following passages in the Niphʻal (Gen. 12:3; 18:18; 28:14) and Hithpaʻel (Gen. 22:18; 26:4; Jer. 4:2) stems; both stems are best translated with a reflexive sense. It occurs two other times in the Hithpaʻel as well, in Deut. 29:19 where it expresses arrogance and Isa. 65:16 where it speaks of the good days to come when everything will be as it should be in the land.

ברכה. The noun ברכה also occurs in Biblical Hebrew: Gen. 27:41; 39:5; 49:28; Deut .33:1; Pss. 3:9; 129:8; Isa. 65:8. It denotes the power coming from Yahweh which brings good fortune and prosperity to man.[53] With regard to the root ברך in the Bible, Christopher Mitchell says,

> The biblical use of the root is quite similar to its use in other NW Semitic languages. However, the biblical authors give no indication that they are drawing upon other literatures for the meaning of *brk*. The root has a greater range of meaning in the Bible than in extra-biblical NW Semitic texts, indicating that the biblical authors developed to some extent the earlier NW Semitic meanings. The biblical contents are quite satisfactory themselves for elucidating the meanings, and there is an abundance of biblical contexts. All things considered, then, the etymology is useful for background information and for comparison to biblical use, but it is not useful for determining the biblical meanings of the *brk* derivatives.[54]

His second sentence is a judgment call which can just as well be argued either way. There is no evidence that the biblical writers do not draw meaning for ברך from other NW Semitic languages. Rather, a comparison with those texts do show a close connection between biblical and other NW Semitic ideas of blessing.

Chapter 2: Semantic Survey and History of Interpretation

The uses of ברך in the Bible can be divided into two overarching categories and then the first of these can be subdivided into several smaller ones. The first use, whether of the noun or verb, is of a person asking/wishing a blessing of a deity (with or without a DN) for another person. The second category involves blessing between a person and a deity either from a person towards a deity, which is praise, or a spoken blessing from a deity (Yahweh) to a person.

Person to person blessings, can be subdivided into several smaller categories of use. BDB suggests the following categories: the blessing of a parent, salutation, or greeting (the last of which may appear in any of the following situations: a) in meeting, b) in departing, c) by messengers, d) in gratitude, e) morning salutation, f) congratulations for prosperity, g) in homage, h) in friendliness.[55] Ritual uses are also included in the Bible: a) declarative human blessings, b) testamental blessings, c) divination blessing pronouncements, d) miscellaneous declarative blessings, e) optative blessings: the priestly blessing, prayers for blessing, thanksgiving benedictions, benedictions of praise and congratulation, psalmodic concluding benedictions, and miscellaneous wishes.[56]

ארר. The verb ארר occurs over sixty times in Biblical Hebrew: over fifty times in Qal "curse, bind with a curse" (most of these as the Qal passive participle ארור); once in Niph'al "cursed"; seven times in Pi'el "lay under a curse, bring a curse"; and once in Hoph'al "be under a curse." God curses (e.g., Gen. 3:14, 17; 4:11; 12:3; Jer. 11:3; 17:5; 48:10; Mal. 1:14; 2:2). Men curse (Gen. 9:25; 27:29; 49:7; Exod. 22:27; Num. 22:6, 12; 23:7; 24:9; Deut. 27:15–26; 28:16–19; Josh. 6:26; 9:23; Judg. 21:18; 1 Sam. 14:24, 28; 26:19; 2 Kgs. 9:34; Jer. 20:14, 15; Ps. 119:21; Job 3:8). The messenger of the Lord curses once (Judg. 5:23).[57] The noun form מארה meaning "curse" occurs five times: Deut. 28:20; Mal. 2:2; 3:9; Prov. 3:33; 28:27.[58] Gevirtz says that in view of the frequency of the curse with ארור "in the Old Testament, and of its restriction to Hebrew sources, this curse form may be recognized as characteristically and specifically Hebrew."[59]

קלל. The verb קלל occurs approximately eighty times in the Old Testament. In Qal it can mean "be slight, trifling" or "be of little account" (e.g., 1 Sam. 2:30; Isa. 40:4). In Niph'al it can mean "show oneself swift" (Isa. 30:16); "humble oneself" (2 Sam. 6:22); with בעיני "be too small a matter to" (1 Sam. 18:23; 2 Kgs. 3:18); "be a light thing to" (Ezek. 8:17). In Pi'el it means "declare cursed, too trifling, of no account" and is used of people in general (e.g., Gen. 12:3; 1 Sam. 17:43); father and mother (Exod. 21:17; Lev. 20:9); the king (2 Sam. 16:5, 7, 10f, 13) or the rich (Qoh. 10:20); God (Exod. 22:27; Lev. 24:15); a servant towards his lord (Prov. 30:10; Qoh. 7:21). It is used of God cursing the field (Gen. 8:21); Bala'am cursing Israel (e.g., Deut. 23:5; Josh. 24:9); cursing the name of the Lord (Lev. 24: 11, 14, 23); and cursing enemies (Pss. 62:5 [4]; 109:28). It also can mean "declare cursed invoking the name of Yahweh" (2 Kgs.

Chapter 2: Semantic Survey and History of Interpretation

2:24); cursing "by his God" (1 Sam. 17:43); call a curse formula upon (1 Kgs. 2:8; Isa. 8:21). The Pu'al means "be declared cursed" (e.g., Isa. 65:20; Ps. 37:22). The Hiph'il means "make light" (e.g., Jonah 1:5; Exod. 18:22; 1 Sam. 6:5); and "treat with contempt" (e.g., 2 Sam. 19:44; Isa. 8:23; 23:9).[60]

קללה. The noun קללה occurs over thirty times in the Old Testament. It means "curse, the formula with which a person, a thing is called ארור" (e.g., Gen. 27:12; Deut. 29:26; 28:15, 45; Judg. 9:57; Jer. 24:9; 25:18; 26:6); become a curse (Jer. 44:22; 49:13); be a formula of curse/cursing (2 Kgs. 22:19; Jer. 42:18; 44:8; Zech. 8:13); take a (formula of) curse (cursing) (e.g., Jer. 29:22). It can refer to a written curse (Deut. 29:26). Several times it is juxtaposed to ברכה (Deut. 11:26, 28f; 23:6; 27:13; 30:1, 19; Josh. 8:34; Zech. 8:13; Ps. 109:17; Neh. 13:2).[61]

אלה. The root אלה occurs in both verb and noun forms in the Old Testament. The verb occurs five times in the Qal, four times it means to "swear an oath before God" (1 Kgs. 8:31||2 Chr. 6:22; Hos. 4:2; 10:4); and once to "curse" (Judg. 17:2). The verb also occurs three times in the Hiph'il with the meaning "adjure, put under oath."[62]

The noun occurs over 35 times and has the meanings "oath" (e.g., Gen. 24:41; Lev. 5:1; Deut. 29:11); "curse" from God (e.g., Num. 5:23; Deut. 29:18); curse from men (Job 31:30; Pss. 10:7; 59:13); and "execration" (Num. 5:27; Jer. 29:18; 42:18; 44:12).[63]

אלה is also used to guarantee fidelity to a treaty, both parties may take the oath or only the weaker (e.g., Gen. 26:28; Ezek. 17:13). It can be used in ratification of a covenant (e.g., Deut. 29:18, 20[19, 21]; Isa. 24:6; Jer. 23:10). Indeed, because such curses are basic components in the ritual transaction of a covenant, אלה can almost be used as a synonym for ברית.[64]

קבב. The verb קבב occurs thirteen times in the Old Testament. It occurs only in the Qal stem and means "utter a curse against, curse" (Num. 22:11, 17; 23:8 (2x), 11, 13, 25 (2x), 27; 24:10; Job 3:8; 5:3; Prov. 11:26). In Num. 23:11, 25; 24:10 it is contrasted to ברך, and in Prov. 11:26 is contrasted to ברכה.[65] In Job 3:8 it is parallel to ארר.

Summary

Blessing and curse have similar emphases in the various literatures. The Akkadian root *karābu*'s dual meaning "prayer" and "blessing" is certainly noteworthy. This can be interpreted as indicating that blessing is prayer, a specific kind of prayer. This would allow then that all blessings may be prayers, invocations of deities to act whether the deities are specified or unspecified. Such an understanding of "bless" makes quite a difference in the way blessing may be understood to operate. This implied call upon the deities to act even when not named could apply to the Bible as well and shed light on the discussion of the next part of this chapter. The Akkadian root *arāru* is important as well: Akkadian and Inscriptional

Hebrew are the only languages which use this root, and use it with apparently the same ideas found in the Bible.[66]

The Ugaritic root *brk* has the same basic meanings as inscriptional and BH ברך, "knee, pool, bless." The blessings of progeny and produce as the intent of *brk* are common to both Ugaritic and BH.[67] General Ugaritic words for "curse" are rare; specific curses predominate.

The root ברכ in Aramaic is not found in the target inscriptions, though it frequently occurs in fifth century inscriptions and later. Likewise, the general word for "curse" אלה occurs only in the Panammu inscription and there is used a bit obscurely in a treaty context. The general root קבב occurs perhaps four times in the Deir 'Alla inscriptions in a context of uttering curses.

The root ברכ is not infrequent in the Phoenician inscriptions. In them it is used much like it is in the Hebrew inscriptions though, unlike the Hebrew inscriptions, it is often accompanied by specific blessings as well. Only אלה occurs as a general word for curse in Phoenician and then only in the debatable context of the first Arslan Tash tablet.[68] General words are quite often found in later Punic inscriptions.

The root ברכ occurs several times in inscriptional Hebrew in much the same contexts that it is used in the Bible and with apparently the same meaning. The root ארר occurs several times in the Hebrew inscriptions and only in them. It is used for the protection of graves and/or inscriptions.

The Bible, containing the latest of the sources consulted, includes more of the general words for blessing and curse found in the various languages than any one of the others (i.e., virtually all of the words found in the languages consulted above also occur in the Bible), and the meanings of the words are quite similar across the board. Though evidence is by no means equal from all of the languages surveyed, general words for blessing and curse are used in the same kinds of literature (letters, votive inscriptions, tomb inscriptions, religious texts-- where available) with what appear to be the same intent: the wish that the deity invoked bestow the desirable things of life (progeny,[69] food, security, etc.) on the favored one or deprive the despised one of the same things. The phrase "general words for blessing and curse" refers to the words surveyed above in contrast to specific wishes for blessing or curse which may or may not use a root like ברכ or שלמ, but have that clear intent nevertheless; such blessings or curses are called "substantive blessings" or "substantive curses" in this study. The next section of this chapter will outline interpretations of blessing and curse in the Bible and trace the development of their interpretation.

The History of Interpretation of Blessing and Curse in Biblical Literature

The views of nine scholars who have had a formative role in the

Chapter 2: Semantic Survey and History of Interpretation 17

interpretation of blessing and/or curse in biblical literature are summarized below. The ideas of these scholars will be presented in chronological order: Johannes Pedersen, Sigmund Mowinckel, Johannes Hempel, Claus Westermann, Gerhard Wehmeier, Josef Scharbert, Herbert Brichto, Anthony Thistleton, and Christopher Mitchell. Their views will be briefly stated concerning the following questions: 1) Of what do blessing and curse consist ? 2) How do blessing and curse operate?[70]

Johannes Pedersen

Pedersen says that blessing is the vital power that no living being can live without.[71] It is the "entire power of life, the strength underlying all progress and self-expansion."[72] Every individual possesses blessing of his or her own which will be stronger or weaker or of a different type according to the individual.[73] Blessing is the strength of life, primarily fertility, but also the power to create wealth.[74]

The act of blessing is the transfer of soul power.[75] God increases the soul power of a person when he blesses him or her. God's power merges with the individual's and becomes indistinguishable. Human blessing is also the transfer of soul power, usually by touch. Humans even strengthen God when they bless him according to Pedersen.[76] Blessing is a self-fulfilling power. It comes from the soul of the one uttering it and that individual can only communicate it according to the power in him or herself.[77] Blessing "consists in a communication of the contents of the soul."[78] An oral blessing comes from the soul power of the one uttering it and so cannot be revoked; "it becomes real."[79]

Curse is "the dissolution of the soul."[80] Curse and sin are so closely tied together that they cannot be parted.[81] Curse (אלה) is a "poisonous substance" which makes the earth infertile, towns to fail, and people to become sterile.[82] Just as one person can utter a blessing into the soul of another, so can one utter a curse into another's soul. "He whose soul creates something evil for another—be it in thought, in word or in deed—he puts the evil into the soul of his neighbour, where it exercises its influence."[83] People who have strong souls also have stronger curses.[84] Curse is also infectious, like blessing, and the cursed person is best removed from his or her community.[85] There are degrees of curse, the worst being ארור, which detaches the cursed individual (the "soul") from his or her community, ultimately bringing about death or expulsion.[86] Curses are by all means to be avoided for they are like a cancer that weakens even the "heavy soul" of the one full of honor threatening to make his soul light (קלל).[87]

Mitchell notes that Pedersen's terminology is mystical and often unclear because he thinks that Israelites conceived of everything in terms of "totalities" or "connected wholes," and that Pedersen runs his definitions together. Further, there is no evidence for Pedersen's idea of "soul power." In every biblical reference, God is either specifically or implicitly the ultimate source of blessing.[88]

Chapter 2: Semantic Survey and History of Interpretation

Sigmund Mowinckel

Mowinckel accepts Pedersen's work and goes on,

> Every living thing has its own particular blessing, which means that life is allowed to expand in "peace and harmony"...It is the mysterious "potency" and power and strength, immanent in life itself; so that the Israelitic *berakha* in many ways corresponds to the power which the phenomenology of religion has called "mana". Blessing is a health-giving power, creating and promoting life, the power of blessing, "blessedness". "The blessed one" (*barukh*) is a person "having in himself blessing."[89]

Indeed, Psalms such as Ps. 29, reflected the old idea that cultic acts and words actually renewed the strength and blessing power of the deity, according to Mowinckel. He thinks such an understanding of "bless Yahweh" was toned down to mean "praise and thank him."[90]

But Israelite understanding did not stay on this level; because blessing was something holy and possessed creative power, the Israelites understood that it must have some connection to the deity and Yahweh became more and more understood as the source of blessing. Though formerly a person uttering a blessing might have been thought to be the source of the blessing (Num. 22:6), in historical times Yahweh himself was understood as the subject of blessing. This does not mean that the Israelites believed in a sort of magical word. Yahweh put his power behind the blessing word of the proper person; thus the power was his.[91]

The center of Mowinckel's work, as might be expected, was to show how blessing had its origin in the cult. He believes that since the transfer of blessing was holy, "a ritual act," deriving its power from "The Holy One," then blessing

> from the very first had its place in the cult itself. To procure, secure and increase "the blessing" that was the the object of the temple services in Israel, put in a nutshell. Through the temple service the congregation got into contact with Yahweh and received his life-creating and life-supporting power, his blessing.[92]

"Curse is the very opposite of blessing, it is blessing with a negative sign."[93] It is a negative power: the cursed one fails in everything, he or she suffers in all of the areas that make life worth living—evil and sudden death, family and name obliterated. Cursedness "is an operative power...spreading from the one who is 'filled with curse' (*'arur*) to his family and all his surroundings." It is contagious and contact with such a person can lead to the possession by demonic powers seen in illness.[94] It can be inflicted on decent people through words and rites. They then must protect themselves by "purifications, and increased blessing and the help of Yahweh. But 'the evil curse' will not affect a person who is full of blessing and under the protection of God, as long as he keeps close to God."[95] The blessing power of an individual and of Yahweh can be turned into a

Chapter 2: Semantic Survey and History of Interpretation

cursing power against the enemy and someone threatening Israel will be placed under the curse of Yahweh (Gen. 12:3).

Cursing had a place in the cults and rites of Israel from ancient days against wrongdoers from among the people (Num. 5:11ff). Mowinckel points out that the self-acting power of the curse is seen in this passage at points, i.e., "water of curse." Referring to incidences where the curse is pronounced on enemies before war (Num. 22-24; Deut. 27:14; Judg. 5:23; etc.), he states,

> the cursing word has usually been replaced by the prayer for Yahweh to crush the enemy. But a prayer like the one in [Ps] 83:10ff with its elaborate description of the disaster imprecated on the enemies of the people is evidently connected with ancient cursing formulas, such as seers and other "divine men" ('iš 'elohim) and possessors of the effectual word would use against the enemy before the battle; with such words Balak expected Balaam to slay the Israelites for him.[96]

He says further, "The fact that Yahweh is not often named as the subject of these wishes for punishment points in the same direction; as a rule the old cursing formula belonging to the idea of the self-acting word has been retained."[97]

Johannes Hempel

Hempel believed that Mowinckel had overemphasized the cult and traced the evolution of Israelite beliefs about blessing and curse throughout the Old Testament.[98] He said that all of blessing is summed up in שלם, a condition of complete prosperity, wholeness, and security.[99] Hempel disagreed with Mowinckel's contention that the idea of "blessing Yahweh" in the Bible carried over the older idea of actually increasing Yahweh's power. Though he agreed that this was the original purpose of the formulas, they no longer had it in the biblical texts.[100]

Hempel distinguished three historical stages in the development of the idea of blessing: 1) in the folk religion stage, blessing and curse were magical, contagious (to people, animals, and objects), and self-fulfilling[101]; 2) in the cultic stage, blessing and curse were no longer regarded as magical but certain ceremonies were required to prompt God to bless;[102] and 3) in the ethical monotheism stage, there was no magical or cultic influence on God, rather blessing and curse came from God and were based on the ethical values proclaimed by the prophets.[103]

There are two problems with Hempel's neat three stage division. First, there is inadequate evidence for the folk religion stage. Hempel must rely on emendations and assumptions to provide examples of this stage. Second, Hempel does not adequately show the development from stage two to three. He assumes that the ethical view is farther removed from primitive folk religion (magic) than is the cultic view but this may not necessarily be the case, cultic usage is based on religious traditions and social customs rather than magic. Also, the cultic and ethical views might well have coexisted.[104]

Claus Westermann

Westermann brings a tradition history approach to the study of blessing and curse. He draws a strong distinction between deliverance and blessing. Deliverance is the saving action of God to rescue his people from some perilous situation. Blessing (a prosperous state of being) is the result of an act of deliverance.[105] The oldest concept of blessing is seen in Gen. 27: the power of life, fertility, and prosperity.[106] The presence of God, success, and peace all belong to the semantic field of blessing.[107]

Westermann sees strong indications of a magical idea of blessing remaining in the Old Testament. He believes pre-Yahwistic passages such as Gen. 27 show this: blessing involves touching, the blessing can only be given once, it is irreversible, and it works unconditionally.[108] The magical nature of blessing was modified by the Yahwist when it was made an historical concept connected with God's promise. In Deuteronomistic theology, blessing was tied to the idea of covenant and so became conditional: blessing was under the control of Yahweh.[109]

According to Westermann, blessing and curse underwent different developmental tracks. While blessing was brought under the power of Yahweh at a very early stage, "curse was never placed in such direct relation to Yahweh's work."[110] He notes that though biblical references to Yahweh blessing someone or something are abundant, "nowhere does it [the Bible] speak of the curse of Yahweh or of Yahweh's putting a curse on someone or something."[111] The Old Testament instead speaks of Yahweh's judgment or punishment. From this he concludes that "in Israel the curse was never theologized the way blessing was."[112] In critique of Westerman's view, it should be noted that though he says the word curse (ארר or קלל etc.) is not put in the mouth of Yahweh, yet passages such as Josh. 6:26 ארור האיש לפני יהוה contradict his statement, Yahweh brings about the curse and is said to have "spoken it" by his servant Joshua (1 Kgs. 16:34). Further, Yahweh does promise judgments against Israel for disobedience, judgments which are best understood in the context of comparison with ancient Near Eastern curses.

Gerhard Wehmeier

Wehmeier notes the pre-Islamic understanding of blessing as a natural force: power manifested in the fertility of humans, animals, and crops, independent of human ritual or actions of gods. Only in the NW Semitic cultures is the power of blessing traced back to the actions of gods.[113] Wehmeier agrees with Pedersen, Mowinckel, and Hempel that originally the power of blessing came from the soul of the blesser and was self-fulfilling but says this idea only survives in three places in the Old Testament: the oldest layers of Gen. 27; 32:22-32; and the Balaʿam narratives. He thinks each of these stories has been edited to circumvent the magical idea of blessing and to show Yahweh himself as the only ultimate source of blessing.[114]

Chapter 2: Semantic Survey and History of Interpretation 21

He disagrees with these other scholars over their understanding of humans blessing God, i.e., that by blessing God they increase his power so that he can bless more in return.[115] Wehmeier not only denies the presence of the magical word in the Old Testament, he also denies that the right conjunction of officially correct person, place, and time will necessarily result in blessing, rather, Yahweh will bless when and whom he wishes.[116]

Wehmeier discusses what he perceives as different emphases of blessing in the works of the different Old Testament writers and editors. For example, blessing is very important for J who identifies the God who blesses with fertility as the same God who acted to bring his people out of Egypt; furthermore, through Israel all humankind will be blessed. Though ברך occurs in E, Wehmeier does not see a particular emphasis. D emphasizes the blessing of the land; the prosperity of the chosen people Israel in the land of promise. P takes the promise of descendants from J and the promise of land from D and applies them to all humankind.[117] Wehmeier does not see blessing as a major theme of the prophets; though they name faithfulness and obedience as requirements for blessing, they do not obligate God to bless anyone. He agrees with Pedersen's statement, "Wisdom is the same as blessing, the power to work and to succeed."[118] Success and prosperity are the result of having wisdom. However, Wehmeier differentiates between prosperity and blessing: material things are "unstable and questionable." Goods accepted as God's gifts are blessings but, by striving to obtain goods on their own and so establishing their lives, people deprive themselves of happiness.[119]

Josef Scharbert

Scharbert agrees with his predecessors that the idea of blessing is rooted in magical beliefs but that few traces of this remain in the Old Testament. "That the oldest texts refer to God...or mention God explicitly, and are used as praise or as a declaration of grateful solidarity rather than incantation, speaks against the magical character of the *barukh*-formulas."[120]

"If man is its subject, in most cases the 'blessing' means a laudatory commendation to the deity to bless someone, or merely a greeting formula."[121] The contexts show the content of the blessing to be a wish for long life, descendants, prosperity, success, and power. The use of the passive participle designates the possession of powers to bestow happiness and promote life, or "the suitability of the blessing from the deity indicated by the commendation." Thus, "apparently the Northwest Semites always understood the deity as the true giver of blessing even when they do not explicitly mention him."[122]

He contends that the *sitz im leben* of blessing was the family and home; not the law where, unlike curse, blessing is a short, later insertion, nor the cult where, contrary to its original spontaneous use in giving thanks to God for specific blessings, it has become a stock phrase in worship.[123]

Scharbert's main contribution to the study of blessing is the attention he draws to the relationship between the one blessing and the recipient of the blessing:

> brk in the Piel always means to express solemn words that show the appreciation, gratitude, respect, joint relationship or goodwill of the speaker, thus promoting respect for the one being blessed and, when a man is the object of brk, the wish that he might receive happiness, success, and increase of earthly possessions.[124]

Also,

> Just as the curse was intended to destroy a man's solidarity with others when he grossly transgressed the basic ethical norms of his clan, so the blessing is intended to strengthen solidarity with individuals and groups with whom he has or seeks paricularly close social, racial and religious relationships...[125]

Scharbert says that blessing can also indicate an eventual or future relationship of this kind. A passage he uses as an example is Isa. 19:25, where ברוך עמי is used by Yahweh to refer to Egypt and Assyria with Israel. Yahweh says that they too will someday be with him in that intimate relationship that blessing presupposes.[126]

On occasions when ברך is used in the Pi'el, the references to God as the source of blessing are oblique or understood. Scharbert notes that many of the old benedictions do not mention God (e.g., Gen. 27; Lev. 9:23; Deut. 33:1; Josh. 22:6; 2 Sam. 19:40). In these cases, the benedictory word of the father, especially if he was the clan or tribe leader, was effective by virtue of his own relationship with the god of the tribe. This allowed him to possess within himself the power and authority to declare a word which imparted blessing.[127]

Similarly, in his discussion of ארר, Scharbert says that though it was "probably" originally thought that the word became effective in and of itself as soon as conditions for its activation were right, yet the early expansion of the ארור-formula seen in Josh. 6:26 to ארור האיש לפני יהוה, argues that the fulfillment of the curse was seen as dependent upon the will of Yahweh from very early times.[128]

In the ancient Near East, curse (especially in the incantation literature) had very definite magical connections which were sometimes only thinly covered by religious ideas.[129] However, in Israelite society the use of curses is closely regulated. This is seen in that the curses are put in the form of prayers to Yahweh.[130] Also the thrust in ancient Israel was that Yahweh was guardian of curse; the curse was a means to see that God's will, divine judgment, etc. were done.[131]

Herbert Chanan Brichto

Brichto is careful to note a distinction between curse as malediction, which anyone can deliver, and curse as a ban or "imposing a spell," a power not given to everyone. God, one who derives the power from God, or a magician who knows occult secrets can bind with a spell. Also, a community or its authorized representatives can ban individuals from relationship with it.[132] Indeed, he says ארר is best rendered "spell," if a spell is seen as a "magic circle, which bars what is within from that which is without." This best describes what the "denotation and connotation of *'rr* is in all of its occurrences."[133] When ארר is applied to the earth it acts as a spell that bars the earth's fertility to men. When it is applied to men or animals, it keeps them from being fertile and associating with their own kind.[134] There are also rules restricting malicious curses.[135] In the Bible, good fortune and misfortune (ברכה and קללה respectively) are traceable to God, and prayers or imprecations invoking these are, even when not made explicit in the text, addressed to the deity.[136]

Anthony C. Thistleton

Thistleton traces the idea that words have power back to two classic studies: O. Grether, "Name und Wort Gottes im Alten Testament" and L. Durr, *Die Wertung des gottlichen Wortes in Alten Testament und in antiken Orient.*[137] Thistleton offers several lucid criticisms of this idea. First, proponents of the idea of magical word say that דבר means both "word" and "thing" and that the word becomes a thing. But Thistleton rightly asks, why can דבר not have two meanings? Why must דבר become a thing when spoken, taking on an existence of its own?

Next, he notes that if the word or blessing has power, that power is clearly derived from the deity who speaks it or backs it according to his will, whether that deity is Yahweh or Marduk or another, not because the spoken word has some innate power of its own.[138]

Thirdly, he notes that contrary to some supposed primitive confusion between "word" and "thing," blessing and curse are best understood as illucutionary or performative utterances. That is to say, word and thing are one but they are so because they are uttered by an acceptable person at an acceptable time and in acceptable manner. For example, the statement "I hereby divorce you" can function as an act of divorce (illocutionary utterance) only in a society where that statement is an acceptable means of divorce.[139]

Finally, he critiques the idea of O. Procksch[140] and G. von Rad[141] which portrays "dianoetic" and "dynamic" views of language as alternatives to each other. The dianoetic element of דבר, for example, is the idea, the thought. The dynamic element is the material force behind דבר. Procksch and von Rad said that the dianoetic function of a word conveyed the intellectual idea behind the word but was insufficient to give the dynamic meaning that the word had for ancient people.[142] Thistleton says, "the functions of words are as diverse as the different functions of a row of tools."[143] No single theory of language is adequate and the theory that says if a dianoetic view of language is inadequate, a dynamic view must be sought, is mistaken: it is not an either-or decision.[144]

Though Thistleton's critique is generally excellent, his third point is still weak. While the concept of illocutionary utterance is helpful at points, it can be understood as magical if it limits either God's power to bless or who may ask a blessing.[145]

Christopher W. Mitchell

Mitchell, says that fertility, dominion, and prosperity are the most common benefits God is seen to bestow because they were the most valued in biblical times.[146]

> God blessed because of his favorable attitude toward a person or a group of people. A blessing is any benefit or utterance which God freely bestows in order to make known to the recipient and to others that he is favorably disposed toward the recipient. The type of benefit God actually bestows when he blesses is of secondary importance.[147]

On the use of ברך from man to God, he says that since man could not wish material benefits for God, "Praise was what man could give to God in lieu of material benefits as an expression of appreciation for God's benefaction."[148] ברוך-formulas with Yahweh as recipient are found in older texts where they are followed by the reason for the praise, as well as in later cultic texts where the reason for praise is either non-specific or not given.[149]

When ברך is used as a greeting the one uttering the benediction did not expect it to be fulfilled in a striking manner. "The greetings and farewells are social customs that usually have little religious value."[150] They express goodwill, friendship, affection, and occasionally religious fellowship; they are primarily expressions of the speaker's sentiments. "Because of their social function, they are usually best translated 'to greet' or 'to say farewell,' rather than 'to bless,' unless they also have a religious function or invoke Yahweh or God."[151] The greetings tell nothing about the relation between God and the addressee, only that the speaker is favorably disposed toward the addressee.[152]

Mitchell's main point is his emphasis on what blessings (not those in greetings) say about the relationship which the people involved have with God. "God is pictured as the ultimate source of blessing throughout the Bible because blessing is a result of a favorable relationship to God."[153] He says "God blessed because of his favorable attitude toward a person or a group of people." Blessings "make known to the recipient and to others that he is favorably disposed toward the recipient."[154] With regard to the Priestly Blessing, Mitchell says the right forms must be observed: the right person (the priest) utters the blessing and at the right time and when the conditions are met, God promises to respond regardless of the identity or character of the persons blessed.[155]

Mitchell critiques the scholars who propounded the idea of the magical

Chapter 2: Semantic Survey and History of Interpretation

word by an expansion of Thistleton's comments on illocutionary utterance. Mitchell reasons that these so-called "magical words" with power to act on their own were actually accepted as powerful by the community because they were illocutionary acts based on "societal conventions."[156] The juxtaposition of time, place, and personnel which may resemble magical blessing is not a manipulation of God to bless. Mitchell rightly notes elsewhere that such manipulation is against Old Testament ideas of blessing.[157] Perhaps Jacob (Gen. 27) was in the right place at the right time, but Yahweh determined who would receive the portion of the first born; this is the central point to the story. In the Old Testament, it is Yahweh in every case concerning Israel who determines when blessing or curse will take place.

Mitchell's contention, that the name of the deity invoked in a greeting blessing says nothing about the speaker's relationship to the deity, would mean that nothing can be learned about personal piety from these greetings since the DN invoked would merely be dictated by the local convention.[158] Apparently, he believes that those ancient greetings had no more significance than a modern "good bye" (originally from "God be with ye") might carry. However, the interpretations of such greetings can never be settled with certainty.

Summary

The views of the scholars surveyed show agreement in the understanding of the ideas of blessing and curse in the Bible. Blessing consists of a wish for someone to receive the things considered good in life: land, numerous progeny, sufficient food, clothing, etc. Curse is the wish that someone be deprived of these same things. On this point there is consensus.

The greatest difference of opinion comes in how blessing and curse are thought to operate. The magical view that an individual's "soul power" enabled him or her to bless or curse another at will without resort to the resources of a deity, gave way to the belief that the Old Testament stories (at least in their current form) reflect a theologized view of blessing and curse, a view that blessings and curses are dependent on the action of the correct person, at the correct time, by means of the correct formula. Perhaps the best understanding is that blessing and curse are invocations of Yahweh and are dependent for fulfillment on Yahweh's will rather than human will. Perhaps the ideas of blessing and curse in the Bible did not actually go through the developmental progression suggested by the scholars above. The inscriptions will be appealed to below for elucidation of these points.

Together, the two parts of this chapter have presented the two overarching ways blessing and curse are seen in the inscriptions: in the use of general words like ברכ and ארר and in the appearance of more specific requests, substantive blessings and substantive curse (i.e., long life, etc.).[159] The next three chapters will present the inscriptions which contain blessings and curses. Better understanding of the meanings of these blessings and curses will be gained by their comparison

Chapter 2: Semantic Survey and History of Interpretation

with each other, other ancient materials, and the Bible.

---

[1] In this study, the following procedure will be followed in the transliteration of texts: alphabetic (non-cuneiform) texts will be provided in Aramaic block letters wherever they are not quoted in a discussion. Where the texts are provided by commentators only in English transliteration, I will transliterate the texts into block letters. Cases where I provide transliteration will, of course, be so noted. Vowels will, of course, not be provided in the transliteration and will be provided in the cited biblical passages only where particularly relevant for the argument. Likewise, final forms will not be used, nor will שׁ and שׂ be differentiated, nor will word dividers be provided (except as found in the originals). All of these steps are taken in the attempt to provide the texts to the reader in a form as near the original as possible. Cuneiform texts quoted herein will be provided in normalized transliteration, i.e., syllable divisions usually will not be shown.

Because this study is not an exhaustive reworking of every occurrence of words for blessing and curse in Akkadian and BH, standard works (i.e., *CAD, AHw,* BDB, and KB) are largely relied upon for general categories of meaning. These works do not, however, determine the final categories of meaning arrived at in this study, nor my conclusions.

[2] The abbreviations DN, PN, and GN are technical abbreviations for the following: DN = deity name, PN = personal name, GN = geographical name, and RN = royal name.

[3] "*birku,*" *CAD,* B vol. 2, pp. 255–57.

[4] "*karābu,*" *CAD,* K vol. 8, p. 192.

[5] Ibid., p. 193.

[6] Ibid., pp. 194–96.

[7] Ibid., p. 196.

[8] Ibid.

[9] Ibid., p. 197.

[10] Ibid., pp. 197–98.

[11] Josef Scharbert, "ברך," *Theological Dictionary of the Old Testament,* vol. 2, ed. G. J. Botterweck and H. Ringgren, trans. J. T. Willis (Grand Rapids: Wm. B. Eerdmans, 1975), p. 281

[12] "*arāru,*" *CAD,* A vol. 1, part 2, p. 234.

[13] Ibid., p. 234.

[14] A *kudurru* is an ancient boundary marker.

[15] "*arāru,*" pp. 234–38. KB, p. 89.

[16] KB, p. 89.

[17] "*arru,*" *CAD,* A vol. 1, part 2, p. 305.

[18] Scharbert, "ארר," *Theological Dictionary of the Old Testament,* vol. 1, ed. G. J. Botterweck and H. Ringgren, trans. J. T. Willis (Grand Rapids: Wm. B. Eerdmans, 1975), p. 406.

[19] "*qallalu,*" *CAD,* Q vol. 13, pp. 55–57. There may be some connection as well to Akkadian *gullulu,* and its cognates which means "to commit a sin" as in *mamman ša ana ilim úgāl* [*1*]*i-lu ul ibašši* "there is no one who committed a sin against the god" ("*gullulu,*" *CAD,* G vol. 5, pp. 131–32; noted in Brichto [*The Problem of "Curse" in the Hebrew Bible,* JBL Monograph Series, vol. 13 {1963; rpt. Philadelphia: Society of Biblical Literature, 1968}, pp. 177–79]).

[20] References to Ugaritic materials in the following notes will usually be dual: Gordon's text designation (a number [*UT*] or perhaps a name) will be given as will the designation in *ARTU* (see the next note for full references). These will be the usual minimum designations, in addition, the translation of Gordon (*UL,* see just below) may be cited as may that of Gibson (*CML,* see below),

Chapter 2: Semantic Survey and History of Interpretation 27

and perhaps that of the first publication of the text in *PRU, Ugaritica V*, etc. These references are complicated, but are given for three reasons: 1) some of the Ugaritic texts bear corrected or multiple designations, 2) some translations are preferable to others, and 3) it is hoped that one or the other resource will be available to each reader.

[21] Cyrus Gordon, *UgariticTextbook*, Analecta Orientalia, vol. 38 (Rome: Pontificium Institutum Biblicum, 1967), pp. 197–98 (hereafter *UT*); *UT* 137: 23, 25, 27, 29; (or Johannes C. De Moor, *An Anthology of Religious Texts from Ugarit*, NISABA, vol. 16, [Leiden: E. J. Brill, 1987], p. 32 [hereafter *ARTU*], Ba'al III:i:23, 25, 27, 29). Other examples are as follows: 1) Kotharu-and-Khasisu, "the skilled craftsman" god, gave a special bow and arrows to Dani'ilu, "he put the arrows on his knees (*lbrkh*)" (*ARTU*, p. 234, Aqhat I:v:27; or Cyrus Gordon, *UgariticLiterature: A Comprehensive Translation of the Poetic and Prose Texts*, [Roma: Pontificium Institutum Biblicum, 1949], p. 89, 2 Aqht V:27 [hereafter *UL*]). After the god has left, Dani'ilu names and blesses ([*yb*]*rk*) the bow to succeed: this is the formal way a child would be received, it has become his child to replace the one he has lost (2 Aqht V:35–39 in *UT*, p. 248; *UL*, p. 89; or Aqhat I:v:36–39 in *ARTU*, pp. 234–35, n. 72). 2) In 3 Aqht 24, [35] (*UT*, p. 249; *UL*, p. 93; Aqhat II:iv:24, [35] in *ARTU*, p. 245), "knees" are the meaning in a less clear context, "Pour out [his] blood like a murderer, like a butcher on to [sic] his knees" (*ARTU*, p. 245). 3) In 'nt II:13, 27 (*UT*, p. 252; *UL*, pp. 17–18; or Ba'al I:ii:13, 27 in *ARTU*, p. 6), the bloody 'Anatu, in the midst of her slaughter revels in her victory, plunging her knees (line 13 *brkm*, line 27 *kbrkm*, probably for "feet") in the blood of those she has slain. Another possible occurrence is *UT* 131, the extant part of line 9 begins with a "*k*" and is followed by a word divider (C. Virolleaud, "Fragments mythologiques de Ras-Shamra," *Syria* 24 [1944–45], 14). This occurrence has been reconstructed as a parallel for 'nt II:13 (*UT*, p. 252; *UL* , pp. 17–18; Ba'al I:ii:13 in *ARTU*, p. 6). Cf. John C. L. Gibson (*Canaanite Myths and Legends*, 2nd ed. [Edinburgh: T. & T. Clark, 1978], p. 131 [afterwards *CML*]) who reads *br*]*k<m>*; but a form of *brk* is probably incorrect as there is not space for the "*m* " and the singular seems unlikely.

[22] Also, Ba'al V:i:16 in *ARTU*, p. 70. "*y* " is a feminine ending (J. C. de Moor "Studies in the New Alphabetic Texts from Ras Shamra I," *UF* 1, ed. K. Bergerhof, M. Dietrich, O. Loretz, J. C. de Moor. Neukirchener-Vluyn [W. Germany: Verlag Butzon & Berker Kevelaer, 1969], pp. 185–86). In the duplicate passage, *UT* 604:6 (after Virolleaud, *Ugaritica V*, pp. 559–61, RS 24.293, variation A, pp. 559–61 and 561, n. 1) it is spelled *brkt*. There is a strange idea that *brkt* /*brky* here should be translated "knee" as elsewhere (M. Dietrich and O. Loretz, "UG, *BŠ, TBŠ, Hebr. *ŠBS [Am 5,11] sowie UG IŠY und ŠBŠ," *UF* 10, ed. K. Bergerhof, M. Dietrich, O. Loretz [Neukirchener-Vluyn, W. Germany: Verlag Butzon & Berker Kevelaer, 1978], pp. 434–35), but this explanation merely contorts the passage in order to avoid another meaning.

[23] *UT* 128:II:18; Keret II:ii:18 in *ARTU*, p. 205.

[24] M. Dietrich; O. Loretz; and J. Sanmartín, "Die Ugaritischen Totengeister RPU(M) und die biblischen Rephaim," *UF* 8, ed. K. Bergerhof, M. Dietrich, O. Loretz, (Neukirchener-Vluyn, W Germany: Verlag Butzon & Berker Kevelaer,1976), pp. 49–50; the text is Aqhat IV, fragment c.iv.7 (translated in *ARTU*, p. 272, note, however, that *ARTU* does not appear to agree with the euphemistic intent Dietrich et al. recommend). KB (p. 154) lists two times in biblical literature where it *may* actually mean "curse" (Pss. 62:5; 109:28) and five times where it undoubtedly does (1 Kgs. 21:10, 13; Job 1:5, 11; Prov. 30:11).

[25] Other examples include: 1) KRT II:v:11 (*ARTU*, pp. 208–09; *UT* 128:v:11), the context is unclear but the guests (perhaps gods) at a pre-mortem funeral banquet (yes, "pre-mortem," the goddess Athiratu tells Keret that he will fall gravely ill and instructs him to arrange the banquet) are requested to bless something, perhaps "the offering of produce of the land." 2) I Aqht 194–95 (*UT*; Aqhat III.iv.32 in *ARTU*, p. 263), a woman calls upon the gods for strength to find and kill the slayer of her brother, "surely you will bless me (*ltbrkn*) and I will go blessed (*brkt*)."

[26] L. Kutler, "A 'Strong' Case for Hebrew *mar*," *UF* 16, ed. K. Bergerhof, M. Dietrich, O. Loretz (Neukirchener-Vluyn, W. Germany: Verlag Butzon & Berker Kevelaer, 1984), p. 118. Most scholars see this meaning of *mrr* "strengthen, fortify" as a development from the stem which means "bitter" which later adopted the meaning "bitterly strong" and still later "generally strong" (so Kutler above and B. Margalit, "Lexicographical Notes on the AQHT Epic [Part I:KTU 1.17–18],"

in *UF* 15, ed. K. Bergerhof, M. Dietrich, O. Loretz [Neukirchener-Vluyn, W. Germany: Verlag Butzon & Berker Kevelaer,1983], p. 70). Pardee sees it as a separate root for "bless," distinct from the root meaning "bitter" (Dennis Pardee, "The Semitic Root *mrr* and the Etymology of Ugaritic *mrr* ||*brk*," *UF* 10, ed. K. Bergerhof, M. Dietrich, O. Loretz [Neukirchener-Vluyn, W. Germany: Verlag Butzon & Berker Kevelaer, 1978], pp. 249–88). Also see de Moor's position in outline *(ARTU,* p. 227).

[27] Krt II:ii:14–16 in *ARTU,* p. 205.

[28] *UL,* p. 74. This would be a good example of a more precise kind of blessing meaning for *mrr*. Further examples are 1) 2 Aqht I:25, 36 where *mrr* is again in parallel to *brk* , this time in asking and receiving the blessing of a son from El, *ltbrknn ltr il aby tmrnn lbny bnwt wykn bnh bbt* "Please bless him, o Bull Ilu, my father, fortify him, o Creator of creatures! And let him have a son in (his) house..." (Aqhat I.i.23–25 in *ARTU,* p. 227). 2) *UT* 6:24, the tablet is very incomplete and the line in question is broken but the context seems to be one of giving birth and so this could be an example of the specialized blessing *mr* (*r*). 3) *UT* 6:26, 27, *šmm tmr zbl mlk* "the heavens defend the princely one the king," defense could be seen as a blessing; lines 27–28 are similar *šmm tlak* [ ]*tl amr* "the heavens send [ ] dew I defend..."(*UL,* p. 52). 4) *UT* 126:iv:2 *il šm' amrk* "Il listen, I will protect thee" (*UL,* p. 80; the translation in *ARTU,* p. 217 [Keret III.iv.2] is quite different). Gibson's translation "I see that you are..." (treats it as from the verb *'mr*, cf. *UT*, p. 361, glossary #229). 5) I Aqht 156–57 (*UT*; in *ARTU*, p. 260 [Aqhat III.iii.50, 52]), *lmrrt tg 'll bnr* is apparently a GN; Gordon says that it is the name of a grape arbor and retains its meaning of "protect" as a "place protected from the sun" (*UL,* p. 98). 6) I Aqht 195 (*UT*; *ARTU,* p. 263 [Aqhat III.iv.33]) *mrr* is in parallel to *brk* in lines 194–95 *ltbrkn alk brkt tmrn alkn mrrt* "Please bless me, (that) I may go blessed, Please fortify me, (that) I may go fortified" (*ARTU,* p. 263), the plea of Pughatu, the daughter of Dani'ilu, who calls on the gods to strengthen her for revenge on her brother's killer. 7) *UT* 1012:13, a very broken text, is identical to that in *UT* 126:iv:2 and may be translated "I shall protect thee." 8) *UT* 2002:3 also possibly contains a form but the text is very short and very broken. 9) The form *amr* occurs translated by Gordon as "I will defend" (*UT* 6:27; *UL,* p. 52). 10) *UT* 51:iv:8, 13, 16–17, Gordon (*UT*, p. 361) suggests this DN *qdš wa 'mrr* Qodesh-and-Amrur (*CML*, p. 59) is connected to the verb *mrr* in its likeness to *brk*. 11) *UT* 150:16 contains a PN in a list of PNs *amrb'l* possibly "I bless Ba'al." 12) Also, in *UT* 2118:12, there is a PN *amri* [*l*] (*UT*, p. 361, glossary #230) possibly "I bless Il"(?). Perhaps *mrr* means "praise" in these last two cases as *brk* sometimes does in BH.

[29] So KB, קלל, p. 839, but John Huehnergard (*Ugaritic Vocabulary in Syllabic Transcription,* Harvard Semitic Studies, vol. 32 [Atlanta: Scholars Press, 1987], p. 174), lists an adjective from this root which is best translated "small, inferior (quality)." And in Aqhat III:i:40 (*ARTU,* p. 250; I Aqht 40 in *UT*) the grieving Dani'ilu casts a spell on the summer heat adjuring the clouds to give the spring rain in the place of the heat. The word is *ysly*, a verb understood as "he curses," which Huehnergard (p. 170), compares to Akkadian *arāru* (referring to *Ugaritica V*, texts 130.[RS 20.149], p. 234; and 137 ii 46', pp. 244–45). De Moor (*ARTU,* p. 250, n. 174), refers to an article he has written ("A Note on CTA 19 [1Aqht]: 1.39–42" in *UF* 6, ed. K. Bergerhof, M. Dietrich, O. Loretz [Neukirchener-Vluyn, W. Germany: Verlag Butzon & Berker Kevelaer, 1974], pp. 495–96) which fully explains why he favors the translation "adjure" over "curse."

[30] For example: Ba'al III:i:5–9 (*ARTU,* p. 30) Ba'al and Naharu exchange curses. Naharu calls for Horonu (god of black magic, master of demons) and Athtartu to break Ba'al's head and for him to go to the place of death, to "fall down at the height of your years." In Ba'al III:iii:13 (*ARTU,* p. 36) 'Athtaru (irrigation god) confronts Yammu who has displaced Ba'al and says "ruin" (a name for the underworld) is Yammu's and threatens to burn down Yammu's palace. In Aqhat I:vi:51–52 (*ARTU,* p. 240; 2 Aqht vi:51–52 in *UT*) 'Anatu maligns (*tlšn*) Aqht and perhaps "curses" him (a word is missing); her intent is certainly malignant. Aqhat III:iii:46–iv:6 (*ARTU,* pp. 259–60) contains three diverse curses against the towns nearest the place where Aqhat was slain (cf. Deut. 21:1–9).

[31] Joseph A. Fitzmyer, "The Padua Aramaic Papyrus Letters," *JNES* 21 (1962), 16; also *TSSI*, vol. 2, pp. 144–45. Dated at 410 B.C., the letter is housed in the Padua, Italy, Civic Museum and is designated Pad I. The letter's precise provenance is unknown.

Chapter 2: Semantic Survey and History of Interpretation 29

[32] Unless otherwise specified, all English Bible texts quoted are from the Revised Standard Version. Several late Aramaic texts are of interest as well: the Genesis Apocryphon (dated 50 B.C. to A.D. 68) uses the root frequently (Joseph A. Fitzmyer, *The Genesis Apocryphon of Qumran Cave 1*, Biblica et Orientalia, vol. 18a, 2nd ed. [Rome: Biblical Institute Press, 1971]). In 21:2–3, Abraham says, "…and I praised the name of God, and I blessed God…" א[ל]הא והללת לשמ אלהא וברכת (pp. 66–67). In 22:15–16, Melchizedek says, "Blessed be Abram by the Most High God" בריכ אברמ לאל עליון (pp. 72–73). Other examples include *KAI* 243.1, 244.1, and 246.1 which are all 1–2 century A.D. inscriptions from Hatra. *KAI* 243 is on a marble statue of a king who is described as blessed of god, בריכ אלהא. *KAI* 244 and 246 were on temple gates and call on the god Ba'alshmyn to bless people who had (apparently) donated money for the building.

[33] See Chapter 4. It also occurs a few times in the first Arslan Tash inscription whose language, whether Phoenician, Aramaic, or other, is debated (see the discussion in Chapter 4).

[34] Should the language of this inscription be called Aramaic? Though generally agreeing that the language is most closely related to Aramaic, scholars are divided whether it should be called Ammonite (Frank M. Cross, "Notes on the Ammonite Inscription from Tell Sîrān," *BASOR* 212 [1973], 13–14), "Gileadite" (P. Kyle McCarter, "The Balaam Texts from Deir 'Allā: The First Combination," BASOR 239 [1980], 50), or an Aramaic dialect with Canaanite isoglosses—unsurprisingly because of neighbors with Canaanite family languages on three sides (Stephen A. Kaufman, "Review Article: The Aramaic Texts from Deir 'Allā," *BASOR* 239 [1980], 73). Because of the general consensus on the origin of the language the root קבב is included here. For a more recent and fuller treatment of this subject see the articles by P. K. McCarter, "The Dialect of the Deir 'Alla Texts"; D. Pardee, "Response: The Linguistic Classification of the Deir 'Alla Text Written on Plaster"; and J. C. Greenfield, "Philological Observations on the Deir 'Alla Inscription"; among others in J. Hoftijzer and G. van der Kooij, eds. *The Balaam Text from Deir 'Alla Reevaluated: Proceedings of the International Symposium Held at Leiden 21-24 August 1989* (Leiden: E.J. Brill, 1991).

In addition, the root לוט/ליט does occur in a couple of late texts: the *Aramaic Proverbs of Aḥiqar* line 151 (A. Cowley, *Aramaic Papyri of the Fifth Century B. C.* [1923; rpt. Osnabrück: Otto Zeller, 1967], p. 225), and in an Aramaic Gnostic text (André Dupont-Sommer, "Le texte araméen gnostique" in *La doctrine gnostique de la lettre "waw" d'apres une lamelle araméene inédite*, Paris, 1946), and *DISO*, p. 136. The root is also found in post-Biblical Hebrew, Gustav H. Dalman, *Aramäisch-Neuhebräisches Handwörterbuch zu Targum, Talmud und Midrasch* (1938; rpt. Hildesheim, W. Germany: Georg Olms Verlag, 1987), p. 215.

[35] So Scharbert ("ברך," p. 281), it may also be read as "the official, steward of Ba'al" (so, W. Randall Garr, *Dialect Geography of Syria-Palestine, 1000–586 B.C.E.* [Philadelphia: University of Pennsylvania, 1985], p. 130). See the discussion in Chapter 3.

[36] *KAI* 10:8; *TSSI*, vol. 3, pp. 94–95. The 341 B.C. inscription from Kition calls for blessing on an altar (?) (*KAI* 32). The 391 B.C. votive inscription from Idalion calls for the god Resheph to bless the giver of a golden mace (*KAI* 38; *TSSI*, vol. 3, pp. 132–33). A 389 B.C. votive inscription from Idalion calls for the blessing of a faithful devotee (*KAI* 39). Further Phoenician and Punic inscriptions down to the first century B.C. call upon Isis, Ashtart, and other unspecified gods for similar blessings (Phoenician: *KAI* 41, 363 B.C.; *KAI* 40, 255 B.C.; *KAI* 58, third century B.C.; *KAI* 47, second century B.C.; *KAI* 48, first century B.C.; Punic: *KAI* 78, third century B.C.; *KAI* 63, third to second century B.C.). There is an extensive group of similar Punic inscriptions from Carthage dating from the early centuries A.D.

[37] Donner and Röllig translate it "verfluchte Brut" but with question marks (*KAI*, vol. 3, pp. 152–53).

[38] Stanley Gevirtz, "West-Semitic Curses and the Problem of the Origins of Hebrew Law," *VT* 11 (1961), 152, n. 7.

Quite recently, both the verb קב and the noun קבת have been found in a Phoenician inscription from southern Turkey, ancient Cilicia (P. G. Mosca and J. Russell, "A Phoenician Inscription from Cebel Ires Daği in Rough Cilicia," *EpigAnat* 9 [1987], 1–27). This inscription is from outside the target area of this study though within the target time. The inscription in question was found in 1980 at the site of Cebel Ires Daği and is dated between 625–600 B.C. It seems to be a record of various land bequests, several (and seemingly unrelated) bequests are mentioned involving several individuals. The section of particular interest is found in lines 5A through 7A:

(5 ואפ.בעל.כר.ישב.בנ.וקב.מחש.קבת.אדרת

(6 .לבל.גזלי.אדמ.שד.אמ.כרמ.בד.שפח.כלש.בכל.אש.יתנ

(7 .ל.מחש

The text is translated in the *editio princeps* thus: "And furthermore, B'L KR he settled (dwelt) in it, and MTŠ cursed a mighty curse so that no one might wrongfully seize it—field or vineyard—from the possession of the family of KLŠ, of all which MTŠ had given to him."

[39] Gibson says these texts seem to be a mixture of Phoenician and Aramaic perhaps purposely mixed to add to their magical efficacy (*TSSI*, vol. 3, pp. 79–80).

[40] Brichto, *The Problem of "Curse,"* p. 40.

[41] See Chapter 4, "Other Words for Curse" for a treatment of the Arslan Tash inscription.

[42] This vague number is because of the unspecified number of occurences of various forms of ארר found at Khirbet Beit Lei. See the discussion in Chapter 4.

[43] See the discusson of the inscription in Chapter 2 for the debate about the language of the inscription.

[44] BDB, pp. 138–39.

[45] Ibid., p. 140.

[46] Christopher W. Mitchell, *The Meaning of BRK "To Bless" in the Old Testament*, SBL Dissertation Series, vol. 95 (Atlanta: Scholars Press, 1987), p. 185, Table 1.

[47] Ibid. Note that KB subscribes to some of the older ideas about blessing. It designates the thrust of the Pi'el as, "gift someone with fortunate power, declare someone to be gifted with fortunate power—declare God to be the source of fortunate power, wish someone to be gifted with fortunate power." (pp. 153–54)

[48] Mitchell, *The Meaning of BRK*, p. 185, Table 1. The root also occurre in various place names (BDB, p. 139).

[49] BDB, p. 140; KB, pp. 154–55. Also the names ברכאל, ברכיה, ברכיהו, and יברכיהו.

[50] Mitchell, *The Meaning of BRK*, p. 149.

[51] Scharbert, "ברך," pp. 288–93. When verse numbers are given as here "68:27(26)" the first verse number refers to the Hebrew text, the number in parenthesis to the English translation.

[52] Scharbert, "ברך," p. 295.

[53] Ibid., pp. 297–98.

[54] Mitchell, *The Meaning of BRK*, p. 10.

[55] BDB, p. 139.

[56] Mitchell, *The Meaning of BRK*, p. 94.

[57] KB, pp. 89–90.

[58] BDB, p. 76.

[59] Gevirtz, "West Semitic Curses," p. 151, n. 2. This is still the case with the published materials thirty years later.

[60] KB, p. 840.

[61] Ibid., pp. 840–41.

[62] BDB, p. 46.

Chapter 2: Semantic Survey and History of Interpretation 31

[63] Ibid.

[64] Josef Scharbert, "אלה," *Theological Dictionary of the Old Testament*, vol. 1, rev., ed. G. J. Botterweck and H. Ringgren, trans. J. T. Willis (Grand Rapids: Wm. B. Eerdmans, 1975), p. 264.

[65] BDB, p. 866. Greenfield says that all sure examples of קבב in BH are from the Bala'am stories (Num. 22-24) which he calls "a repository of dialect words" ("Some Phoenician Words," *Semitica* 38 [1990], 157).

[66] For example, it is used to protect a Neo-Assyrian inscription (*CAD*, A vol. 1, part 1, p. 235) and in Hebrew to protect the En gedi inscription (treated in "ברך and ארר" in Chapter 5).

[67] These may sometimes be the intent in the Hebrew inscriptions as well though the contexts are not always clear.

[68] Except for the appearance of the root קבב, verb and noun, in the recently discovered Cebel Ires Daği inscription, see note 38 for this chapter.

[69] It is interesting to note that the biblical verb בָּכַר ("bear early, first") and the noun בכר ("first born") from the same root are simple rearrangments (metathesis?) of the radicals in ברך. The physical similarity of the words and the fact that children were one of the choicest of blessings suggests that this connection is more than chance.

[70] These questions are modified versions of those asked by Mitchell (*The Meaning of BRK*, p. 17). The use of the first six scholars below was suggested by his study of blessing but where possible, the views of these same scholars about curse have been incorporated and the ideas of others have been added as well as a summary of the contribution of Mitchell himself.

[71] Johannes Pedersen, *Israel: Its Life and Culture*, 4 vols in 2 (Copenhagen: Branner Og Korch, 1926), vol. 1, p. 182.

[72] Ibid., p. 212.

[73] Ibid., pp. 182–92.

[74] Ibid., p. 209.

[75] Ibid., p. 200. Dr. E. W. Nicholson is to be thanked for pointing out that Pedersen draws many of his ideas about ancient understanding of the "soul" from Vilhelm Grönbech whose ideas may be found in his work: *The Culture of the Teutons*, 3 vols., trans. W. Worster (Copenhagen: Jespersen Og Pios Forlag), 1931. Grönbech's highly readable work, though very romantic, has extensive chapters on the ancient idea of the soul (chapters VII "Life and Soul," IX "The Soul of Man," X "The Soul of Man is the Soul of the Clan"), ideas which he does not hesitate to attribute to other ancient peoples as well, including the Hebrews (vol. 1, p. 284).

[76] Pedersen, *Israel*, vol. 1, p. 204.

[77] Ibid., p. 200.

[78] Ibid.

[79] Ibid.

[80] Ibid., p. 437.

[81] Ibid., p. 441.

[82] Ibid., p. 437.

[83] Ibid., p. 441.

[84] Ibid., p. 442.

[85] Ibid.

[86] Ibid., p. 451.

[87] Ibid., p. 452.

[88] Mitchell, *The Meaning of BRK*, p. 19.

[89] Sigmund Mowinckel, *The Psalms in Israel's Worship*, trans. D. R. Ap-Thomas (Oxford: Basil Blackwell, 1962), vol. 1, pp. 44–45.

[90] Ibid., p. 46.

[91] Ibid.

[92] Ibid.

[93] Mowinckel, *The Psalms*, vol. 2, p. 48.

[94] Ibid.

[95] Ibid.

[96] Ibid., pp. 51–52.

[97] Ibid., vol. 1, pp. 236–37.

[98] J. Hempel,"Die israelitische Anschauungen von Segen und Fluch im Lichte alt-orientalisher Parallelen," *BZAW* 81 (1961), 30–31. (This was originally published in *ZDMG* 79 [1925], 20–110).

[99] Ibid., pp. 58–61.

[100] Ibid., pp. 96–97.

[101] Ibid., pp. 31–35.

[102] Ibid., pp. 67–100.

[103] Ibid., pp. 100–113. Claus Westermann says that Hempel's three stages actually correspond to Mowinckel's two because Mowinckel says ethical monotheism developed within the cult (*Blessing in the Bible and the Life of the Church*, trans. Keith Crim [Philadelphia: Fortress Press, 1978], pp. 22–23).

[104] Mitchell, *The Meaning of BRK*, p. 22.

[105] Westermann, *Blessing in the Bible*, p. 5.

[106] Ibid., p. 54.

[107] Ibid., p. 28.

[108] Ibid., pp. 56–58.

[109] Ibid., p. 58.

[110] Ibid., p. 23, n. 12.

[111] Ibid.

[112] Ibid. Though, technically, Yahweh is never said to curse in the inscriptions (see the possible exceptions of Lachish 5 in the treatment by Torczyner under "ארר" in Chapter 4, and the treatment of the Khirbet el-Qôm inscription by Dever in Chapter 5, each of these readings are questionable and held only by these scholars), ארר does occur in six definite examples (Siloam Tomb; Khirbet Beit Lei 4, 5, 6, 7; and En Gedi; all are discussed in Chapter 4), all of which are Hebrew.

[113] G. Wehmeier, "Deliverance and Blessing in the Old and New Testaments," *IJT* 20 (1971), 32–33. For his fuller treatment see his *Der Segen im Alten Testament*, Theologisches Disertationen, no. 6 (Basel: Friedrich Reinhardt Kommissionsverlag, 1970).

[114] Wehmeier, "Deliverance and Blessing," p. 35.

[115] Ibid., p. 34.

[116] Ibid., pp. 34–35.

[117] Ibid., pp. 37–39.

[118] Ibid., p. 39, n. 25.

[119] Ibid., p. 40.

[120] Scharbert, "ברך," p. 286.

[121] Ibid., p. 283.

[122] Ibid.

[123] Ibid., pp. 304–06.

[124] Ibid., p. 293.

[125] Ibid., p. 303.

[126] Ibid., p. 285.

[127] Ibid., p. 288.

[128] J. Scharbert, "ארר," *Theological Dictionary of the Old Testament*, vol. 1, rev., ed. G. J. Botterweck and H. Ringgren, trans. J. T. Willis (Grand Rapids: Eerdmans, 1975), p. 412. This applies also to the case of Num. 5 "the water of bitterness that brings the curse"; though Scharbert says the magical origins are "quite clear," they have been weakened by the fact that the ritual is ordered by Yahweh who is seen as the real judge of the woman and the one who causes the misfortune named in the curse (p. 413).

[129] Ibid., p. 416.

[130] Scharbert says that אלה is related to word magic: attributing "to the spoken and written word the power to actually bring about that which is stated in the word, whether it be good or evil" ("אלה," p. 265). But usually the curses designated אלה were prayers to Yahweh to bring the curse about. Further, the words of the priest in Num. 5:21 "leave no doubt that the Israelites knew they could not force Yahweh to act by word and ritual. This understanding is made even clearer in 1 Kgs. 8:31∥2 Chr. 6:22." He furthers his argument by saying that only after אלה was no longer understood as magic could it be associated with the covenant idea. Yahweh's covenant with Israel did have curses associated with it but they went into effect only at Yahweh's discretion and could be stopped by him at any time (p. 266).

[131] Scharbert, "ארר," pp. 417–18.

[132] Brichto, *The Problem of "Curse*," JBL Monograph Series, vol. 13 (1963; rpt. Philadelphia: Society of Biblical Literature, 1968), p. 111.

[133] Ibid., p. 114.

[134] Ibid., pp. 114–15.

[135] Brichto suggests that Ps. 24:4 (and perhaps the prohibition of Ex. 20:7) was an attempt to put an end to the abuse of curses which call upon Yahweh (*The Problem of "Curse*," pp. 59–67). Cf. Job 31:29–30.

[136] Ibid., p. 218.

[137] Anthony C. Thistleton,"The Supposed Power of Words in the Biblical Writings," *JTS* 25 (1974), 283–99. The complete citation for the former is *BZAW* 64 (1934), 103–7, and the place of publication and the date of the later is Leipzig, 1938.

[138] Ibid., pp. 290–91.

[139] Ibid., pp. 293–94.

[140] O. Procksch, "λεγω," *Theological Dictionary of the New Testament*, vol. 4, ed. G. Kittel, trans. G. W. Bromiley (Grand Rapids: Wm. B. Eerdmans, 1967), pp. 92–93.

[141] G. von Rad, *Old Testament Theology*, vol. 2, trans. D. M. G. Stalker (New York: Harper

& Row, 1965), p. 80.

[142] Also compare Pedersen, *Israel*, vol. 1, pp. 182, 195.

[143] Thistleton, "The Supposed Power of Words," p. 298.

[144] Ibid., pp. 297–98.

[145] See the critique of Mitchell below.

[146] Mitchell, *The Meaning of BRK*, p. 165.

[147] Ibid.

[148] Ibid., pp. 146–47.

[149] Ibid., p. 149.

[150] Ibid., p. 168.

[151] Ibid., pp. 106–07.

[152] Ibid., p. 168.

[153] Ibid., p. 169.

[154] Ibid., p. 165.

[155] Ibid., p. 97.

[156] Ibid., p. 174.

[157] Ibid., pp. 96–98.

[158] Compare parts of Olyan's critique of Tigay (Chapter 1, note 3).

[159] The combination of these two ways also occurs.

Chapter 3

BLESSING

This first chapter dealing with the inscriptions will treat those that either have a form of ברכ or contain some sort of blessing without a curse. The first section will examine inscriptions where ברכ and שלמ occur together. The second section will be a treatment of inscriptions having ברכ without שלמ. The third will present inscriptions where a form of ברכ is found without a context. The fourth section will discuss the inscriptions where שלמ occurs without ברכ. The last section will treat "Substantive Blessings," blessings which do not use a form of ברכ.

ברכ and שלמ

Several inscriptions from the target area and time use both ברכ and שלמ in greetings and religious texts. These inscriptions are presented below in the following order. Inscriptions which include a DN will be considered first, beginning with those including Yahweh and followed by one with another DN. Last in this section will be a text which includes both words but no DN.

With the DN Yahweh

Of the texts which contain both ברכ and שלמ without a curse, most include the DN Yahweh (Arad 16, 21, 40, Kuntillet ʿAjrud 4, Ketef Hinnom 1, 2). These, predominantly Judean inscriptions, will be treated first.

Arad 16.1–3 Tel Arad is located approximately 25 miles south of Jerusalem. The Iron Age citadel of Arad (where the Arad letters were found) is located on a mound on the edge of the Bronze Age site. Arad letters 16 and 21 were all found in the stratum VI level, 605–595 B.C.[1]

1) אחכ.חנניהו.שלח.לשל
2) מ.אלישב.ולשלמביתכברֿ
3) כתבֿליהוה.[2]

Aharoni's division of the first two and a half lines is as follows:

1) אחכ חנניהו שלח לשל
2) מ אלישב ולשלמ ביתכ בר
3) כתכ ליהוה[3]

And he translates these lines, "Your brother Ḥananiahu greets Eliashib and your house, I have said a blessing to the Lord for you."[4]

Second Samuel 8:10 is a strong parallel, וישלח תעי את־יורם־בנו אל־ המלך־דוד לשאל־לו לשלום ולברכו "And Toi sent Joram his son to King David to ask after his welfare and to bless him."[5] Another interesting parallel comes from one of the Elephantine documents אל.מרי.מיכיה.עבדכ גדל.שלמ.וחיו. שלחתלכ.ברכתכ.ליהה.ולחנ[וב] "A mon seigneur Michée ton serviteur Gadol. Salut et vie je t'envoie. Je te bénis par Yahô et par Ḥnu[b]."[6]

In 2 Sam. 18:28, שלום is used to indicate completion of a successful military endeavor. This is of interest in comparison to the military *sitz im leben* of the Arad inscriptions and points out that שלמ is not used only to speak of someone's physical condition.[7]

Aharoni says that "your brother" is an address not necessarily reserved for a relative but also used for addressing another of the same rank (cf. Num. 20:14 and 1 Kgs. 9:13).[8] Pardee disagrees. He thinks that this is correspondence between family members, noting that where this phrase occurs elsewhere (the letters from Hermopolis and Saqqara, *KAI* 50) there is a family relation.[9] The letter follows a typical opening formula "X inquires after the welfare of Y, I have blessed you to the Lord." Pardee says that the formula of

kinship term + *šlḥ* + *lšlm* PN + *wšlm bytk* + *brktk lyhwh*

is a form of address used between family members.[10] Furthermore, שלח שלמ occurs in Hebrew only in these letters from Arad (16, 21, 40) and Wadi Murraba'ât.[11] The extension of greetings to "your house" is not frequently found in letters.[12] However, in light of the BH examples, Pardee's contention that the letter must be between family members does not seem sustainable.

Foresti thinks that the phrase "PN *šlḥ šlm* PN" is an abbreviation of the similar phrase in the Hermopolis letters PN *šlḥ lš'l lšlm* PN (also compare 1 Sam. 25:5; 2 Sam. 8:10//1 Chr. 18:10) and has a similar interpretation "PN sent (the letter) for the welfare of PN."[13] He also draws a firm distinction between the inscriptional expression ברכתכ ליהוה and the biblical ברוך אתה ליהוה saying that the two are independent of each other; the first being active and best translated "before Yahweh" and the latter being passive and best translated "by Yahweh."[14] He does not say why this difference is important. Some commentators might draw the distinction based on ideas of how blessing and curse are thought to operate in the Bible, i.e., that some people have the power to bless in themselves (thus active "I bless you...") but that texts implying this power have been adapted to reflect more orthodox ideas (thus "blessed be thou to Yahweh").[15] But Foresti does not give this reason and, indeed, gives a biblical example quite close to the inscription's form, Gen. 27:7 ואברככה לפני יהוה "And let me bless you before the Lord."[16] There is insufficient evidence to distinguish between these two phrases on the basis of the active or passive form of the verb or the use of different forms of the pronoun. The two phrases are best interpreted as two ways of saying the same thing; in both cases Yahweh is the one called upon for blessing, in each case the

Chapter 3: Blessing

recipient is being called to Yahweh's attention for blessing.[17]

Aharoni thinks that the use of the historic spelling ברכתכ rather than ברכתיכ is another mark of the formulaic character of ברכתכ ליהוה;[18] this may seem remarkable in light of the fuller spelling שלחתי in this same inscription (line 4), but medial *matres lectionis* are much rarer at this early date than final *matres lectionis*. ברכתכ is a perfect (probably Pi'el) and should be interpreted as an epistolary perfect, "I hereby bless/pronounce a blessing in favor of someone to a deity."[19] The formula *brk* PN (or second person pronoun) *l*-DN also occurs in the Edomite Ḥorvat 'Uza inscription and the post-period letters from Saqqara and Hermopolis.[20]

The force of ברכ in this and other target letters is debatable. Is there in each case a sincere, conscious wish on the part of the sender for the welfare of the recipient? It will be argued that the uses of ברכ in the inscriptions was a sort of short-hand method for asking numerous specific blessings (e.g., offspring, land, food, safety, etc.). Did the letter writers wish these blessings from the deities or are these phrases so formulaic as to have lost any religious force? Do they really tell us anything about the religious ideas of blessing? Based on an analogy with the present day letter address "dear" and its ubiquitous, superficial usage, it may be argued that ברכ has little value. The question remains, though, how much ancient uage can be compared to such modern conventions. McCarter suggests the following

> I think it would be fair to distinguish at least three degrees of blessing: (1) blessings like those found in literary texts, where the idea of blessing is self-consciously entertained by the author, (2) formulaic blessings like those found in treaty texts and on gravestones, and (3) simple greetings, such as are routinely found at the beginnings of letters. It seems to me that we learn most about the religious idea of blessing from type (1) and least from type (3).[21]

This summary is certainly true. Unfortunately for this study, the types occur in reverse order to their value for discovering religious significance.

Arad 21.1–4 .[22]

1) בנכ.יהוכל.שלח.לשלמ.גדליהו[ן
2) אליאר.ולשלמ.ביתכ.ברכתכל[
3) ה.ועת.הנ.עשה.אדני.[ ].
4) ישלמ.יהוה.לאדנ[ן

Aharoni translates, "Your son Yehocal sends greetings to Gedalyahu son of Elyair and greets your house, I have blessed you to the Lord, and now: Behold my master has done [ ] the Lord give my master his due..."[23]

While the use of a verb form of שלמ (line 4) is not found elsewhere in the Hebrew letters, it does occur numerous times in Ugaritic letters.[24] Pardee translates line 4, "may Yahweh restore to my lord."[25] Though the literal translation of line 4, "may Yahweh give peace to my lord," is attractive, it seems probable that the translation of Aharoni or Pardee is correct. It is not likely that the substance of the letter would be introduced ("And now") and then further greetings given.[26] This inscription does, nevertheless, contain both ברכ and שלמ (lines 1–2) and they appear in the same formula found in Arad 16.

Arad 40.1–3. This inscription was found in the stratum VIII level (destroyed 701 B.C.) of the Iron Age Citadel at Arad.[27]

1) [ונח]ם בנכמ.גמר[
2)]מיהו.שלח[
3)]ה [מלכיהו.ברכתכ]

Aharoni translates these lines, "Your son Gemar[yahu] and Neḥamyahu gre[et] Malkiyahu; I have blessed [you to the Lor]d."[28]

Aharoni's reconstructions are based on the examples of letters 16 and 21, supplying [מ.לשלמ] at the end of line 2 and [ליהו.] in the gap on line 3. He restores line 2 thus because he points [שלחמ] as an active particle masculine plural, pointing consistent with his conjectural pointing of letters 16 and 21.[29] Pardee prefers to restore it as [שלח.ו.שלמ] treating the verb as a perfect.[30] There is a consistent confusion of the plural forms: in line 1 בנכמ should be בנכ (or בנכ), שלח in line 2 should be שלחו (it might have actually been correct before the text was broken[31]), and in line 3 ברכתכ should be ברכנכ. In each case the changes are needed because two people rather than one are sending the letter. While the first of these bespeaks true confusion on the part of the scribe—he has made the pronominal suffix rather than the noun plural—and the second is uncertain, the third might be attributed to standard formulaic letter form elsewhere found in the singular.[32]

Kuntillet ʻAjrud 4. This mid-ninth to mid-eighth century inscription is found on the second pithos (Pithos B) near the top of the vessel not far from a handle.[33] The inscription is written on several short lines in a narrow column and is partially over-written by sections of three abecedaries.[34]

[אמר]
1) [א]אמריו
2) מר ל אד[נ]י[
3) השלמ.א[ת]
4) ברכתכ.ל[י]
5) הוה [תמנ]
6) ולאשרתה.יב
7) רכ.וישמרכ
8) ויהי.עמ.אד[נ]

Chapter 3: Blessing

(9 י
(10 כ ³⁵

Hadley translates the inscription, "Amaryau says: say to my lord: Is it well with you? I bless you by Yahweh of Teman and by his asherah. May he bless you and keep you and be with my lord."³⁶

The expression השלם pointed הֲשָׁלוֹם is quite common in biblical literature, e.g., Gen. 29:6; 43:27; 2 Sam. 18:32; 20:9; 2 Kgs. 4:26; 5:21. It appears to be a fairly standard greeting used to inquire about someone the speaker knows, whether that person be a relative, a special aquaintance, or fellow military officer, etc. Especially interesting is 2 Kgs. 9:17–19, 22, 31 הֲשָׁלוֹם "is it peace?" or "is it well?" In this passage the questioners were asking a similar question, the context is like that of the Lachish letters where the sender wished for his superior to hear "peace" in a war context.³⁷

Wiseman has recently argued convincingly (at least as regards communications between leaders or groups) that in the Bible asking the peace of someone (השלום or שאל לשלום) often caries the "nuances of diplomatic usage" found in Akkadian documents.³⁸ By comparing the language and personnel in the Akkadian and biblical materials, Wiseman has determined that שלם often indicates the attitude of non-hostility necessary for persons or groups to develop relationships necessary for treaties or covenants.³⁹

Interesting parallels are found in several of the Aramaic papyri from Elephantine. The solicitude of the sender after the health of the recipient is expressed by the phrase שלם מראן אלה שמיא ישאל שגיא בכל עדן עדן "The health of your lordship may the God of Heaven seek after at all times."⁴⁰ The Saqqara papyrus (*KAI* 50:2–3) uses a similar formula to ask after the recipient's health, after which request the sender gives assurance of her own health and follows with a blessing very similar to Kuntillet 'Ajrud 4. ושלמ את אפ אנכ שלמ.ברכתכ. לבעל צפנ.ולכל אל.תחפנחס.יפעלכ.שלמ "And are you in good health? I, too, am in good health. I bless you by Ba'al-Ṣaphon and by all the gods of Taḥpanḥes. May they make you to be in good health!"⁴¹

Meshel speaks of the Phoenician charateristics of the script of the Kuntillet 'Ajrud inscriptions and suggests that they may date from the time of the "half-Phoenician" queen Athaliah notorious for her encouragment of the worship of Ba'al and Ashtarte. A Phoenician-influenced worship site in southern-most Judean territory is intriguing.

Peckham emphasizes the high degree of religious sharing between Israel and the Phoenicians. He also points out areas where Israel appears to have deliberately broken with Phoenician religious traditions, either claiming other deities' characteristics as the true prerogative of Yahweh or repudiating those characteristics which could not be correctly appropriated for Yahweh. He notes several examples of these contacts between the peoples both of sharing and repudiation, including: the use of מצבת;⁴² and kingship ideas. In contrasting the ideas of kingship, he notes that the position of the king in the Phoenician texts, that of provider of food, clothing, secure boundaries, and peace (Kilamuwa 1 and Azitawada texts), is the reserve of Yahweh in the biblical texts (Deut. 8:3–4; 10:8).[43] Likewise, he notes what may be the biblical writers' parody of the Phoenician concern for the tranquility of the dead (cf. for example Isa. 14:3–23 where the prophet, in the form of a dirge for the king of Babylon actually ridicules the elements of beatific afterlife venerated in Phoenician inscriptions).[44] He also briefly compares beliefs about the cycle of life and death;[45] and the cult of Astarte.[46]

Ketef Hinnom 1 and 2. In the summer of 1979, two small amulets were found in a burial cave in Jerusalem's Hinnom Valley. The late seventh century amulets are of almost pure silver and were tightly rolled.[47] After being carefully unrolled, the scrolls each eventually revealed forms of the Priestly Blessing contained in Num. 6:24–26. A full *editio princeps* of the amulets has yet to be published, but some photographs which at least give an idea of the amulets' appearance are available along with a drawing of the smaller of the two.[48] The larger of the plaques shows seventeen of what are thought to have been an original nineteen lines. Ninety-three letters are discernible; seventy-one can be read with a high degree of probability.[49] The smaller of the two which is described as "less well-preserved than the larger one," probably also contained nineteen lines of writing.[50] This, with very incomplete readings of the the two amulets,[51] was all of the information available on the inscriptions until a recent article by Ada Yardeni. Yardeni, originally responsible for most of the deciphering of the plaques, now offers more complete readings of both with comparisons to biblical passages. Her readings and comparisons are included below (Table 1).[52]

The first plaque has only the first verse and the beginning of the second verse of the Numbers blessing but may have originally been virtually identical to it. The second plaque has an omission at the end of the second verse and the beginning of the third verse of the MT that may be explained as homoioteleuton between the occurrences of אליך or as a shorter version of the blessing.[53] That other, shorter, versions of the Priestly Blessing were used in Israel's worship is shown by Psalm 67:2: אלהים יחננו ויברכנו יאר פניו אתנו.[54] Gerhard Wehmeier notes (regarding the Numbers passages) that the priestly pronouncement of Yahweh three times is not necessary grammatically. It is repeated in order to emphasize that Yahweh himself is the blesser, not the priest of himself and not the ceremony.[55]

Yardeni notes that the "shining face" of God and God "lifting eyes upon" someone to show favor are also known from second millennium Ugarit and Mesopotamia.[56] She speaks both of the power of illocutionary utterance and magical word but, I think, does not draw a sharp enough distinction between them. Yardeni writes, "The priestly blessing belongs to the group of strictly formulated prayers which are part of certain ceremonies. In prayers of this kind the formula had an *automatic* validity as securing the blessing."[57] She gives no

Chapter 3: Blessing   41

evidence for this position, however, except for late (Mishnaic) examples of magic amulets. She does not satisfactorily show that these centuries earlier silver amulets also were considered to have magical force.[58] Two other objections may be made to this interpretation: 1) the fact that it was Yahweh who was specifically called upon to bless and who may not be forced into blessing;[59] and 2) that this kind of magical talisman is specifically prohibited in the Bible (see below). Whether or not this biblical text was used in an unacceptable way by the owners of the plaques cannot, of course, be decided.[60]

Table 1

Comparison of the Plaques with the Massoretic Text

Plaque 1	Plaque 2	MT
1) [ ].הׄוׄ		
2) [ ]		
3) [ ]		
4) [ ] הברית [ו]		שמר הברית
5) [ה]חסד לאה		והחסד לאהביו
6) [ ] ב שמרי|ן		ולשמרי מצותיו
7) [ ]		
8) ה על משׄ|כ]		
9) בה [ ] א/המכל		
10) [ ] ומהר/דע	1) [ ] א/ת/זבו ו	
11) כי בו גאל	2) [ ]לניהו	
12) ה כי יהוה	3) [א/ד/ר] [ ] יה]	
13) [י]שיבנו [ ]	4) [ ] ו ע/טר/ד]	
14) [ ]ור יבר]כ]	5) [ ] יבר]כ]	יברכך
15) כ יהוה ו	6) כ יהוה ו	יהוה
16) [י]שמרכ [י]	7) [י]שמרכ	וישמרך
17) [א]ר יהוה	8) יאר יה	יאר יהוה
18) [פ]נ]יו	9) [ו]ה פניו	פניו
19) [lines missing]	10) [א]לי]כ ו]י]	אליך ויחנך ישא יהוה פניו אליך
	11) שמ לכ ש	וישם לך
	12) ל]מׄ [	שלום
	13) [ ]	
	14) [ ]	
	15) [מ]	
	16) [דל]	
	17) [ ] ל]ל [	

As the plaques stand, the introductory, general request for blessing using ברך is followed by two other rather general requests. The request that Yahweh "keep" (שמר) someone is surely the blessing of safety from enemies as well as, perhaps, the petition for necessary provisions. Based on biblical parallels, the supplication that Yahweh "shine" (יאר) his face upon someone indicates his favor. The verb אור occurs in the Hiph'il eight times (Num. 6:25; Pss. 31:17[16]; 67:2[1]; 80:4[3], 8[7], 20[19]; 119:135; and Dan. 9:17 also cf. Ps. 13:4-5[3-4]). Of these, only Num. 6:25; Pss. 67:2 and 119:135 are not requests that Yahweh deliver the speaker from enemies (the Daniel passage is a request that Yahweh have regard for his destroyed sanctuary). Predominantly, the verb is used to request Yahweh's favorable attentions and subsequent deliverance from enemies. The three occurrences of the verb in Ps. 80 are particularly interesting. Three times the same refrain ends sections of the psalm and two of the sections particularly refer adversaries: אלהים השיבנו והאר פניך ונושעה "Restore us, O God; let thy face shine that we may be saved." These verses are significant because the word ישיבנו appears to occur in line 13 of plaque 1 as well.

The first of the passages in the MT column of the table is a proposed parallel to plaque 1:4-6 from Deut. 7:9 "...who keeps covenant and steadfast love with those who love him and keep his commandments..." Yardeni does not comment on these words in the article nor does she say anything about the other words which do not parallel the Priestly Blessing. However, it appears that the phrase עֹל מֹשׁ[כ]בה "upon his resting place/bed" occurs in lines 8-9. משכב is also used to refer to a final resting place in Is. 57:2; Ezek. 32:25; 2 Chr. 16:14 and the early fifth century Phoenician Tabnit inscription, line 8.[61] Based on her readings of 1:11-13—{ כי בו גאלה כי יהוה [י]שיבנו]—the possibilities are intriguing. Even a perfunctory attempt at translation seems to show that Yahweh is expected to come as the redeemer who will bring about restoration (cf. Gen. 48:16; Hos. 13:14; Jer. 50:34; Pss. 19:14[15]; 69:19; 72:14; 106:9[10]).[62] Lines 1-4 of plaque 2 are much more difficult to read.

There are several interesting parallels to this text. Certain Old Babylonian letters found at Tell al Rimah contain the wish that gods guard the recipient's life, ^{d}UTU ù $^{d}AMAR.UTU$ liballiṭúki "May Šamaš and Marduk grant you long life," and the wish that news of the recipient's welfare not be denied the sender šulumki ù šulum LÚ.TUR-ri la iparás "May news of you and news of the workmen not be denied me."[63] In the first parallel, the wish for the recipient to have long life is substantially a blessing and, what is so prosaically translated as "news," is better translated "your welfare" šulumki. Some Akkadian letters from Ugarit contain the phrase lú šulmu ana muḫḫika ilânu^{mes} ana šulmani liṣṣuru^{ru}-ka "May there be peace to you, may the gods guard your health."[64] Ugaritic letters often bear the parallel formula yšlm lk ilm tǵrk tšlm "May there be peace to you, may the gods preserve you and give you peace."[65] An Aramaic letter from late fifth century B.C. Egypt ends with the prayer that אלהיא שלם ישמו לך "the gods

Chapter 3: Blessing

may grant thee peace."[66]

Barkay says that the fact that the text was written on silver shows the special place of honor it held. He compares this to Second Temple law scrolls of Alexandrinians living in Jerusalem which used gold letters for occurrences of "Yahweh."[67] The custom of wearing amulets containing verses from the Pentateuch is known from the time of the Late Roman and Byzantine Samaritans. Roman period lead and bronze amulets have also been found.[68] Barkay suggests that these were like the tefillin described in Exod. 13:9, 16 and Deut. 6:8–9 (perhaps Prov. 1:9 and 6:21 as well) and notes the phrase "tefillin of silver" תפלה זי כסף which occurs in an Aramaic papyrus from Edfu dating from ca. 300 B.C. (*AP* 81:30).[69]

Perhaps these amulets or plaques were used as magical charms and thus condemned by the prophet Isaiah (3:20) and destined for removal with other adornments on the day of judgment. The Arlsan Tash texts contain incantations and were designed to be hung up to protect a room or a dwelling from demonic mischief.[70] Cross and Saley note the occurrence of the word מזזת "doorposts" in line 26 of Arslan Tash 1 and suggest that this text was "a pagan prototype of the *mezuzah*, the Israelite portal inscription."[71] Perhaps though, these amulets with their biblical supplications found near the endangered holy city and dating from close to the end of the kingdom of Judah were rather meant as prayers of the faithful to the God who had repeatedly delivered his people in the past.

With Other DN

The site of Ḥorvat ʿUza (Khirbet Ghazza, ancient Ramoth Negev[72]) is in the Eastern Negev about 5 miles SSW of modern Arad. The site was occupied at three different periods: the seventh to sixth centuries when it was Judean fortress, as well as in the fifth, and the third centuries B.C.[73] The ostracon is complete in four pieces found together in the gatehouse of the fortress. The pieces were found with Hebrew ostraca in a seventh to sixth century Israelite stratum.[74] More recently, the text has been dated to ca. 586 B.C.[75]

1) אמר.למלך.אמר.לבלבל.
2) השלמ.את.והברכתכ
3) לקוס.[76]

Beit-Arieh and Cressen translate these lines, "(Thus) said Lumalak (or <E>limelek): Say to Blbl! Are you well? I bless you by Qaus."[77] They suggest a comparison to 2 Sam 20:9 השלום אתה אחי. Chase's reading of Kuntillet ʿAjrud 4 השלמ את "Is it well with you?" is another valuable parallel.[78] The form of ברכ used is apparently in the Hiphʿil stem unlike Hebrew or Phoenician. The epistolary form is virtually the same as that of Kuntillet ʿAjrud inscriptions 3 and 4. Zwickel suggests that the identification of the word והברכתכ as a Hiphʿil is incorrect and that the ה should be more properly identified as a ה interrogative;

the phrase והברכתך לקוס. then being interpreted as an "ironische Frage."⁷⁹ Zwickel is quite right to point out that Hiphʻil ברכ is not used with this meaning in BH (only once in Hiphʻil and there with the meaning "kneel") and orthographically his interpretation is possible, but it makes little sense in the context and he is unable to give any examples. While it may not be definitely said, with Beit-Arieh and Cresson, that this form "is an Edomite trait,"⁸⁰ that explanation coincides best with the known formulas.⁸¹ The fifth century letter from Saqqara, Egypt (*KAI* 50) also contains a first person blessing calling upon another deity, Baʻal Ṣaphon, to bless an equal.⁸²

With No DN

The ostracon which bears inscription Samaria C1101 is a section of the outer rim of a shallow bowl covered with a red slip and burnished on the outside. It dates from the second half of the eighth century. Birnbaum thinks that each of the three lines was written by a different person.⁸³ The first of the transliterations below is that of Sukenik in the *editio princeps* and the second is that of Albright.

(1 ברכ שלג/מ
(2 ברכ 2 הרד/עמה/הרשבו
(3 ימנהשערמ 10 מ̇ ⁸⁴
―――――――――
(1 ברכ של[מ]
(2 ברכ הפעמ הק̇שב ו]
(3 ימנה שערמ 13⁸⁵

Albright translates the text, "(O) Baruch, greet[ings]! (O) Baruch, now pay attention and [give X son of] Imnah barley, 13 (measures)."⁸⁶

Sukenik suggests בְּרָךְ "blessing" or בָּרוּךְ "blessed" for ברכ and שָׁלֹם "peace" or שָׁלֵם "healthy" for שלמ. Alternatively, he suggests the proper names בָּרָךְ, שָׁלֹם, or שלמ.⁸⁷ Albright's translation takes ברכ as the PN בָּרוּךְ, also the name of several people in the Bible. The reading of מ as the last letter of line 1 is virtually certain.

Albright interprets the ostracon as a letter to an underling ordering him to pay the given amount to the person specified. The second word of the first line is then the briefest of greetings.⁸⁸ Moscati concurs with Albright.⁸⁹ Donner and Röllig translate it "Baruch, zahle [ ] Baruch..."⁹⁰ This may be compared to Arad inscription 21:4]לאדנ.ישלמ.יהוה which Aharoni translates "the Lord will give my master his due," as well as biblical examples 2 Sam. 3:39b; Prov. 25:22b; and Ruth 2:21.⁹¹

Birnbaum translates the first line as "Greetings, Shallum!" and then sees the first word of the next line as "Baruch" (PN). He bases this interpretation on a comparison with Lachish letters 4, 5, 8, and 9 which all begin with a greeting in contrast to letters 2 and 6 which begin with the PN, howbeit preceded by אל.⁹² He thinks it unlikely that the two words are both greetings ("be blessed, peace!")

Chapter 3: Blessing

but allows that both may be PNs, i.e., "Baruch, Shallum."[93]

Lemaire translates the text "Baruk a accompli[." He says that different hands can be distinguished by the writing and that the first line was actually written last by Baruch himself.[94] Gibson translates the text "Baruch (son of) Shallum," an interpretation allowed by other scholars.[95] Two Aramaic letters from Elephantine are of interest at this point, שלם אוריה "Shalom 'Uryah" (76, no. 1, A.1) and [ש]לם אחוטב "Shalom 'Aḥitub" (78, no. 2, A.1) both use שלם at the start of a letter as a greeting and are followed by a PN. In the second case some word or words preceded שלם.[96]

Albright's interpretation is very probably correct. If the note is to an underling, extensive salutations may have been deemed unnecessary (contrast greetings in the Arad and Lachish letters sent from superior to inferior with those sent from inferior to superior or from equal to equal). However, Ruth 2:4 bears an example of a very civil greeting exchanged between a landowner and his workers, indeed initiated by the landowner.[97]

Summary

Arad 16 and Kuntillet ʿAjrud 4 are the clearest examples of ברך and שלם occurring with יהוה in the same inscriptions. In Arad 21, יהוה is partially erased and in Arad 40 שלם is completely missing. שלם is missing from the first of the Ketef Hinnom amulets and both words are partially damaged and restored largely on the basis of a comparison with the blessing of Numbers 6 in the second amulet. The "Edomite" Ḥorvat ʿUza ostracon and the later examples noted (Saqqara, Padua I, and the Hermopolis letters) testify to the popularity of this combination in Edomite, Phoenician, and Aramaic inscriptions (all after the target period except Ḥorvat ʿUza) accompanied by other DNs. ברך and שלם are each also found frequently in biblical greetings. Of the inscriptions found in Israelite territory, Kuntillet ʿAjrud 4 asks blessing of Yahweh and perhaps the deity "Asherah" or a cult image which may have represented a consort for Yahweh.[98] The only example where ברך and שלם might be found without a DN (Samaria C1101) is probably a short greeting preceding or following a PN. The earliest of these texts is Kuntillet ʿAjrud 4 predating Arad 40 by perhaps a century (ca. 800 versus ca. 701 B.C.). The latest texts in this category are Arad 16 and 21 (ca. 595 B.C.). This combination of DN with ברך and שלם is found in Phoenician, Hebrew, and Edomite inscriptions of the target period. ברך in these inscriptions might well be a compact way of expressing all of the more specific blessing ideas seen in the longer inscriptions. The numerous contacts between the Hebrew kingdoms and Phoenicia lend weight to the idea proposed here, that the use of ברך for blessing at this early period is further evidence of Hebrew-Phoenician sharing of religious concepts.[99]

ברכ Alone

ברכ also occurs without שלמ in the inscriptions. This section will present texts where a) ברכ occurs with Yahweh, b) occurs with another DN, and c) appears to occur without a DN.

With Yahweh

ברכ (without שלמ) occurs with Yahweh in the Kuntillet 'Ajrud 5, 3, 2, and Khirbet el-Qôm inscriptions.

Kuntillet 'Ajrud 5. This mid-ninth to mid-eighth century inscription is written on the rim of a massive stone bowl (weighing about five hundred pounds).[100] Meshel thinks that its presence at Kuntillet 'Ajrud is testimony to the sacredness of the site.[101]

לעבדיו בנ עדנה ברכ הא ליהו

The text is translated, "Belonging to 'Obadyahu son of 'Adnah, may he be blessed by Yahweh." This text is probably best interpreted as a dedicatory inscription, the bowl being consecrated for use in the cult center by the man whose name appears thereon.[102] The word ברכ is best interpreted as a passive participle. The phrase ברוך יהוה is not uncommon in the Bible (e.g., 1 Sam. 25:32; 2 Sam. 18:28; 1 Kgs. 1:48; 5:21; 8:15, 56; Zech. 11:5; Pss. 28:6; 31:22; 41:14) but the phrase with a personal pronoun inserted and the ל is much rarer (Judg. 17:2; 1 Sam. 15:13; 23:21; 2 Sam. 2:5; Ruth 2:20; 3:10; Ps. 115:15). Of these, Ruth 2:20 provides an almost perfect match, ברוך הוא ליהוה "blessed be he to the Lord."[103] The other passages provide a form of the second person pronoun (את, אתה, or אתמ) except for Judg. 17:2 which has ברוך בני ליהוה "blessed be my son to the Lord." "Yahweh" is written in the abbreviated form יהו here on the bowl and elsewhere only on Kuntillet 'Ajrud 6.[104]

Kuntillet 'Ajrud 3.[105] This inscription was found on sherds of the reconstructed vessel designated as "the first pithos" or "pithos A." The two pithoi bearing inscriptions also bear numerous drawings. This inscription was written just above and partially overlapping the most famous of the drawings, those of two similarly clothed "Bes-figures" with a seated harpist to their right and in the background. This same pithos also holds drawings of a cow nursing a calf and of some sort of cat-like beast, both to the left of the three figures mentioned above.[106] Beck suggests that the inscriptions are "route prayers" written by caravaneers and other travellers and dedicated to deities in order to secure their protective watch over the wayfarers' journeys.[107] She further concludes that there is no relationship between the drawings and the inscriptions.[108] Like the other inscriptions from this site, this text is dated to the mid-ninth to mid-eighth century.[109]

Chapter 3: Blessing

(1 אמר א[ ]ה[ ]כ[אמר ליהל[לאל] וליועשה ו   ברכת אתכם
(2 ליהוה שמרן ולאשרתה[110]

Hadley translates the inscription, "X says: say to Yehal[lel'l] and to Yo'asah and [to Z]: I bless you by Yahweh of Samaria and by his asherah."[111]

The form ברכת is best identified as a (Pi'el) perfect first common singular and is followed by the second masculine plural form of the personal pronoun. In biblical texts this combination does not occur. However, in two cases the perfect first common singular of the verb does occur with the second person singular pronoun, both times as a pronominal suffix (וברכתיך Gen. 26:24 and Exod. 20:24). In each of these cases God is the subject. Further, when the verb form used is the imperfect first common plural, the subject is usually either God blessing men or men blessing God; only rarely is it used of men blessing men (Gen. 27:7, 33; 48:9) and then never with ליהוה.

Kuntillet 'Ajrud 2. This black ink on plaster inscription "probably dropped off the west wall of the bench room."[112] The inscription is dated from the mid-ninth to mid-eighth century B.C.[113]

(1 [ברכ יממ וישבעו]
(2 [היטב יהוה]

Meshel translates these lines, "blessed be their day...God favoured"[114]

Another translation may be "blessed be the days and may they swear...Yahweh will deal well..."[115] For a biblical passage concerning a day being blessed, see Exod. 20:11 על־כן ברך יהוה את־יום השבת ויקדשהו "therefore the Lord blessed the sabbath day and sanctified it," or not blessed, Jer. 20:14b־אשר יום ילדתני אמי אל־יהי ברוך: "the day which my mother bore me, let it not be blessed." Perhaps the last word of the first line is best read with a ש thus וישבעו "and may they be full." Though the parallel to this inscription is not precise, being "filled with days" is a biblical way to describe a long and good life: Gen. 35:29; Job 42:17; 1 Chr. 23:1; 29:28. It is also used in other blessings, Jer. 31:14b ועמי את־טובי ישבעו נאם־יהוה "and my people shall be satisfied with my goodness, says the Lord" and Deut. 33:23 נפתלי שבע רצון ומלא ברכת יהוה "Naphtali shall be satisfied with favor and filled with the blessing of the Lord."

The first word of the second line is problematic. The form of the third letter, transliterated as ט, is actually anachronistic for these inscriptions. A ט with this shape would be the norm for Aramaic documents of the fifth or later centuries but an eighth century Hebrew or Phoenician ט would be shaped like a circle with an "x" in it turned so that one of the strokes is vertical.[116] If the reading is accepted comparable biblical passages include Lev. 10:19 הייטב בעיני יהוה "would it have been acceptable in the eyes of the Lord?" (Note that the ה is a ה interrogative.) Yahweh is also the subject of the Hiph'il form in the following passages Gen. 32:12; Num. 10:32; Judg. 17:13; 1 Sam. 25:31; Zeph. 1:1; Ps.

125:4. If the reading is not accepted, and this seems the better choice, the interpretation is even more problematic.[117]

Khirbet el-Qôm. This inscription was found on a wall between chambers one and two of Tomb II at a site now identified as biblical Makkedah six and one-half miles ESE of Lachish. The heavily-scarred condition of the wall, the varying force with which the letters were inscribed, the presence of erasures, and the double writing of some letters makes this inscription, particularly line 2, very difficult to read and thus many readings are available. Dever dates the inscription to the mid-eighth century B.C.[118]

1) לאריהו.הקשב.כתבה
2) ברכ.אריהו.ליהוה
3) ומארר.יד לאשר תההוש עלה
4) לאניהו

Dever translates the inscription, "(Belonging to) 'Uriyahu. Be careful of his inscription! Blessed be 'Uriyahu by Yahweh. And cursed shall be the hand of whoever (defaces it)! (Written by) 'Oniyahu."[119] Dever cites the examples of the Khirbet Beit Lei and Siloam tomb inscriptions for the form of ארר he reads in line 3.[120] However, though מארר occurs in BH (though as a Pi'el participle and only in Gen. 5:29 and Num. 5:18–27, not as the Pu'al he reads here) and as a noun מארה, neither occurs in the inscriptions and specifically not in the warnings to grave robbers which he cites as examples. For the word which he reads "defaces" from the root תוה, he cites 1 Sam. 21:4 as an example but the spelling is problematic and this is not the word used in other inscriptions of this type.[121] Indeed, Dever is the only scholar to read blessing *and* curse in this inscription but his treatment (the *editio princeps*) is included here, rather than in Chapter 5, for organizational purposes.

Many scholars have offered readings of this inscription. The most controversial point of the inscription is the identity of "asherah" in lines 3 and following.[122] Apart from this, and the meaning of the adjective describing 'Uriyahu in line 1, opinions about the intent of the inscription are remarkably uniform. While all are agreed that 'Uriyahu has been blessed, commentators differ as to perspective of the inscription: is 'Uriyahu blessed because he has been delivered (past tense) or is past blessing being considered a valid basis to ask for Yahweh's assistance in a current time of trouble?[123] All of the scholars hold one of these two views except William Shea. His reading and interpretation will be dealt with separately below.

The first opinion, that the inscription is simply praise for past deliverance, is held by Lemaire, Miller, Mittmann, Jaroš, Hadley, and Margalit.

1) אריהו.העשר.כתבה
2) בּרכ.אריהו.ליהוה

Chapter 3: Blessing

3) ומצריה.לאשרת[.][הושעלה
4) לאניהו
5) ולאשרת
6) רה

    This is André Lemaire's reading. His translation is, "Uriyahu *le riche l'a fait ecrire*: Béni soit Uryahu par Yhwh et <*par son ashérah*>; de ses ennemis {*par son ashérah*}, (*il l'a sauvé*) (Par) Onyahu *Et par son ashérah*."[124]

    Lemaire transposes the word לאשרת in line 3 to a position before the first word of the same line. He says that the transposition of these words on the stone is the mistake of an engraver working by oil lamp; it was forgotten at its proper place and had to be added in its current location.[125] By this transposition, Lemaire has obtained an antecedent for "his asherah" and found a parallel to the formula seen in Kuntillet 'Ajrud inscriptions 3 and 4. Though Lemaire's suggestion of the accidentally transposed words may be questionable, his reading of line 3 seems the best accepted of those given to date. The adjective העשר replaces the patronym and is found with the article as here in Exod. 30:15 and 2 Sam. 12:4. The idea that riches were a sign of God's blessing is seen in such biblical passages as Zech. 11:5 and Ps. 112:3.[126]

    Miller's translation is "(For) Uriyahu the rich: his inscription. [Or: 'has written it'] Blessed is Uriyahu by Yahweh; Yea from his adversaries by his asherah he has saved him. (written) by Oniyahu (...?) and by his asherah."[127] He agrees with Lemaire's reading and allows that the scribal mistake Lemaire postulates may be correct.

    Miller's general interpretation of the inscription is either that it is a prayer of 'Uriyahu hiding in the tombs and asking Yahweh's deliverance from an external threat, or, as he thinks more likely, a prayer or psalm for the deceased 'Uriyahu buried or expected to be buried in the tomb.[128] As an alternative to Lemaire's transposition in line 3, he sees in the extant form, "two essentially poetic lines creating a psalm of thanksgiving."[129] He says that ליהוה in line 2 is paralleled to לאשרתה in line 3 like parallels in the Psalms where "Yahweh" is the word in the "A" phrase and one of his characteristics ("his word," "his soul," "his hand") is the word in the "B" phrase (e.g., Pss. 11:5; 28:5; 33:11; 34:4; 130:5).[130] Miller recognizes a further parallel between ברכ in the line 2 and הושע in line 3, a parallel also found in the Bible in Ps. 3:9 (where both forms are nouns) and Ps. 28:9 (where both are verbs).[131] He also suggests that lines 2 and 3 are a type of parallelism which says "A is so, and what's more B," i.e., God is not only the God of salvation but also the provider of blessing.[132]

    Siegfried Mittmann's translates the text, "Uriahu, der Sänger, hat geschreiben ein Gesegneter Jahwes ist Uriahu und aus Bedrängnis heraus preist er den Gott seines Dienstes, der ihm hilft."[133] He thinks that the inscription is actually two inscriptions intentionally written together, the inscriptions of two men ('Uriyahu and 'Oniyahu) buried in adjacent burial chambers.[134] The small human hand engraved in the stone below the inscription, then, accompanies 'Oniyahu's incomplete inscription. The hand with his name served together as his memorial.[135]

The only difference in reading Mittmann proposes in the first two lines is to delete the ע that other scholars accept as the seventh letter of line one. Subsequently, he translates the word as "Sänger" and understands Uriyahu to have been involved in liturgical service. As is seen in his translation of line 3, this affects his interpretation of the inscription. This reading is not to be preferred, however, since some letter has been intentionally written where Mittmann leaves out the ע.[136]

Karl Jaroš's translation is, "Urjahu, der Reiche, hat dies schreiben lassen: Gesegnet sei Urjahu durch JHWH, denn von seinen Feinden hat er ihn durch seine Aschera gerettet. (Grab) des Onjahu durch seine Aschera [ ] w[ ]h."[137] His interpretation of the text as a votive inscription matches a suggestion rejected by Miller.

Judith Hadley's reading of the inscription is virtually the same as that of Lemaire. She translates the text, "Uriyahu the rich wrote it. Blessed be Uriyahu by Yahweh For from his enemies by his (YHWH's) asherah he (YHWH) has saved him.[   ] by Oniyahu [   ] and by his asherah [   ] his a[sh]erah."[138]

Hadley notes that "Yahweh and his asherah" may be regarded as a compound subject and therefore not require a plural verb form, cf. Prov. 27:9 and 29:15. She repeats and summarizes earlier suggestions made concerning the hand below the inscription: 1) it may be a commemorative marker (cf. 1 Sam. 15:12, 2 Sam. 18:18; Isa. 56:5, etc.) and serve as a suppliant's enduring request for remembrance before God or, 2) the hand may be a sacred symbol intended to guard the tomb.[139]

In an article dedicated to the meaning of the hand in relation to the inscription, Schroer makes a detailed comparison of tomb inscription components and concludes that the hand is apotropaic in nature. The hand functions here as the explicit curses function in other tomb inscriptions—to warn away would-be violators. In this case the warning is pictoral rather than literary.[140] Schroer's example-rich article is persuasive; a warning against violators is expected. However, I know of no other examples where this symbolism occurs.

Baruch Margalit has recently proposed a new reading and interpretation of the inscription. Though his reading largely agrees with that of the above scholars and the relevant part of his interpretation coincides with theirs, his reading varies just enough to solve one of the problems which has given so much trouble. He reads the second and third lines as:

(2 ברכ.אריהו.ליהוה [כי הצל(ה)ו.מ(כפ.)איביה]

(3 ומצריה [...] הושע.לה לאנידו

Translated as "Ur(i)yahu the rich composed it: 'Blessed be Uri(i)yahu unto YHWH— <For he rescued him from (the hands of) his enemies>, And from his foes [...] he

Chapter 3: Blessing

saved him.' Inscribed by On(i)yahu." He says that the lower part of the inscription is actually a second text which he reads as: ולא<ש>רתה [.ליהוה] and translates: "[(Dedicated) to YHWH] and to his consort (Asherah)." By leaving out לאשרתה in line 2 he avoids the awkwardness that Lemaire attempts to alleviate but without the equally awkward rewrite of the inscription. Margalit says לאשרתה

> "may be understood either as a scribal error anticipating the lower inscription...or as the result of the word having been written initially (prior to the upper iscription) in connection with, and in reference to, the tree design, where it eventually 'collided' with the inscription written above.[141]

The interpretation that the inscription is a request for Yahweh's future blessing of 'Uriyahu is held by Naveh, Zevit, and O'Connor.

1) אריהו.השר.כתבה
2) ברכ.אריהו.ליהוה
3) נצרי.ולאשרת.הושע.לה
4) לאריהו

This is Joseph Naveh's reading. His translation is, "Uriyahu the governor wrote it. May Uriyahu be blessed by Yahweh my guardian and by his Asherah. Save him (save) Uriyahu."[142]

Table 2

Graffito Inscriptions Comparison

El-Qôm graffito:	ולאשרתה	אריהו ליהוה נצרי	ברכ
'Ajrud Pithos I:	ולאשרתה	אתכמ ליהוה שמרנ	ברכת
'Ajrud Pithos II:	ולאשרתה	ליהוה [שמרנ]	ברכתכ

Naveh notes that there is a close linkage between the formulas found in graffiti and those found in votive or dedicatory inscriptions. The dedicatory inscriptions have two parts: the first part names the object dedicated and the second part has a call for the deity to hear the offerer's prayer and to bless him. In graffiti, which he considers this inscription to be, only the second part is found.[143] Naveh compares this inscription to two of the Kuntillet 'Ajrud inscriptions (in this study numbers 3 and 4). Table 2 is a comparative chart of the relevant portions of the inscriptions.[144]

Ziony Zevit's translation is, "Uryahu, the prosperous, his inscription: I blessed Uryahu to YHWH And from his enemies, O Asherata, save him. [   ] by Abiyahu [   ] ?? and to Asherata [   ] A[she]rata."[145]

Zevit's reading of the first three lines agrees with that of Lemaire in every point except for Zevit's inclusion of a ת at the end of the first word of line 2, ברכת. Some obscure mark is at this point but all of the other scholars consulted read it as a damaged word divider. The addition of the ת changes the form to a perfect first common singular and thus "I blessed."[146] He notes the absence of the direct object marker before אריהו but cites numerous biblical precedents, virtually all of which are a) poetic texts, b) imperative forms, and c) have either יהוה as subject with an object that is not a proper noun or have יהוה as the object of blessing with the meaning praise. For these reasons the texts he cites are not particularly valuable parallels.[147] In Arad 16:2–3; 21:2–3; 40:3; and Kuntillet ʿAjrud 4:4, a pronominal suffix is attached to the first person singular form of the verb (as is the case in the Ḥorvat ʿUza Edomite ostracon). In Kuntillet ʿAjrud 3:1 the verb is followed by the sign of the direct object with the appended pronominal suffix. In the Ketef Hinnom blessings the suffix is appended to the imperfect form of the verb.[148] In biblical examples, the first person form of the verb is always either used by Yahweh to pronounce blessing on someone, and if so has a pronominal suffix attached; or it is used by a person with regard to Yahweh having the meaning "praise" and may or may not have את preceding Yahweh.[149] On the other hand, the participial forms may (Kuntillet ʿAjrud 5) or may not (Kuntillet ʿAjrud 1, 2) have a pronoun between them and the DN. The absence of the expected sign of the direct object argues against Zevit's reading.

O'Connor accepts Zevit's reading of the text. His translation is like that of Naveh, either, "May you bless Uriah, O Yahweh And from his enemies, O Asherata, save him," or "You have blessed Uriah, O Yahweh O Asherata, may you save him from his enemies."[150] O'Connor accepts the ת that Zevit reads as an additional letter at the end of ברכ in line 2, but translates the resulting word as a perfect second person singular form. That form occurs seven times in the Bible but not with this emphasis. The second person singular imperfect also occurs seven times with the nearest parallel being Ps. 5:13a כי־אתה תברך צדיק יהוה "For thou dost bless the righteous, O Lord." Note that no vocative ל appears in this example, contra Zevit's interpretation of ליהוה in line 2.

In a recent article, William Shea has proposed a remarkably different reading and subsequent translation.

1) אריהו.האשר.כתבה
2) ברכ.אריהו.ליוהו
3) ומצריה.לאשרתה.ושעלה
4) (hand) לאניהו
5) לאשרתה
6) ולא]רתה

Uriyahu was the one who wrote it. Blessed be Uriyahu by
Yahweh, And his Egyptian (servant) by his asherah, and here
is his hand print: (hand sunk in relief) for Oniyahu. By his
asherah. And by his a( )erah.[151]

First, Shea sees ʾUriyahu as the writer rather than as the one who had

Chapter 3: Blessing

the inscription written. Second, Shea sees no indication of deliverance from enemies either past or present. Instead, he understands a reference to 'Uriyahu's Egyptian slave Oniyahu (the ה on ומצריה is thus a third person masculine pronominal suffix). This slave's devotion to Asherah, notwithstanding his Yahwistic name, Shea explains plausibly.[152] He goes on to explain the over-written letters as the attempt of the illiterate slave to contribute his efforts to the inscription—the "ghost letters" occur only in the third line where the slave is mentioned—though the hand print would be interpreted as his "X," the signature of the illiterate devotee. The inscription should then be understood as a simple general call for blessing by the Yahwistic 'Uriyahu to Yahweh and by the foreign slave to his goddess Asherah.[153]

All of the commentators (except Shea) treat this inscription as either a votive inscription (in which case Uriyahu's wealth/position or the deliverance from harm he has experienced may be seen as a sign of blessing Lemaire, Miller, Mittmann, Jaroš, Hadley) or as a plea for help in present danger (based on Yahweh's favor shown in past blessings Naveh, Zevit, O'Connor).[154] Like Arad 16, 21, 40 and Kuntillet 'Ajrud 3, 4, the Khirbet el-Qôm inscription contains a variation of the formula ברך ליהוה.

With Other DN

In the following three texts ברך (without שלמ) occurs with a DN other than Yahweh. In Karatepe A.i.1 and Kuntillet 'Ajrud 1, it occurs with the DN Ba'al. On the Ivory Box from Ur it occurs with Astarte.

Karatepe A.i.1 . Only the first line of Karatepe A will be treated here, the rest will be treated later. At that point general matters concerning the inscription will be noted.[155]

אנכ אזתוד הברכ בעל

This may be translated, "I am Azitiwada, the blessed of Ba'al."[156] Scharbert agrees with this translation.[157] Segert lists this form as a passive participle and compares the Punic PN BURUCBAL.[158] Rosenthal translates the text "the blessed of Ba'l" but includes "or perhaps: chief official (habarakku) of Ba'l" in a note.[159] Donner and Röllig translate it, "der ein Gesegneter des Ba'al ist" but seem no more certain than Rosenthal.[160] Harris gives the form ברכבעל as a PN found in two Phoenican texts CIS I 860 and 3526.[161] Jean and Hoftijzer list הברכ as a Qal passive participle.[162] Bron makes a lengthy study of the different commentators' views and decides that הברכ is indeed a participle functioning as a verb which he translates "le béni de Ba'al."[163] He notes the problem is not so much the presence of the article as it is the absence of some preposition before בעל.[164] Gesenius §127f gives numerous examples of the article attached to the first word of a construct chain in BH grammar, especially cases where the absolute in the chain is a proper noun, rendering the article redundant.

Gibson suggests two interpretations. First, and in his opinion less likely, is "The blessed one of Baal" with the article on the first word of the construct chain. He gives the example ברוך יהוה found in Gen. 24:31 and 26:29 (which each have a titular flavor). For the unusual use of the article on the first word of the construct, he gives a Phoenician example from Kilamuwa 1.5 הלפניהם and biblical examples Jer. 38:6 הבור מלכיהו "the pit of Malchiah" and Lam. 2:13 הבת ירושלם "the daughter of Jerusalem." In each of the last two cases the second word of the pair is, like the Karatepe inscription, a PN.[165] Second, "Grand Vizier of Baal," in which the ה equates to the א found in a title of ultimate Sumerian origin (Akkadian *abarakku*) applied to Joseph (Gen 41:43) אברך. Gibson admits it is not clear why a chief minister should be called "vizier of Baal" rather than "vizier of the king" but notes that Azitiwada and Joseph do have some interesting similarities.[166] Garr, under his treatment of the Qal passive participle, says the form is most likely a noun "official or steward."[167] Though the evidence is ambiguous, "the blessed of Baʻal" seems the marginally better choice.

Kuntillet ʻAjrud 1. This inscription was found in the debris of the entrance to the bench room (like inscription 2). Meshel says that parts of five lines of writing have survived and then says "words which could be deciphered include" those given below. He dates the Kuntillet ʻAjrud inscriptions from the mid-ninth to mid-eighth centuries B.C.[168]

ובארח אל ב[
ברך בעל בים מל[
שם אל בים מל

Meshel offers the translation, "and in the (just) ways of God...blessed be Baʻal in the day of...the name of God in the day of..."[169]

Moshe Weinfeld's reading and translation differ significantly:

ובזרח...אל וימסן הרם
ברך בעל בים מלח[מה]
לשם אל בים מלח[מה]

"When God shines forth (=appears) the mountains melt...Baal on the day of w[ar] ...for the name of God on the day of war."[170]

Though no really close parallels have been found, the following possibilities are suggested. Perhaps the letters ending the second and third lines may begin the word מלא whose cognate appears in Ugaritic but is not apparently applicable.[171] In BH, 1 Kgs. 8:15 might have the same intent, "Blessed be the Lord God of Israel who with his hand has fulfilled what he promised with his mouth to David my father" ברוך יהוה אלהי ישראל אשר דבר בפיו את דויד אבי ובידו מלא.

These letters may begin מלאכה referring to the "work" or "business" of God, cf. Gen. 2:2. Or, the letters may start מלכ and the line be translated "Blessed be Baʻal in the day, king of [," compare 1 Kgs 5:21(7) "Blessed be the

Chapter 3: Blessing 55

Lord this day who has given to David a wise son" ברוך יהוה היום. Or, compare Ps. 31:21 which begins "blessed be the Lord" and has a warfare simile. Also, the phrase ליום מלחמה occurs in Prov. 21:31 "The horse is made ready for the day of battle, but the victory belongs to the Lord." Ba'al is so often pictured as a warrior that one might expect him to be sought as a deliverer. In this vein, compare Pss. 68:35 and especially 124:6 and 8 ברוך יהוה שלא נתננו טרף לשניהם: עזרנו בשם יהוה עשה שמים וארץ: "Blessed be the Lord, who has not given us as prey to their teeth! ...Our help is in the name of the Lord, who has made heaven and earth." These verses use יהוה and בשם יהוה in a sort of parallelism much like that seen in Kuntillet 'Ajrud 1, line 2 ברך בעל "blessed be Ba'al," line 3 שם אל "name of El" and perhaps both have a warfare context.[172]

This inscription is remarkable because it is the only Hebrew inscription to use ברך with a DN other than Yahweh in a preserved context.[173] Though the time of the inscription and the biblical evidence for polytheism make this understanding of the inscription quite plausible, there is another possibility. It is possible that בעל is here being used to refer to Yahweh as it seems clear was sometimes the practice in the Northern Kingdom (cf. the eighth century prophet-Hosea, 2:16). If this were the case then there would be no Hebrew inscriptional uses of ברך with a DN other than Yahweh (except for Kuntillet 'Ajrud 3 and Khirbet el-Qôm where the troublesome pairing of Yahweh with Asherah/asherah occurs). This possibility may be diminished by the fact that בעל is paired with אל in this inscription. Indeed, a better translation might hace rendered "El" for אל. Since we are not certain what constituted "orthodox" Hebrew address of Yahweh at the time (or if there was a standard), this question must remain open. On the other hand, eighth century Israelite syncretistic polytheism is well-attested by the Bible and probably by such inscriptions as Kuntillet 'Ajrud 3 and Khirbet el-Qôm.

Ivory Box from Ur.[174] This inscription is on the lid of an ivory box found under Nebuchadrezzar's pavement of the NE chamber of the E-nun-mah sanctuary at Ur.[175] Presence of objects placed beneath the floor of the temple may be evidence for the practice of dedicating them in this way.[176] This Tyro-Sidonian text is dated from the early seventh century.[177]

(1) ארנ.[ז].מגנ.אמתבעל.בת.פטאס.אמת.אדננ

(2) מתת.לעשתרת.אדתי.תברכי.בימי.אדננ.[---].א.בנ.יסד

"Box which (or This box) Amat-Ba'al daughter of Paṭ-Esi, handmaid of our lord (?), offered (as) a gift to Astarte on his behalf. Mayst thou bless in his days our lord..., the son of Yasad (?)."[178]

Barnett's translation of line 2 starting at תברכי is "may she bless me in the days of our master."[179] He treats תברכי as an imperfect second feminine singular with a first person singular suffix. Dussaud agrees with this interpretation.[180] Gibson, however, translates the same portion as "May you bless her in her day(s)!

Our master is...son of YSD." He also interprets תברכי as a second feminine imperfect of the verb but with a third feminine singular suffix attached. But Segert interprets this form as an imperfect third feminine singular with a third feminine singular suffix, "she may bless her."[181] Harris also identifies it this way.[182] Indeed, neither Harris nor Segert recognize a second feminine singular imperfect verb form at all.[183] Donner and Röllig's translation agrees.[184] Savignac apparently agrees with this identification, but sees the suffix as refering to the box, the gift, not the giver.[185] Though any of the translations seem plausible (except Savignac's), that of Harris and Segert seems most like that seen in other votive inscriptions and is therefore to be preferred.

Without DN

Samaria C1220 is an ostracon from a bowl covered with a red slip inside and out with a burnished exterior and rim. The bowl's quality and shape suggests that it was a high foot bowl. The inscription is dated to ca. 735 B.C.[186] Just below the rim on the inner surface, the seven letters of the inscription were inscribed in the red slip.[187] The inscription is ברכ אחזי.[188]

Sukenik does not decide on a translation but explores two possiblities. First, he notes that the inscription appears to begin with the word ברכ but seems to continue after the last word and says it might be a letter or note beginning "Blessed by Aḥazia" or "Be blessed, Aḥizia." He rejects this proposal saying that the potsherd is too small and the letters written too large for a note to fit on such a small sherd.[189] His second and preferred possibility is that this is a votive inscription and אחזי is the name of the offerer, the breaking off of the inscription then prevents one from reading the expected ל followed by the DN of the god to whom the vessel was dedicated, he compares Gen. 14:19 ברוך אברם לאל עליון.[190]

Birnbaum, apparently responding to Sukenik's rejection of the sherd as a letter on the basis of its size, says that the writer realized the space problem after proceding thus far and began again on another sherd. He suggests two translations based on treating the sherd as a letter "May Ahaz be blessed" or "Mayst thou be blessed, Ahaz." He rejects the option that these are the names of the person to whom the vessel belonged (i.e., "belonging to Baruch son of Aḥaz") saying, correctly, that ל would thus be expected before ברכ. However, his denial of Sukenik's votive suggestion is weak, saying only that "it does not seem right" and, "We know of no Hebrew votive inscriptions."[191] But now Hebrew votive inscriptions are known (cf. Kuntillet 'Ajrud 3 and En-Gedi).

A problem with Birnbaum's letter interpretation is that I find no other letters that begin thus, though 2 Kgs. 4:29 and other passages do use an imperative form of ברכ as a greeting. Also, it does not seem likely that the scribe would have been so short-sighted concerning the size of the sherd or the size of the letters that he was making, though the note might have been quite short like

Chapter 3: Blessing

Kuntillet 'Ajrud 5 and have fit around the rim of a vessel.

Another interesting parallel is found in the En Gedi text.[192] The context of the inscription is quite obscure and the text is fragmentary, but it is clear that in lines 4, 6 and 7, the word ברכ is followed by what are either PNs and/or DNs/titles.

Lemaire suggests that the inscription was written not on an ostracon but on an entire vessel and designated the owner "Baruch (son of) Aḥaz."[193] Problems with this interpretaion are the absence of the ל indicating ownership before ברכ, for which there is plenty of room, and the absence of the expected בנ between the words if the second is the patronym. Neither the absence of the ל preceding or בנ between the names is too great a barrier as seals are often found lacking both though clearly indicating the relationship between the two names.[194] And, as a seal and a name label on a vessel serve the same purpose, it would not be surprizing to find that they share the same abbreviated forms.[195]

Sukenik's preferred interpretation is more likely. Though it seems more probable that a dedicated vessel would have its inscription on the outside near the rim, rather than on the inside as is the case with C1220, the huge bowl found at Kuntillet 'Ajrud (number 5 from that site in this study) has its inscription written on the top of its edge. Most other sacred vessels found at Arad (number 104),[196] Beersheba (3985/1),[197] and Hazor[198] were inscribed with the word קדש on the outside of the vessel.[199] It seems probable that the inscription is some sort of votive inscription on a donated vessel whose presence in the sacred place would continually call for blessing on its owner. The stone bowl from Kuntillet 'Ajrud (Kuntillet 'Ajrud 5) provides an interesting possible parallel לעבדיו בנ עדנה ברכ הא ליהו "Belonging to 'Obadyau son of 'Adnah, blessed be he to Yahu." Since C1220 seems to continue, it could very plausibly be reconstructed as ברכ אחז[יהו] ליהו "Blessed be Aḥaziahu to Yahu."[200]

Summary

When used with Yahweh, ברכ alone only occurs in the inscriptions from Kuntillet 'Ajrud and Khirbet el-Qôm, sites where the troublesome combination of "Yahweh and his asherah" occurs (in Kuntillet 'Ajrud 3 and Khirbet el-Qôm). Kuntillet 'Ajrud 5 contains a formula (ברכ הא ליהו) very similar to that found in BH (ברוך אתה יהוה), and Samaria C1220, while ambiguous, seems very close. The Kuntillet 'Ajrud 2 formula does not have ברכ and יהוה together, the two words are on separate lines the ends of which are broken. Though the overall text has a strong blessing flavor, the parallel to the other formulas is not precise. Biblical parallels or near parallels are extant as noted. With regard to the use of ברכ with other DNs: while the Karatepe A.i.1 reference is ambiguous, the Ivory Box from Ur and Kuntillet 'Ajrud 1 clearly contain appeals to Astarte and Ba'al respectively. Only Kuntillet 'Ajrud 1, of the inscriptions in this category, is not a request for blessing on someone. Kuntillet ʿAjrud 1 is a "blessing" on Baʿal, best interpreted as "praise" for Baʿal.

ברכ Without Context

In three instances, in two inscriptions, forms of ברכ are read in the inscriptions without any decipherable context. These words did not originally appear alone. Little can be determined from these letters except that they are in Hebrew and are found in a Judean context.

In Arad 28:1, 7, a very fragmentary text written on a broken ostracon, Aharoni proposes to read ברכה "blessing" twice. The text dates to stratum VI like letters 16 and 21, 605–595 B.C.²⁰¹ In the first line of the text the three letters ב כה are visible and in the seventh (last) line all four letters are clear.²⁰² No comprehensive translation of the text is possible.

Lachish 31:2 was found in the make-up of the floor of Room 4084 near the inner city gate (Area G) in a early sixth century level (Stratum II).²⁰³ Parts of five short lines remain and the roughly rectangular ostracon is broken on all four sides. The letters יברכ in the second line are the clearest of the text. These letters could compose a verbal form "may he bless" or perhaps be part of a PN.²⁰⁴

שלמ Alone

Just as ברכ can be found in inscriptions without שלמ, the opposite is also true. The two words occur so often together in letter greetings and שלמ is so often a blessing asked that it should be considered on its own in this study of the period inscriptions. It is found both with the DN Yahweh and without a DN.

With Yahweh

This "blessing" occurs only with the DN Yahweh. While it is found once in the Arad letters (no. 18), it is predominantly found in the Lachish letters (2, 3, 5, 6, 7, 9).

Arad 18:1–3, 8. The inscription, like 16 and 21, is from stratum VI, 605–595 B.C.²⁰⁵

(1   אל אדני אלי
(2   שב יהוהיש
(3   אל לשלמכ.ועת
(4   תנ.לשמריהו
(5   ? .ולקרסי

Chapter 3: Blessing   59

6) תתנ. ? ולד
7) בר.אשר.צ
8) ותני.שלמ.
9) בית.יהוה.
reverse
10) הא.ישב

Aharoni translates the inscription, "To my lord Eliashib, May the Lord seek your welfare, and now: Give to Shemaryahu a lethech (?), and to the Kerosi give a homer (?), and as to the matter which you commanded me—it is well; he is in the house of God."[206] Aharoni says the absence of the sender's name from the address is evidence that the letter was sent to Eliashib by an inferior. He says that the number of examples is too small to determine whether the choice of address "was accidental or dependent on the taste of the writer" but thinks that certain forms were used in military dispatches like those of Lachish as against forms used in civil dispatches like the Arad letters.[207]

Pardee translates the lines, "To my lord Elyashib May Yahweh concern himself with your well-being..." He thinks that this is the only certain example of correct letter form from an inferior to a superior. He says it represents the Jerusalem tradition since the בית יהוה is within the writer's sphere of contacts.[208] Pardee disagrees with Aharoni's supposed distinction between military and civilian forms of address saying it will not stand scrutiny.[209]

For the phrase שאל לשלמ compare 2 Sam. 11:7b וישאל דוד לשלום יואב ולשלום העם ולשלום המלחמה:. Examples are also found in Akkadian letters, e.g., Tanaach 1:5–6 *ilani^nu lišalu šulumka šulum bîtika* "May the gods take note of thy welfare, the welfare of thy house"[210] and EA 96:4–6 *ilanu^nu šulumka šulum bîtika lišal* "May the gods be concerned for thy welfare (and) the welf(a)re of thy dynasty."[211] Some Old Babylonian letters from Tell al Rimah contain the wish that gods guard the recipient's life (*^dUTU ù ^dAMAR.UTU liballiṭúki* "May Šamaš and Marduk grant you long life") and that news of the recipient's welfare not be denied the sender (*šulumki ù šulum LÚ.TUR-ri la iparás* "May news of you and news of the workmen not be denied me").[212] Note that what is translated "news of you" is actually *šulumki*, perhaps better, "your welfare." This is the way the second occurrence of שלמ is used in the letter, line 8. With regard to the concern of a previous letter, the inferior replies that the situation is indeed שלמ. In 1 Sam. 30:21, David asks after the welfare of his men וישאל להם שלום. Later Aramaic parallels are not rare, e.g., "The health of your lordship may the God of Heaven seek after exceedingly at all times" שלמ מראן אלה שמיא ישאל שגיא בכל עדן.[213]

Foresti translates the phrase, "May Yahweh obtain for you prosperity."[214] He notes other biblical examples as well, Gen. 43:27; Exod. 18:7; Jud. 18:15; 1 Sam. 10:4; 17:22; 25:5; 30:21; 2 Sam 8:10; 11:7, etc. In all of the biblical passages the subject is a person and the expression is used in greetings (e.g., Exod. 18:7; Jud. 18:15), as a question concerning health (e.g., Gen. 43:27; 2 Sam. 11:7), or in wishing someone well (e.g., 1 Sam. 25:5; 2 Sam. 8:10). Foresti thinks the formula found in this letter came from one person wishing another health and God is invoked "as the supreme guarantor of the prosperity of the addressee."[215]

Lachish. The Lachish documents included in this part of the study date from the early sixth century B.C., were found in one of the gate complex rooms, and were all discovered over a period of a few months in early 1935.[216] Lachish (Tell ed-Duweir), one of the largest and best known biblical sites, is located approximately 25 miles SW of Jerusalem. The Lachish letters show several variations of a formula in which Yahweh is invoked for peace or tidings thereof as a greeting to a superior. The variations can be divided into four groups as follows.

Lachish 2:1–3 and 9:1–2 are good examples of the first group which also includes the poorly preserved 7:1–2. The phrase found here is translated "May Yahweh cause my lord to hear tidings of peace." (Note that letter 2 is provided with an address while 7 and 9 are not.)

(1 אל אדני.יאוש ישמע
(2 יהוה את אדני.שמעת של
(3 מ.עת כימ עת כימ

Torczyner translates lines 1–3, "To my Lord Ya'ush: May Yhwh let hear my Lord tidings of peace (well-being) even now, even now."[217] Lemaire notes that the form found here is not found in the Bible, but the phrase משמיע שלם does occur in Nah. 2:1(1:15). He says that the phrase עת כימ עת כימ signifies that the situation had so deteriorated in the face of Nebuchadrezzar's invasion that the arrival of "good news" was imperative.[218] Michaud agrees, saying that the style is "nerveuses" and "d'une précision extrême."[219] Pardee points out that the Hiphʻil of ישמע is used as a greeting only these six times and only at Lachish. It is "unquestionably a formulaic feature of the scribal tradition represented by these letters."[220] In this context, the wish for someone to have שלמ, or tidings thereof, is certainly the wish not just for health but for safety and here the safety of hearing that the enemy has retreated, the threat is over (cf. 2 Chr. 19:1, Jehoshaphat returned to Jerusalem בשלום after joining Ahab in battle against the Arameans; also Yahweh included among Solomon's unasked blessings "the life of your enemies," and promised, "I will lengthen your days" as a reward for obedience, 1 Kgs. 3:10–14).

Lachish 7:1–2

(1 [ישמע יה]וֹה [את אד]
(2 [ני שמעת שלמ] עת כ]

"[May Yh]wh [let hear my lord tidings of peace, even] now:..."[221]

Chapter 3: Blessing

Lachish 9:1–2

1) ישמע יהוה.את [א]ד
2) ני שמעת שלמ

"May Yhwh let hear my lord tidings of peace."²²²

Lachish 3:1–3. This ostracon is the only example of the second type and varies slightly from the above category in that it is more formal; the latter part of the greeting, exactly like 2, 7, and 9, is preceded by the name of the sender "who has sent to tell" (שלח.להג[ד]) the recipient Ya'ush, "May Yahweh cause my lord to hear tidings of peace."

obverse
1) עבדכ הושעיהו.שלח.ל
2) הג[ד] לאדני יאו[ש] ישמע.
3) יהוה [את] אדני שמעת.שלמ

Torczyner translates these lines, "Thy slave Hosha'yahu has sent to tell my lord Ya'ush: May Yhwh let hear my lord tidings of peace..."²²³

Lachish 5:1–2. The third variation is represented by only one text and that broken and a poor basis on which to determine a variant. Letter 5, according to Torczyner, includes the phrase common to the first and second categories but also includes וטב. In two letters (4 and 8) which comprise the fourth category, טב is substituted for שלמ,²²⁴ however, none of the letters of שלמ are dependable in this inscription.

1) ישמע יהוה [א]ת [אד]ני
2) [שמ]עת.[שלמ] וטב [עת]
3) [כימ עת כימ]

"May Yhwh let hear my [lord] [tid]ings of [peace] and good, [even now, even now.]"²²⁵

W. L. Moran mentions an interesting point in his treatment of the Sefire inscriptions. He notes that in Akkadian treaties, the combination *ṭūbtu u sulummû* is only used of peace effected by a treaty.²²⁶ While the Lachish letters are not treaty documents, nor are they Akkadian; they are not prohibitively separated either in time (he notes such treaties were used until the seventh-sixth centuries²²⁷) or geography from those documents and a peace treaty would be good news indeed to soldiers manning a garrison at some distance from the capital.²²⁸

Lachish 6:1–2 differs significantly enough from the others so as not to be seen as simply a variation of that type.

1) אל אדני יאוש.ירא.יהוה א
2) ת אדני אתהעתהזה.שלמ

Torczyner translates these lines, "To my lord Ya'ush. May Yhwh let see (us) my lord, (while) thou (art) even now in peace."²²⁹

Torczyner divides the third group of letters in line 2 thus: אתה עתה זה. He treats ירא as a Hiph'il imperfect, "may Yahweh cause us to see," being the emphasis. He says that this phrase has the same intent as the phrase written with שמע.[230]

Joüon disagrees with Torczyner's word division in line 2 citing major grammatical difficulties.[231] Based on his own word division, he translates the beginning of the inscription, "Que Yhwh fasse que mon seigneur voie le moment-ci (à l'état de) pacifique" (i.e., "bon, parfait, etc.").[232] The intent of the greeting is that the superior (recipient) will receive the letter and it will find him in a good state, or in a good disposition to be informed of the letter's contents.[233] Albright divides these words in the same way and his translation is similar, "May the Lord cause my lord to see this season in good health!"[234] Ginsberg agrees, he says Torczyner's exegesis is "forced" and subsequently "graphically impracticable."[235]

Gordon's interpretation varies significantly because of his division and translation of these words, "May YHWH make my lord see this present signal! Peace!"[236] The context of the inscription is the desperate condition of the Judean army and the "pious hope" that the addressee will receive the signal.[237]

Without DN

Only the papyrus from Wadi Murabba'ât (designated papMur 17) has the combination of שלמ without ברכ and without a DN.[238] The nearly inaccessible site is 25 kilometers SE of Jerusalem and eighteen kilometers south of Qumran Cave 1.[239] The circa 675 B.C. text is a palimpsest; the lines given below are from the original message written on the papyrus.[240]

1) אָמֹר.-.---יהו.לכ].[שׁלֹ֯ח.שלחת.אֵ֯ת שלמ ביתכ
2) וְעֵת.אל.תשמע לכ[ל] דְּבֹר אֹשֶׁר יֹדבר.אֵליכ.

The text is translated in the *editio princeps*, "...yahu te dit: 'j'envoie mes salutations à ta famille. Et maintenant, n'écoute pas toute parole que te raconte.'"[241]

In Num. 22:37, the infinitive absolute is used with the first person perfect form of שלח as it is here, but not with שלמ. Aramaic inscriptions that use the phrase שלח שלמ include *AP* 41:3 אשלח שלמכ and some of the Hermopolis letters both in this order and paired with וחין and with a different twist, לשלמכי preceding שלחת.[242] Some of the late fifth century Aramaic documents treated by G. R. Driver use a synonym הושרת in a parallel phrase.[243] Instances of greetings being sent to the recipient's house have been noted above in texts ranging from the El-Amarna letters to the Arad letters.[244] In Hebrew it is only elsewhere found in Arad letters 16, 21, and 40.

Summary

שלמ without ברכ does not, certainly, occur with a DN other than Yahweh in the target inscriptions.[245] This type of greeting is found, within the temporal and geographical limits of this study, only in letters which are in

Chapter 3: Blessing

Hebrew and of Judean provenance. It has been noted in a late seventh century document from Assyria (though the lacunae of the text may obliterate an original, fuller greeting and/or a DN) as well as in a couple of fifth century Aramaic letters from Egypt. This sort of greeting seems to be a later development of the fuller greeting form. שלמ also occurs frequently in the Bible both as a blessing and as a greeting.

Substantive Blessing

This category includes inscriptions which contain a prayer or wish for what can be termed blessings even though they do not actually include the word ברכ. These inscriptions ask for those same life characteristics that are seen in other texts as the boons which make life full and enjoyable. Such substantive blessings are seen in biblical texts like Ruth 1:8–9, "May the Lord deal kindly with you...the Lord grant you husbands...," and Ruth 4:11–12, "May the Lord make the woman who is coming into your house like Rachel and Leah..."[246] These substantive blessings are found with Yahweh, with another DN, and without a DN.

With Yahweh

Only in the Kuntillet 'Ajrud 6 inscription is substantive blessing found with Yahweh. This inscription dates from the same time as the other texts from this site (mid-ninth to mid-eighth century)[247] and was apparently discovered at the same time but was not published in the *editio princeps*.

כל אשר ישאל מאת חננ...ונתנ לה יהו כלבבה

Hadley translates the text, "Whatever he asks from a man, may it be favored...and let Yahweh give unto him as he wishes (according to his heart)."[248] Weinfeld notes two elements contained here: grace in the eyes of men and favor in the eyes of God. He compares this text to Prov. 3:4 "So you will find favor and good repute in the eyes of God and man": ומצא־חן ושׂכל־טוב בעיני אלהים ואדם. He also notes Exod. 3:21; 11:2–3; and 12:35–36 (the exiting Israelites' "plunder" of the Egyptians) 3:21 ונתתי את־חן העם־הזה בעיני מצרים "And I will give this people favor in the eyes of the Egyptians," (cf. 11:2 וישאלו איש מאת רעהו "each man shall ask his neighbor."[249]

Further, Weinfeld suggests several parallels with late Ramesside Egyptian letters.[250] Edward Wente translates the first letter he treats, "I tell Arsaphes, Lord of Heracleopolis, Thoth, Lord of Hermopolis, and every god and [every] goddess by whom I pass to give you life, prosperity, health, and a long lifetime, and a good ripe old age; and to give you favor before gods and men."[251]

Two very strong parallels are found in the Hadad inscription lines 4 and 12–13. Line 4 reads ומז.אשׁאל.מנ.אלהי.יתנו.לי, "and whatever I asked from the gods they used to give to me."²⁵² Lines 12–13 read לי֯[.]י֯ת֯[נ֯]ן֯.מת.אלהי.מנ.אשאל.ומה "and what I asked from the gods, they used always to give to me."²⁵³ The second instance is preceded and succeeded by statements that Panammu enjoyed the favor of his gods: they accepted his offerings (line 12) and he had their "favor" (וארקן).

In other biblical texts the same meaning is found. In Judg. 9:9, 13 the olive tree and the vine each refuse to become king of the trees because they "honor" (v. 9 יכבדו) or "cheer" (v. 13 המשׂמח) "gods and men" (אלהים ואנשים). First Samuel 2:26 contains the same sentiment if not identical language והנער שמואל הלך וגדל וטוב גם עם־יהוה וגם עם־אנשים: "Now the boy Samuel continued to grow both in stature and in favor with the Lord and with men." Likewise, Gen. 33:10 is interesting, ויאמר יעקב אל־נא אם־נא מצאתי חן בעיניך ולקחת מנחתי מידי כי על־כן ראיתי פניך כראת פני אלהים ותרצני: "Jacob said, 'No I pray you, if I have found favor in your sight, then accept my present from my hand; for truly to see your face is like seeing the face of God, with such favor have you received me'." Qoheleth 2:10 begins similarly וכל אשר שאלו עיני לא אצלתי מהם "And whatever my eyes desired I did not keep from them." First Kings 3:5 has a similar content ויאמר אלהים שאל מה אתן־לך "And God said, 'Ask what I shall give you'."

Strong parallels also come from other inscriptions. One of the reasons for the erection of the statue on which the Tell Fekherye text is inscribed (line 14) is "And to lengthen his life and so that his words may please gods and men" ולמארכ.חיוה.ולמען.אמרת.פמה אל.אלהנ.לאל.אנשנ. Lines 9–10 of the same inscription perhaps have the same meaning though not the same words. תצלותה. ול.מלקח.אמרת.פמה.ולמשמע "and so that his prayers may be heard and so that the words of his mouth may be accepted."²⁵⁴ Perhaps line 23 of the Panammu inscription had a similar meaning, the preceding line is broken and so the context is not certain. קדמ.אלהי.וקדמ.אנש..²⁵⁵ The fifth century Yeḥaumilk inscription from Byblos also includes this element in lines 9–11, ותתנ [לו הרבת ב]עלת גבל חן לענ אלנמ ולענ עמ ארצ ז וחנ עמ ארצ "And may [the lady], Mistress of Byblos, give [to him] favour in the sight of the gods and favour in the sight of the people of this land!"²⁵⁶

With Other DN

Four tenth century Phoenician dedicatory inscriptions from Byblos have been found which request substantive blessings from other gods. The oldest of them, the Yeḥimilk inscription, is also the longest. The reasons that Yeḥimilk gives for deserving the blessings are also the most developed of the four. The other three (Abibaʿal, Elibaʿal, and Shipiṭbaʿal), with minor variations, are in the same mold set by Yeḥimilk. The last inscription in this category is a dedicatory inscription from Zendjirli, Kilamuwa 2. These dedicatory inscriptions differ little

Chapter 3: Blessing

except for the object dedicated in each case.

Yeḥimilk. This mid-tenth century inscription was discovered in the eighth season of excavations at Byblos.[257] The beginnings and endings of certain lines are superimposed over a Pseudo-Hieroglyphic inscription.[258]

1) בת.ז בני.יחמלכ.מלכ גבל
2) הֹאת.חוי.כל.מפלת.הבתמ
3) אל.יארכ.בעל שממ.ובעל‹ת›
4) גבל.ומפחרת.אל גבל
5) קדשמ.ימת.יחמלכ.ושנתו
6) על גבל.כ מלכ.צדק.ומלכ
7) ישר.לפנ.אל גבל.קדׁשׁמֹ [הֹאֹ][259]

The temple which Yeḥimilk king of Byblus built; it was he who restored the ruins of these temples. May Baal-shamem and Baal(ath)-Gebal and the assembly of the holy gods of Byblus prolong the days of Yeḥimilk and his years over Byblus as a rightful king and a true king before the h[oly] gods of Byblus![260]

Gibson says Yeḥimilk's failure to mention his parentage and emphasis on his right to the throne suggest that he was founder of a new dynasty, perhaps an usurper.[261] The statement that he has restored the temples is doubtless meant to gain Yeḥimilk favor with the mentioned gods, though it is not explicitly stated as a reason why he should be blessed (contrast Solomon's fantastic sacrifices at Gibeon and Yahweh's subsequent offer of blessings—wealth and safety—before the Temple was built, 1 Kgs. 3:3–14).

The feminine plural ימת is also seen (ימות)[262] in Deut. 32:7 and Ps. 90:15, in both of these passages it also parallels שנות. The feminine plural of שנה is more common in BH, occurring ten times including the two passages mentioned above. ימ with the feminine plural ending is found also in Shipiṭbaʻal line 5 and the Tell Siran Bottle inscription (ימת line 7), whereas ימם is found in Hebrew Kuntillet ʻAjrud 2 and Phoenician Karatepe A.iii.5. Likewise, שנת occurs in the Phoenician inscriptions Elibaʻal (line 3) and Shipiṭbaʻal (line 5) and the Ammonite Tell Siran Bottle inscription.

The wish for long life is one of the most common greetings/blessings in the inscriptions. The desire for long life is one of the reasons why King Hadad-yisʻi erected his statue, ולמארכ.חיוה (Tell Fekherye line 14).[263] In the Bible, king (Deut. 17:20 למען יאריך ימים על־ממלכתו) and people (Deut. 5:33 והארכתם ימים בארץ אשר תירשון) are promised longer life if they keep Yahweh's ways (cf. Ps. 91:16 "With long life I will satisfy him" ארך ימים אשביעהו).[264] Ugaritic parallels include RS 16.265, lines 2–6 *ilm tǵrk. tšlmk tʻzzk. alp ymm w rbt šnt bʻd ʻlm* "that the gods may guard you and give you peace and strengthen you a thousand days and myriad year(s) forever; and *UT* 125:14–15, 98–99 (*CTA* 16 I:14–15; II:98–99) "In thy life, O our father, would we rejoice, Thine immortality we would be glad therein!" *bḥyk abn n!šmh blmtk ngln*.²⁶⁵ More specifically, this wish is for the long reign of the ruler in each of the target inscriptions.²⁶⁶

Abibaʿal. This inscription was the first Old Byblian text to be published (1905) but was not deciphered until much later. The inscription was written on the base of a statue of Pharaoh Sheshonk (935–914 B.C.), perhaps given as a gift to Abibaʿal on a visit to Egypt late in that Pharaoh's reign, and is dated about 925. The Phoenician inscription may have then been added upon Abibaʿal's return when the statue was placed in the temple of the goddess.²⁶⁷ The text is heavily damaged but is included here because the restorations are well-accepted.²⁶⁸

(1 [מש](?)ז י[בא.אבבעל מלך [גבל ביחמלכ](?).
(2 מלך] גבל.במצרמ.לבעל[ת.גבל.אדתו.תארכ.בעלת.גבל.ימת
(3 אבבעל.ושנתו.[על גבל ²⁶⁹

Albright translates the text, "[The statue (?) which] Abibaal king of [Byblus son of Yeḥimilk (?) king] of Byblus brought from Egypt for Baal[ath-Gebal, his lady. May Baalath-Gebal prolong the days of Abibaal and his years] over Byblus!"²⁷⁰ The restored content is virtually identical to that of the Yeḥimilk inscription.

Elibaʿal. The three line inscription of Elibaʿal, son of Yeḥimilk, is written on a bust of Pharaoh Osorkon I (914–874 B.C.) and is dated about 914. Three fragments bought by the Louvre from a private source were published by Dussaud in 1925,²⁷¹ subsequently other fragments were found enabling other letters to be read though the inscription is still incomplete.²⁷²

(1 מש.ז פעל.אלבעל.מלך.בי[ח]מלך.מלך.גבל]
(2 [לב]עלת.גבל.אדתו.תארך.בעלת[.גבל]
(3 [ימת.א]לבעל.ושנתו.על [גבל] ²⁷³

Albright translates this text, "The statue which Elibaal king of Byblus, son of [Yeḥimilk king of Byblus], made [for Ba]alath-Gebal, his lady. May Baalath[-Gebal] prolong [the days of E]libaal and his years over [Byblus]!"²⁷⁴ This inscription has the same substantive blessing as Yeḥimilk.

Shipiṭbaʿal. This inscription was discovered in 1935 on a part of the Byblos acropolis associated with the temples of Hathor (Egyptian equivalent for Baʿalat-Gebal) and Hershef and is dated ca. 900 B.C. The inscription was found near the remains of a wall, perhaps the one built by Shipiṭbaʿal and mentioned in the text.²⁷⁵ The surface of the stone is pitted but the text is complete and not difficult to read.

(1 קר.ז בני.שפטבעל.מלך
(2 גבל.בן אלבעל.מלך.גבל

Chapter 3: Blessing

3) ביחמלכ.מלכ.גבל.לבעלת
4) גבל.אדתו.תארכ.בעלת גבל
5) ימת.שפטבעל.ושנתו.על.גבל [276]

Albright translates the inscription, "The wall which Shipiṭ-Baʻal, king of Byblus, son of Elibaal, king of Byblus, son of Yeḥimilk, king of Byblus, built for Baalath-Gebal, his lady. May Baalath-Gebal prolong the days of Shipiṭ-Baʻal and his years over Byblus."[277] The substantive blessing in this text is the same as that in the Yeḥimilk inscription.

Kilamuwa 2. This small ornament (overall length 67 mm.) is made up of eight rectangular gold plates which are joined together in groups of four and whose long sides are fastened together so that they form an open ended box (22 mm. in diameter). The two sets of four plates are joined at their ends to make a hollow tube, one end of which is closed. The plates are attached to each other and decorated by means of soldered gold wire. The letters of the seven short and complete lines of the text are pressed into two of the plates at the open end of the ornament. The inscription was found in the royal palace at Zendjirli. The inscription is unusual in that though the orthography is Phoenician, the majority of scholars think the language is Aramaic.[278] The text is dated circa 825 B.C.[279]

1) סמר ז קנ (2) כלמו (3) בר חי (4) לרכבאל (5) יתנ לה ר(6)כבאל
(7) ארכ חי [280]

"A *smr* fashioned by Kilamuwa, son of Hayya, for Rakkabel. May Rakkabel give him a long life."[281]

The object was apparently a sheath for a wooden pole, perhaps the handle of a staff or scepter[282] or perhaps it was affixed to the pole of a ceremonial chariot.[283] Gibson has the most basic and safest translation of *smr*, "nailed ornament."[284]

The form of חי at the end of the inscription would not be unusual if read as a Zendjirli Aramaic oblique plural ending, however, if the text is Phoenician it must be assumed that the final מ is left off because of lack of space (cf. Karatepe A.iii.3).[285] Dupont-Sommer believes that חי is surely the intended form since the writer could have inserted the מ had he wished. Though he says it may be used "primitively" with a plural meaning, he notes the use of the singular in this formula from Elephantine ליהה חי.[286] He suggests that the form in this inscription is either a fixed formula or indicative of the Zendjirli dialect which has case endings similar to Akkadian and Arabic.[287]

This type of substantive blessing is also seen in biblical passages like Ps. 21:5(4) (in the context of ברך, verses 4[3] and 6–7[5–6]) "He asked life of thee; thou gavest it to him, length of days forever and ever" חיים שאל ממך נתתה לו ארך ימים עולם ועד:. Other biblical parallels include Deut. 30:20; Ps. 23:6; Lam. 5:20; and Job 12:12. Dupont-Sommer suggests a fine late Aramaic example וחינ

אריכן ינתן לך "and may he grant you long life."²⁸⁸

Dupont-Sommer points out that the name of the god רכבאל also occurs in the Phoenician inscription Kilamuwa 1:18 and the Aramaic inscriptions Hadad (lines 2, 3, 11, 18); Panammu (line 22); and Barrakkab 1:5.²⁸⁹ The name means "chariot of El" and, he believes, the object on which the inscription occurs is a sheath or cap for the pole of a chariot. He suggests that this was a ceremonial chariot used in a procession to carry the images of the gods and even compares 2 Kgs. 2:11–12, the chariot and horses of fire that carried off Elijah (note the exclamation of Elisha, "My father, my father, the chariot of Israel and his horsemen" [also 13:14]); 2 Kgs. 23:11, the removal of the horses of the sun and the chariots of the sun (מרכבות השמש) dedicated by the kings of Judah; and finally Ezek. 1 which, he thinks, pictures Yahweh borne in a divine chariot.²⁹⁰ Dupont-Sommer thinks that little by little the chariot bearing the deity became an object of veneration in its own right.²⁹¹

Without DN

Two inscriptions have been found which may be interpreted as having substantive blessings without a DN (the Gezer Calendar and the Tell Siran Bottle inscription). These are included for the sake of completeness, though their use for blessing is at best implicit and is very debatable.

Gezer Calender .²⁹² This small tablet was found in trench number 8 in what Macalister termed "Fourth Semitic débris." Because of its association with pottery and other finds that he dated to the sixth century, he assigned the tablet a like date. This date was disputed immediately by scholars who interpreted the tablet to be two centuries older.²⁹³ Subsequent reexamination of the pottery types has revealed that Macalister's dates were incorrect and that materials he dated to the sixth century (including the inscription) actually date to the eleventh and tenth centuries.²⁹⁴

(1) ירחו אספ.ירחו ז(?)
(2) רע.ירחו לקש
(3) ירח עצר פשת
(4) ירח קצר שערמ
(5) ירח קצר וכל
(6) ירחו זמר
(7) ירח קצ.
אבי[ה] ²⁹⁵

Months of vintage and olive harvest; months of sowing; months of spring pasture; months of flax pulling; months of barley harvest; months of wheat harvest and measuring; months of pruning; months of summer fruit. (margin) Abijah.²⁹⁶

Chapter 3: Blessing   69

Macalister agrees with Mark Lidzbarski's first comments on the text,[297] that the tablet was probably the work of someone, perhaps a peasant, showing off his unusual writing skill. Macalister's further comments are important for this study, "There is nothing historical, votive, or epistolary in the inscription. It contains nothing talismanic or magical—unless, indeed, the missing base of the stone bore an invocation to some Superior Power for a blessing on the labors enumerated; but this hardly seems likely."[298]

There have been many scholarly opinions about the Gezer tablet and most agree that it had no religious significance; there have also been a few notable exceptions. Ronzevalle said that far from being the fancy of an idle peasant, the facts that the tablet was prepared for mounting and bore a signature show its direct and official value.[299] He considers it to be a copy or extract of an administrative document describing the periodic work demands placed by a ruler on the local population. The writing, poor for a professional scribe, he attributes to the text being inscribed on stone rather than written with pen and ink.[300] Cooke says the fact that it had no official value is seen in that it was thrown away with the broken pottery.[301] Daiches identifies the text as a calendar, based on comparison with the Babylonian calendar, even though that calendar includes twelve or thirteen months; the fewer months listed in the Gezer inscription being selected for their agricultural import.[302] Others, led by Albright, say that the tablet is a school exercise, pointing to Egyptian and Mesopotamian parallels and the fact that the soft limestone tablet was scraped and reused.[303] Albright's view seems to be generally accepted now.

It is interesting to note, however, that even Albright's reading of the text and his translations open the possibility for another interpretation of the tablet's function in its ancient community. Albright reads line 5 as ירח קצר וגל which he translates "his month is harvest and festivity."[304] In his discussion of this line, he notes the connection between religious celebration and harvest in the Bible (e.g. Deut. 16:9–12 and Isa. 9:2).[305] From this point, it is not difficult to see the need for a treatment of the Gezer tablet in this study. Inscriptions, as well as the Bible, interpret the bounty of the land as a sign of the deity's blessing: Karatepe A.iii.7–9 "May this city be owner of plenty (of grain) and new wine; and may this people who dwell in her be owners of sheep, and owners of plenty (of grain) and new wine; and may they bear many (children)"[306] (cf. Lev. 26:4–5); and in contrast, the absence of food is used as a curse (e.g., Tell Fekherye lines 19 and 22; Sefire 1.A.24; Lev. 26:26; Deut. 28:38; Mic. 6:14–15; Hag. 1:6).

Henri Michaud represents a sort of bridge position in his interpretation. The text is not precise enough for a calendar but is the school excercise of a student, Abiyahu, whose life the agricultural aspects of the year govern.[307] Michaud almost recognizes the tablet as a sort of blessing in itself by saying that it is an acknowledgement of Yahweh's blessings in nature,

La succession régulière des travaux agricoles est une bénédiction, car chaque homme d'Israël peut vivre heureux sous sa vigne et sous son figuier lorsque la terre produit son fruit et que le culte dû à Yahvé est assuré. Mais quelle malédiction lorsque la terre ne répond pas à l'espérance des hommes...Le petit Abiyahu de Gézer vivait en un temps où Yahvé accordait la fertilité au sol que cultivaient ses enfants.[308]

Wolf Wirgin has been much more vigorous in his advocacy of the idea that the Gezer text is a blessing tablet.[309] Life, for the ancients, centered on the climatic changes and depended on them; rain and heat were necessary, too much or too little would prove equally disastrous. When the cycle was disrupted and prayers of regular individuals were unavailing, the prayers of a holy individual, someone with a reputation for being heard in heaven, were sought. Balaʿam was such an individual.[310] Such an effect may have also been achieved by inscribed stones. In 1900, Frederick Bliss found 51 limestone tablets at Tell Sandaḥannah, mostly in Greek, many of which were identified as cursing tablets by R. Wünsch.[311] According to Wünsch, a person with a grievance would write a prayer against the author of his or her grievance on a limestone tablet and deposit it in a place where the plaintant believed the god would read it. He then "departs assured that he will be avenged in the manner he suggests."[312] Wirgin says that the Gezer tablet is approximately the same size as the cursing tablets,[313] the material, limestone, is significant,[314] and the fact that it was reused indicates that this stone had magical powers and had proven effective.[315] In view of the importance of securing the regular change of the seasons, he says, it is doubtful that a tablet enumerating the agricultural seasons in their true order would be just a writing excercise to be followed by peasants who had done their work in this order since childhood. Further, it is doubtful that a peasant would be able to write at all.[316] Rather, the tablet is a silent prayer that nothing might happen to disrupt the all important agricultural cycle, and the hole served that it might be mounted in a "public holy place" for everyone to see and thus be reminded of the divine commandments in order that their prayer be heard.[317] Victor Sasson, in his treatment of the Amman Citadel inscription, also accepts this purpose for the Gezer calendar and compares it in its function as "blessing tablet" to the Amman Citadel inscription.[318]

Tell Siran Bottle. The second longest of the known Ammonite inscriptions was found by archaeology students excavating the site of Tell Siran (on the campus of the University of Jordan in Amman).[319] The remarkably clear text, dating around 600 B.C., is inscribed in the surface of the bronze bottle.[320] Kent Jackson terms lines 6–8, "a wish or blessing formula."[321]

1) מעבד עמנדב מלך בן עמן
2) בן הצלאל.מלך בן עמן
3) בן עמנדב מלך בן עמן
4) הכרמ.וה.גנת.והאתחר

Chapter 3: Blessing

5) ואשחת
6) יגל וישמח
7) ביומת רבם ושנת
8) רחקת

    Thompson and Zayadine translate the text, "The works of Amminadab, king of the Ammonites, the son of Hiṣṣal'el, king of the Ammonites, the son of Amminadab, king of the Ammonites, a vineyard and the gardens and the *thr* and cisterns. May he rejoice and be glad for many days and long years."[322]

    Zayadine and Thompson draw many parallels between the inscription and the Bible. They note that יומת in line 7 is unusual and say they would expect the much more frequent ימם or ימ, though the feminine form is seen in two poetic biblical texts, Deut. 32:7 and Ps. 90:15.[323] The feminine plural of שנה is more common in BH, they note, occurring ten times including the two passages mentioned above. They believe that the end of the inscription "shows a strong relationship with the Psalms" and say, "It is possible that the biblical usage reflects cultic influence. This in turn may underline the dedicatory and cultic significance of the Tell Siran inscription."[324] However, the form in the Bible is ימות rather than יומת as here. Also, ימ with this plural ending (feminine? note the masculine adjective רבם which follows) is found in other inscriptions, cf. Yeḥimilk line 5 and Shipiṭbaʻal line 5 (whereas ימם is found in Hebrew Kuntillet ʻAjrud 2 and Phoenician Karatepe A.iii.5). Likewise, שנת occurs in the Phoenician inscriptions Yeḥimilk (line 5), Elibaʻal (line 3), Shipiṭbaʻal (line 5).

    Zayadine and Thompson do an extensive study of the occurrences of שמח and גיל in the Bible. They say that of the 125 times that שמח occurs as a verb, 49 times have cultic or religious significance; of the 94 times it occurs as a substantive, 36 have cultic or religious connections; and of the 21 times it occurs as an adjective, one has cultic or religious significance. Hosea 9:1, is, they say a reference to a Canaanite religious festival that the prophet is warning the people away from; a festival which Israelites later celebrated as the Feast of Booths. Likewise, Gen. 31:27 implies that שמח was a feast "consisting of dances, songs, and music."[325]

    The word גיל is found in the Bible 55 times of which 36 have religious or cultic significance. It is not found in any texts earlier than Hosea 9:1 and is found in parallel with שמח only in poetic texts. Zayadine and Thompson note that while the text of the inscription does not refer to any festival, yet barley and wheat were found in the bottle. The bottle and its contents might have been part of a foundation desposit, the grain being a symbol of prosperity offered as a sacrifice to the deity.[326] Compare Albright's reading of the Gezer text, line 5 ירח קצר וגל which he translates "his month is harvest and festivity."[327] He also draws attention to the connection between the feast of weeks and the harvest celebration to be celebrated with "rejoicing" (גיל, Deut. 16:9–12 and Isa. 9:2).[328] The words occur frequently in parallel in biblical passages, always in poetic contexts, and almost exclusively to speak of the rejoicing of the righteous or creation over something that the Lord has done (e.g., Pss. 14:7; 16:9; 31:7; 32:11; 48:12 [11]).

Charles Krahmalkov calls the text "an Ammonite lyric poem" and translates lines 4–8, "'To the vineyard and the orchard! Or shall I be left behind and destroyed?' He (who says this) rejoice and be happy that life is long and there are years yet unlived."[329] The bottle is an ancient *objet d'art*, a model of a wine storage vessel intended for a connoisseur's collection. The author reflects on future days of enjoyment to encourage himself and draw his mind away from death as the end (cf. Cant. 5:1).[330]

Shea thinks that the bottle is actually a drinking vessel and that the inscription is a sort of injunction to responsible imbibing. His translation, minus the genealogy of Amminadab, is "(The wine in this vessel comes) From the cultivation of Amminadab...of the vineyard and of the garden, and shall I inflame myself (with it) and be ruined? (No!) It shall make me glad and bring joy for many days and long years."[331] He notes that שמח is twice used in Pi'el with wine as the subject (Ps. 104:15 and Qoh. 10:19), as well as a comparison of the heart of being glad as with wine (Zech. 10:7), and rejoicing in someone's love which is better than wine (Cant. 1:4).[332]

Coote interprets the first word of the inscription מעבד as "work" or "product" and thinks that it referred to the contents of the bottle, perhaps scented oil. His translation reflects his interpretation that the things mentioned in the preceding lines are the product that it is wished may "give pleasure for many days and for years far off."[333]

Baldacci translates the verbs, in lines 6–8, as Hoph'als rather than jussives or optative Qal forms, "May he (viz. '*mndb*) be gratified and may he be congratulated after many days and long years."[334]

Cross notes an interesting Ugaritic parallel, "Thy grace in the midst of Ugarit as long as the days of the Sun and moon and the pleasance of the years of 'El."[335] Another Ugaritic parallel is 16.265, lines 2–6 *ilm tġrk tšlmk t'zzk alp ymm w rbt šnt b'd 'lm* "that the gods may guard you and give you peace and strengthen you a thousand days and myriad year(s) forever." Another Ugaritic parallel is *UT* 125:14–15, 98–99 (*CTA* 16 I:14–15; II:98–99) "In thy life, O our father, would we rejoice, Thine immortality we would be glad therein!" *bḥyk abn n!šmh blmtk ngln.*[336]

Smit ties the text of the inscription most closely to Qoh. 2:4–6, "I made great works; I built houses and planted vineyards for myself; I made myself gardens and parks, and planted in them all kinds of fruit trees. I made myself pools from which to water the forest of growing trees." While the biblical text goes on to talk about the speaker's more exotic acquisitions (including slaves and precious metals), Smit suggests that the Ammonite text is simpler, "praising exactly those things which are necessary to sustain everyday life."[337]

Chapter 3: Blessing 73

Summary

Of the inscriptions in this section, only Kuntillet 'Ajrud 6 (the only "substantive blessing" with Yahweh) is not among the very oldest included in this study. The inscriptions having a substantive blessing with another DN are all Phoenician from the tenth century and all ask for the same blessing, long life. The two inscriptions which include substantive blessing without DNs (the Gezer Calendar and Tell Siran Bottle) both have agricultural overtones. The Gezer Calendar (debatable as to its function and thus its meaning) has been interpreted as understanding the successfully completed agricultural cycle as a sign of divine beneficence. Its function, according to some, was to serve as a blessing tablet, a perpetual prayer for the continuation of the vital natural cycle. The Ammonite Tell Siran Bottle asks the blessing of a happy long life for the king of the text.

Conclusions

As was shown above, the combination of ברכ and שלמ with a DN was been found in target period letters and blessing texts from Arad (16, 21, 40), Kuntillet 'Ajrud (4), Ketef Hinnom, and Ḥorvat 'Uza. Later examples noted (e.g., Saqqara, Padua I, and the Hermopolis letters) testify to the use of this combination in Edomite, Phoenician, and Aramaic inscriptions accompanied by other DNs. The only example where ברכ and שלמ might be found without a DN (Samaria C1101) is probably a short greeting preceding or following a PN. The earliest of these texts is Kuntillet 'Ajrud 4 (ca. 800 B.C.) predating Arad 40 by perhaps a century. The latest texts in this category are Arad 16 and 21 (ca. 595 B.C.). Because target inscriptions in Hebrew, Phoenician,[338] and Edomite contain the combination and post-period inscriptions in Aramaic and Phoenician contain it, this combination of DN with ברכ and שלמ cannot be said to belong to any single nationality. ברכ in these inscriptions might well be a compact way of expressing all of the more specific blessing ideas seen in the longer inscriptions.

No distinct conclusions should be drawn between usages of ברכ with שלמ and ברכ without שלמ. Most of these differences may be attributed to either the type of literature (the combination is found exclusively in letters and votive texts), or to the broken condition of the inscription in question. The few others do not provide an adequate basis for drawing distinctions, though the difference appears to be simply a matter of style rather than meaning. All of the inscriptions in this category are either Hebrew, Phoenician, or the Phoenician-influenced Hebrew seen at Kuntillet 'Ajrud. The root ברכ is not seen in Aramaic inscriptions until after the target period.

שלמ without ברכ does not occur with a DN other than Yahweh. This type of greeting is found in target inscription only in letters which are Hebrew and Judean. This sort of greeting seems to be a later development of the fuller greeting form.

The earlier inscriptions (tenth to mid-eighth century) are more prone to use specific blessings/invocations for life, health, fertility, food,[339] while the later inscriptions (all mid-eighth century or later) tend to use the general term ברך. The inscriptions of the eighth century may use either style or use a combination of the two. Note that the Israelite and Judean inscriptions, primarily belonging to the later group, tend to use the more inclusive blessing with ברך, while the oldest of the Israelite inscriptions, Gezer (if interpreted as a blessing tablet), holds to the older tendency to use substantive blessing, however, this could be due to the type of literature of the Gezer tablet, not to a chronological tendency.

---

[1] Yohanan Aharoni, *Arad Inscriptions*, ed. Anson F. Rainey, trans. Judith Ben-Or (Jerusalem: Israel Exploration Society, 1981), pp. 8–9, 134. As noted above, the form in which the inscriptions are shown follows these guidelines: where the inscriptions referred to are alphabetic and non-cuneiform (i.e., Hebrew, Aramaic, Phoenician, Moabite, Edomite, or Ammonite), they are transcribed in Aramaic block letters; word dividers and spaces are given as they appear in the inscriptions themselves unless otherwise stated; final forms are not included since those languages during the relevant period had none; שׁ and שׂ are not distinguished; and vowel points are not provided either in the inscriptions or in the quoted biblical texts unless they are particularly necessary for the current discussion. Cuneiform texts will be presented in italicized transliteration.

[2] Ibid., p. 30

[3] Ibid. Aharoni also provides pointing.

[4] Ibid. I agree with Pardee (Dennis Pardee, et al., *Handbook of Ancient Hebrew Letters*, SBL Sources for Biblical Study, vol. 15 [Chico, CA.: Scholars Press, 1982], p. 311) against Aharoni, that שׁלח (line 1) should be pointed as a perfect rather than an active participle. This is the form typically seen in letters (e.g., papMur 17:1 which is discussed later in this chapter).

[5] The translations of biblical texts are from the RSV unless otherwise specified; this translation is my own.

[6] A. Dupont-Sommer, "Le syncrétisme religieux des Juifs d'Éléphantine d'après un ostracon araméen inédit," *RHR* 130 (1945), 18–20. Another parallel is found in the letter from Elephantine designated Pad I (Joseph A. Fitzmyer, "The Padua Aramaic Papyrus Letters," *JNES* 21 (1962), 15–24; and more recently in Joseph A Fitzmyer, *A Wandering Aramean, Collected Aramaic Essays*, SBL Monographs Series, no. 25 [Chico, CA.: Scholars Press, 1979], pp. 219–26. The *editio princeps* is Edda Bresciani, "Papiri aramaici egiziani di epoca persiana presso il Museo Civico di Padova," *Rivista degli studi orientali* 35 [1960], pp. 11–24).

1) [שלמ ב]ית יהו ביב אל ברי שלממ [מ]נ אחוכ אושע שלמ ושררת [שגיא הושרת לכ]
2) [כעת מ]נ יומ זי אזלת בארחא זכ בר[י] לי טיב אפ אמכ כעת ברכ אנת [ליהו אלהא]
3) [זי יח]וני אנפיכ בשלמ

[Greetings to the temp]le of Yahu in Elephantine! To my son, Shelomam; from your brother, Osea. [I send you] greetings and prosperity [in abundance!]. [Sin]ce the day when you went on that caravan, my son, all goes well with me and your mother too. May you be blessed [by Yahu, the God, who may gr]ant me to see your face (again) in peace!

Bresciani compares [ברכ אנת [ליהו.אלהא] of the suggested reconstruction (line 2) to the blessings found in 1 Sam. 15:13 and 23:21 ברוך אתה ליהוה. Fitzmyer (p. 18), following the reading of the *editio princeps*, treats ברכ as a defective spelling for בריכ which might be expected here. (Compare the fourth century Aramaic Carpentras Stele lines 1 and 3 where the feminine form of the participle occurs בריכה [*TSSI*, vol. 2, pp. 120–21]).

Chapter 3: Blessing

[7]Mitchell includes a good brief account of שלם in the Bible as a close, but not complete, synonym for ברך (Christopher W. Mitchell, *The Meaning of BRK "To Bless" in the Old Testament*, SBL Dissertation Series, vol. 95 [Atlanta: Scholars Press, 1987], pp. 181–83.

[8]Aharoni, *Arad Inscriptions*, p. 30. Cf. "your son" in Arad 21, which he says is used of an inferior addressing a superior, cf. 2 Kgs. 5:13, ויגשו עבדיו וידברו אליו ויאמרו אבי "But his servants came near and said to him, 'My father...'" (Ibid., p. 71). Notice also, that in the inscription cited in note 6, the sender addresses himself to "my son" and identifies himself as "your brother."

[9]Dennis Pardee, "Letters from Tel Arad," in *UF*, vol. 10, ed. K. Bergerhof, M. Dietrich, and O. Loretz (Neukirchener-Vluyn, W. Germany: Verlag Butzon & Bercker Kevelear, 1978), p. 310.

[10]Pardee, et al., *Handbook*, pp. 49–50.

[11]Ibid., p. 50.

[12]It, also, is found only in the seventh century Wadi Murabbaʿât papyrus (papMur 17:1 discussed below) besides the three Arad letters discussed in this section (nos. 16, 21, and 40). Examples are also found in earlier cuneiform letters, e.g., Tanaach 1:5–6 *ilanu lišalu šulumka šulum bîtika* "May the gods take note of thy welfare, the welfare of thy house" (following the reading of W. F. Albright, "A Prince of Taanach in the Fifteenth Century B.C.," *BASOR* 94 [1944], 12; and the translation of W. F. Albright and G. E. Mendenhall, "Taanach 1," in *ANET*, p. 490) and *EA* 96:4–6 *ilanu šulumka šulum bîtika lišal* "May the gods be concerned for thy welfare (and) the welf(a)re of thy dynasty" (following the reading of J. A. Knudtzon, ed., *Die El-Amarna-Tafeln mit Einleitung und Erläuterungen*, vol. 1 [1964; rpt. Aalen: Otto Zeller Verlagsbuchhandlung, 1915], pp. 442–43; and the translation of R. Youngblood, "Amorite Influence in a Canaanite Amarna Letter [EA 96]" *BASOR* 168 [1962], 24). The Akkadian/Aramaic bilingual from Tell Fekherye, Aramaic section line 8, (discussed in Chapter 5) includes a very similar phrase ולשלם ביתה, "and for the welfare of his house"; Akkadian section line 11 *šulum bitišu*. (that is the editors' "Lecture développée" their transcription is *SILIM É-šú*) A. Abou-Assaf; P. Bordrueil; and A. R. Millard, *La statue de Tell Fekherye et son inscription bilingue assyro-arameene*, (Paris: Editions Recherche sur les civilisations, 1982), pp. 23, 15, 13.

[13]F. Foresti, "Characteristic Literary Expressions in the Arad Inscriptions Compared with the Language of the Hebrew Bible," *EphCar* 32 (1981), 333.

[14]Ibid., pp. 333–34. Donner and Röllig (*KAI*, vol. 2, p. 68), in their comments on the similar phrase in the Saqqara papyrus ברכתך לבעל צפנ (*KAI* 50), while they do not contrast this meaning with the biblical formula, do interpret the phrase as originally meaning "Ich nannte deinen Namen vor dem Gotte, um ihn zu veranlassen, dich zu segnen."

[15]See the summary of Johannes Pedersen's views in the "History of Interpretation" of blessing and curse in Chapter 2.

[16]The translation is mine.

[17]In an appendix to an article by Dupont-Sommer ("Note on a Phoenician Papyrus from Saqqara," *PEQ* 81 [1949], 52–57), Sidney Smith says that two distinct religious ideas must be reckoned with in this kind of blessing: 1) blessing someone by a/the god/God—this "implies that gods/God have (has) given you special powers" or 2) one may mention a person to his/her god in the god's presence in order to "make the god bless him." This second idea is represented by *brk l* -DN of which he says, "the use with ל seems only to occur in texts implying polytheism, which is not surprising" (p. 57). His first idea reflects an earlier understanding of blessing and curse (see the arguments presented in Chapter 2); the second idea, that *brk l* -DN is only used in polytheistic contexts, can no longer be held if the writers of the Arad letters are understood as monotheistic Yahwists.

[18]Aharoni, *Arad Inscriptions*, p. 31.

[19]Pardee, *Handbook*, p. 311.

[20]Cf. the treatment of the Ḥorvat ʿUza inscription later in this chapter.

[21] P. Kyle McCarter, personal communication.

[22] See the introduction to Arad 16 for information on the Arad inscriptions.

[23] Aharoni, *Arad Inscriptions*, pp. 42–43. The readings and translation are his. The only different division Aharoni provides in his modified version (p. 42) is a space between ל and כ at the end of line 2.

[24] Otto Kaiser, "Zum Formular der in Ugarit gefundenen Briefe," *ZDPV* 86 (1970), 15, n. 29. Kaiser lists several Ugaritic letters that use the form *yšlm lk* which he translates "Möge es dir wohl ergehen," e.g., RS 15.08, 4–6 (also designated *PRU* II, 15:4–6, or *UT* 1015), RS 15.174, 4–6 (*PRU* II, 16:4–6; *UT* 1016), RS 17.139, 3 (*PRU* V, 9:3–4 ; *UT* 2009), et al. In RS 15.08, 4–6 (*PRU* II, 15:4–6; *UT* 1015) this is followed by the phrase *ily ugrt tġrk tšlmk* "may the gods of Ugarit guard and preserve thee." (This will be further discussed in the treatment of the Ketef Hinnom inscriptions.)

[25] Pardee, "Letters from Tel Arad," p. 318. Lemaire agrees, "que Yhwh rende à mon maître," *Inscriptions Hébraïques, Tome I: Les ostraca*, (Paris: Les Éditions du Cerf, 1977), p. 186.

[26] Aharoni, *Arad Inscriptions*, p. 42. He suggests comparison with 2 Sam. 3:39; Prov. 25:22; Job 21:31; and Ruth 2:21.

[27] Ibid., pp. 8–9, 136.

[28] Ibid., p. 71. The readings and translation are his. The final כ of ברכתכ is clear and its translation "you" need not be in the author's parenthesis.

[29] Ibid.

[30] Pardee, "Letters from Tel Arad," p. 324.

[31] Lemaire's translation shows that he treats it as if it were originally correct, "envoi(ent saluer)," *Inscriptions Hébraïques*, p. 207.

[32] I.e., the two letters discussed above and the Kuntillet 'Ajrud inscription just below. However this might just as well be attributed to the poor grammar of the scribe.

[33] Ze'ev Meshel and Carol Meyers, "Did Yahweh Have a Consort? The New Religious Inscriptions from the Sinai," *BAR* 5/2 (1979), 32. They give the inscriptions this very general date range suggesting that the texts may be from the time of the "half-Phoenician" queen Athaliah (p. 34).

[34] The best photograph available is illustration 11 in Meshel (*Kuntillet 'Ajrud: A Religious Centre from the Time of the Judean Monarchy on the Border of Sinai*, Israel Museum Catalogue No. 175, [Jerusalem: Israel Museum, 1978], no page numbers). Lemaire's drawing simplifies too much and leaves out several letters which actually belong to the inscription (André Lemaire, *Les écoles et la formation de la Bible dans l'ancien Israël*, Orbis Biblicus et Orientalis, vol. 39 [Göttingen: Éditions Universitaires Fribourg Suisse Vandenhoeck & Ruprecht, 1981], p. 27, figure 11).

[35] This reading is based on those given by Meshel (*Kuntillet 'Ajrud*, no page numbers) as well as that given in M. Weinfeld, "Kuntillet 'Ajrud Inscriptions and Their Significance," *SEL* 1 (1984), 125; and J. Hadley, "Some Drawings and Inscriptions on Two Pithoi from Kuntillet 'Ajrud," *VT* 37 (1987), p. 185. Obtaining dependable readings of the Kuntillet 'Ajrud inscriptions is problematic because of their incomplete publication in the catalogue and the fact that no *editio princeps* has been published. Chase ("A Note on an Inscription from Kuntillet 'Ajrud," *BASOR* 246 [1982], 65) contributes the reading of the second person singular masculine pronoun את after השלם, cf. 2 Sam 20:9. The line numbers are from Chase's article (p. 63), also see her footnote 3.

[36] Hadley, "Kuntillet 'Ajrud," p. 185. For discussion of the identity of the much disputed "Asherah" or "asherah" in this passage, see any of the studies mentioned in the discussions of Kuntillet 'Ajrud inscriptions 3, 4, or the Khirbet el-Qôm inscription.

[37] The Lachish texts in question are 2:1–3; 3:2–3; 4:1–2; 5:1–2; 7:1–2; 8:1–2; 9:1–2. The phrase is

Chapter 3: Blessing 77

1) אל אדני.יאוש ינשמע
2) יהוה את אדני.שמעת של
3) מ.עת כים עת כים

"To my Lord Ya'ush: May Yahweh let hear my Lord tidings of peace (well-being) even now, even now." (2:1–3) This reading and translation (p. 37) as well as a treatment of the other texts may be found in Harry Torczyner, et al., *Lachish I (Tell ed Duweir): The Lachish Letters* (London: Oxford University Press, 1938).

[38] D. J. Wiseman, "'Is it Peace?'—Covenant and Diplomacy," *VT* 32 (1982), 317.

[39] Ibid., p. 326.

[40] A. Cowley, *Aramaic Papyri of the Fifth Century B.C.*, (1923; rpt. Osnabrück: Otto Zeller, 1967), pp. 111, 113. (Hereafter designated *AP*.) This text is *AP* 30. Other examples include numbers 39:1 and 56:1. Relevant, partially restored texts include 17:1; 37:1–2; 38:2; 21:2; 40:1; and 41:1.

[41] Dupont-Sommer, "Note on a Phoenician Papyrus," p. 53.

[42] Brian Peckham, "Phoenicia and the Religion of Israel," in *Ancient Israelite Religion*, ed. P. D. Miller, Jr., P. D. Hanson, and S. D. McBride (Philadelphia: Fortress Press, 1987), pp. 80–81.

[43] Ibid., pp. 81–82.

[44] Ibid., pp. 82–83.

[45] Ibid., p. 84.

[46] Ibid., pp. 84–85.

[47] Gabriel Barkay, *Ketef Hinnom: A Treasure Facing Jerusalem's Walls*, (Jerusalem: The Israel Museum, 1986), pp. 22, 29. The specific site is designated "repository of cave 25" and was used at three different periods between the seventh and first centuries B.C. The scrolls are dated on the basis of epigraphy.

[48] Ibid., pp. 35–36. Actually, very little can be determined from the photographs; that of the larger "plaque" (p. 36) shows only two or three recognizable letters and that of the smaller (p. 35), though much better preserved, yields few recognizable words. A helpful drawing is placed beside the photograph of the smaller text.

[49] Ibid., p. 29.

[50] Ibid.

[51] The transcriptions in *Ketef Hinnom* give only four lines of the "longer plaque" and only eight lines of the "shorter plaque" (pp. 29-30). Even though the catalogue says the first plaque is longer and better preserved, very few lines of the text are actually given and the place of these lines in the overall text is not specified. Barkay does say (*Ketef Hinnom*, p. 29) "The ending of the owner's name '...yahu' may have been preserved in the first line. It is followed by the words יהוה, [א]הבה, חסד, שמרי, ומהרע..., which are legible, but which are very difficult to interpret in a consecutive sense."

Barkay says (Ibid.) that the owner's name perhaps "Benayahu" or "Shebanyahu" was originally in the first line of the shorter plaque and further on the word רעה occurs. Though Barkay does not speculate on the interpretation of רעה or its relationship to the proper name, perhaps the owner was a priest "shepherd" during his life. There are many biblical references to the priests as shepherds in Jeremiah (e.g., 2:8; 3:15; 10:21; 12:10; 22:22; 23:1) and Ezekiel (32:4ff.). In these passages they are condemned and rebuked as "false" shepherds. Perhaps this term was current in the day of the plaques' owner(s)—the plaques are dated between 650 and 600—about the time of Jeremiah and Ezekiel and this was an accepted epitaph for the priests before it became a cudgel of prophetic scorn. Might a priest also be more likely to own such a blessing text?

[52] Ada Yardeni, "Remarks on the Priestly Blessing on Two Ancient Amulets from Jerusalem," *VT* 41 (1991), 178. She offers no translation of the plaques. As the text of the Priestly Blessing of

Numbers is quite familiar and the rest of the text of the plaques is uncertain, I will also offer no comprehensive translation, except for a few comments. Table 1 is an edited version of Yardeni's comparison. I have made the transliteration into block letters. Yardeni describes how very difficult the amulets were to decipher: between the time of their discovery in 1979 and the winter of 1986 very little of the plaques could be interpreted. The facts that the scrolls were rolled up for so long and that they were written carelessly (Ibid.; she also says the two are the work of separate scribes, p. 180) add to the difficulty of reading them.

[53] Ibid. p. 180.

[54] Barkay, *Ketef Hinnom*, p. 30.

[55] Gerhard Wehmeier, "Deliverance and Blessing in the Old and New Testaments," *IJT* 20 (1971), 34.

[56] Yardeni, "Two Ancient Amulets," p. 181.

[57] Ibid., p. 183. The italics are mine.

[58] Ibid., pp. 184-85.

[59] See the discussion of how blessing in thought to operate, in Chapter 2. One example that especially comes to mind is the Jerusalemites' talismanic mantra "The temple of the Lord," which Jeremiah condemned (ch. 7) and Micah's even earlier warning that the sacred city was not inviolate (3:12).

[60] E.g., in Jeremiah's time the people's incorrect understanding of Yahweh's deliverance a century earlier (Isa. 36-37) lead to believe that Yahweh would *always* deliver the city.

[61] For the Tabnit inscription see TSSI, vol. 3, p. 103.

[62] For Yahweh as redeemer, though the texts are from a slightly later time see Isa. 41:14; 43:14; 44:6, 24; 48:17; 49:7, 26; etc. also K. Randolph Joines, *The Incomparable Divine Kinsman of Second Isaiah* (Haddonfield, NJ: Haddonfield House, 1976).

[63] S. Dalley; C. B. F. Walker; and J. D. Hawkins, *The Old Babylonian Tablets from Tell al Rimah* (Baghdad: The British School of Archaeology in Iraq, 1976), p. 52, text 40, lines 4–5 and 12–13. Texts 41:4–6, 8–9 (p. 52) and 43:4–6, 15 (pp. 53–54) also contain these wishes.

[64] Texts containing this wish include RSL 1, 5–7, *Ugaritica V*, p. 85; RS 20.255 A, 1–3, *Ugaritica V*, p. 101; and RS 20.172, 4–6. See Kaiser ("Zum Formular," p. 16 and n. 33) for other examples and variations.

[65] Gordon, *UT*, p. 5*, text 2010:4–6. Other letters contain a shorter version *ilm tġrk tšlmk* "May the gods guard you, may they give you peace" *UT* 95:7–9, p. 185; *UT* 101:1–2, p. 186; *UT* 117:7–9, p. 189; *UT* 138:5–6, p. 198 (translations are from Gordon, *UL*, pp. 116–19).

[66] G. R. Driver, *Aramaic Documents of the Fifth Century B.C.*, (Oxford: Clarendon, 1954), p. 35 (hereafter *AD*).

[67] Barkay, *Ketef Hinnom*, p. 30.

[68] Ibid.

[69] Ibid. *AP*, pp. 192, 196. The English translation is Barkay's including the plural translation of the singular תפלה.

[70] See "Other Words for Curse" in Chapter 4.

[71] Frank M. Cross, Jr. and Richard J. Saley, "Phoenician Incantations on a Plaque of the Seventh Century B.C. from Arslan Tash in Upper Syria," *BASOR* 197 (1970), 49.

Though the first Padua Aramaic papyrus is too late for this study (late fifth century) it does contain some interesting parallels to the target inscriptions in this section (the first lines of that inscription are quoted in a note to the discussion of Arad 16 above). Fitzmyer notes that, like several of the Hermopolis papyri, the Padua I letter begins with a salutation to the temple of the god whose blessing was to be invoked on the recipient. (This is what Fitzmyer believed because

Chapter 3: Blessing 79

the Hermopolis letters had not been published at the time of his study of the Padua letter. But the first four Hermopolis letters begin with greetings to the temples of deities [those of Nebo in i, Banit in ii and iii, and Bethel and the Queen of Heaven in iv] who are *different* from the one invoked for blessing [in each case Ptah]. So the Hermopolis letters cannot confirm Fitzmyer's reconstruction at the end of line two. Despite this, the reconstructions are fairly certain on the basis of other Aramaic letters edited by Kraeling [*The Brooklyn Museum Aramaic Papyri*, {1969; rpt. New Haven, Conn.: Yale University Press, 1953}, letters 12:2; 1:2; 9:2; {also 2:2; 4:2; 10:2; 3:3; 4:10}]; Cowley [*AP*, nos. 6:4; 25:6; 30:6, 24–26; etc.]; and Driver [*AD* 3:1; 5:1; 13:].)

[72] Y. Aharoni, *The Land of the Bible, A Historical Geography*, 2nd ed. (Philadelphia: Westminster Press, 1979), p. 441.

[73] I. Beit-Arieh and B. Cresson, "Notes and News: Horvat 'Uza, 1982," *IEJ* 32 (1982), 262–63. For information on other Edomite epigraphic finds see, Nelson Glueck, "Ostraca from Elath," *BASOR* 80 (1940), 3–10; and with the same title in *BASOR* 82 (1941), 3–11.

[74] I. Beit-Arieh and B. Cresson, "An Edomite Ostracon from Horvat 'Uza," *TA* 12 (1985), 96. These "Hebrew ostraca" were unavailable for examination at the time of this study. A recent, brief report on the excavation notes that 26 inscriptions have now been found at Horvat 'Uza. The most recent are reportedly Iron II economic texts bearing PNs (I. Beit-Arieh and B. Cresson, "Horvat 'Uza," *Excavations and Surveys in Israel* 7-8 [1988/89], 181).

[75] I. Beit-Arieh, "New Light on the Edomites," *BAR* 14/2 (1988), 34.

[76] Ibid., pp. 96–97. I have added a word divider (after למלכ in the first line) left out of the authors' block letter transliteration but clearly visible in the photograph (plate 12, no. 2), their drawing, and their English transliteration.

[77] Ibid. Ernest Knauf suggests that the letter was addressed to an Edomite king (למלכ) named BLBL, a name of Babylonian extraction. His translation of lines 2b-3a preserves the presence of the preposition on the DN Qaus, "Ich empfehle dich dem Segen von Qaus" ("Supplementa Ismaelitica," BN 45 [1988], 79).

[78] Chase ("A Note on an Inscription," pp. 63, 65). She compares this to the beginning of the Saqqara Papyrus (*KAI* 50) ושלמ את.

[79] Wolfgang Zwickel, "Das 'edomitische' Ostrakon aus Hirbet Gazza (Horvat 'Uza)," *BN* 41 (1988), 37–38.

[80] Beit-Arieh and Cresson, "An Edomite Ostracon," p. 98.

[81] Even more basically, Zwickel vigorously opposes the certainty with which the language of the inscription is called "Edomite" ("Das 'edomitische' Ostrakon," p. 39). But Knauf agrees with Beit-Arieh and Cresson and identifies other edomite traits in the letter ("Supplementa Ismael-itica," p. 79).

[82] This papyrus letter dates from the middle to late sixth century. Noël Aimé-Giron ("Adversaria Semitica (III): VII—Ba'al Saphon et les dieux de Tahpanhès dans un nouvneau papyrus phénicien," *ASAE* 40 [1940], 435), says the inscription cannot be later than the beginning of the fifth century and suggests dates between 568–525 B.C. It was found, with several demotic papyri, in the shaft of a *mastaba* in Saqqara, Egypt. Being from Egypt, and after the time-frame for this study, the inscription is included here only as an example. (Only the first two-and-a-half lines are included here; the rest of the inscription pertains to a business matter.)

(1 אל ארשת בת אשׄמנׄ[תנ]

(2 אמר.לאחתי.ארשת.אמר.אחתכ.בשא.ושלמ את.אף אנכ.שלמ.ברכתכ.לב

(3 על צפנ.ולכל אל.תחפנחס.יפעלכ.שלמ.

Reading from *KAI*, vol. 1, p. 12, text 50.)

> To Arišut daughter of 'ŠMNY[TN]. Say to my sister *Arišut*, your sister *Baš'u* says: 'And are you in good health? I, too, am in good health. I bless you by *Ba'al-Saphon* and by all the gods of *Tahpanhes*. May they make you to be in good health!'

(The translation is by Dupont-Sommer ["Note on a Phoenician Papyrus," p. 53] except for the address, line 1, which he does not provide [it is my translation]. The italics are his.)

There is also a collection of fifth century Aramaic letters from Hermopolis, Egypt which contain the formula PN *brk* PN *l*DN (cf. Kuntillet 'Ajrud 4 and, with slight variation, 3). The Hermopolis papyri are too late to be included as target inscriptions in this study but are important for purposes of comparison.

i:1–3

1) שלם בית נבו אל אחתי רעיה מן אחכי מכבנת
2) ברכתכי לפתח זי יחוני אפיך בשלם שלם בנתסרל וארג
3) ואסרשת ושרדר חרוץ שאל שלמהן...

Greetings to the temple of Nebo. To my sister R'YH from your brother Makkibanit. I have blessed you by Ptah, that he may let me see your face in peace. Greetings to BNTSRL and 'RG and 'SRŠT and ŠRDR ḤRWṢ asks after their welfare...

(*TSSI*, vol. 2, pp. 129–30. Only the first few lines of the first letter is given here for an example. The rest of the text and letters ii–vii may be found in *TSSI*, vol. 2, pp. 132–43). Letters i–iv all begin with the wish of peace to the temple of a god: i to the temple of Nebo; ii and iii to the temple of Banit; and iv to the temples of Bethel and the "Queen of Heaven." This greeting is followed in each case by a call for the blessing of the recipient by the god Ptah. Letters v and vi do not contain the greeting to a temple but do have the invocation to Ptah to bless in order that Ptah might allow them a safe and healthy reunion.

[83] S. A. Birnbaum, "The Sherds," in *The Objects from Samaria*, J. W. Crowfoot, G. M. Crowfoot, K. M. Kenyon (London: Palestine Exploration Fund, 1957), p. 11.

[84] E. L. Sukenik, "Inscribed Hebrew and Aramaic Potsherds from Samaria," *PEQ* 65 (1933), 153.

[85] William F. Albright, "Ostracon C 1101 of Samaria," *PEQ* 68 (1936), 212. The block letter transliteration is mine. The letter Albright reads as פ in the second word of the second line הפעם is the most contested. Other scholars read it as ד, ו, or ר.

[86] Albright, "Ostracon C 1101," p. 212. Albright's translation is given instead of that in Sukenik, the *editio princeps*, because Sukenik was unable to decide on a single translation though he suggested several ("Potsherds from Samaria," pp. 153–54).

[87] Sukenik, "Potsherds from Samaria," p. 153. I think that there was a mistake made in the printing of these articles and a holem should be present in each of his suggestions for the reading of שלם. The printed vowels are as he gives them.

[88] Albright, "Ostracon C 1101," p. 212.

[89] Sabatino Moscati, *L'epigrafia ebraica antica, 1935–1950*, Biblica et Orientalia, vol. 15 (Roma: Pontificio Istituto Biblico, 1951), p. 38.

[90] *KAI*, vol. 2, p. 185.

[91] Aharoni, *Arad Inscriptions*, pp. 42–43.

[92] Birnbaum, "The Sherds," p. 12. He actually includes 4 (IV) in both groups but the second "IV" should have been 6 (VI).

[93] Ibid.

[94] André Lemaire, *Les ostraca*, pp. 246, 248.

[95] *TSSI*, vol. 1, pp. 14–15.

[96] Arthur Ungnad, *Aramäische Papyrus aus Elephantine* (Leipzig: J. C. Hinrichs'sche Buchhandlung, 1911), pp. 103, 105.

[97] This is realizing, of course, that the written greetings form may have been shorter than the

Chapter 3: Blessing 81

spoken ones as a matter of style. For a very different interpretation of this text, see Kurt Galling's treatment ("Ein Ostrakon aus Samaria als Rechtsurkunde," *ZDPV* 77 (1961), 173–85). He interprets the ostrakon as a "justice document" which describes a theft and the penalty (translation, p. 185).

[98] See the discussions of Walter Maier, III (*'Ašerah:ExtrabiblicalEvidence*, Harvard Semitic Monographs, no. 37 [Atlanta: Scholars Press, 1986], pp. 168–73) and Saul Olyan (*Asherah and the Cult of Yahweh in Israel*, SBL Monograph Series, no. 34 [Atlanta: Scholars Press, 1988], pp. 23–37) and their bibliographies for the arguments.

[99] See Brian Peckham's comments cited on pages 53–54 of this study and his article "Phoenicia and the Religion of Israel," in *AIR*, pp. 80–81.

[100] See the discussion of Kuntillet 'Ajrud 4 in the section "ברך and שלם" above for a note on dating.

[101] Meshel, *Kuntillet 'Ajrud*, no page numbers. The block letter transliteration and translation are mine. Good photographs are found in Meshel, *Kuntillet 'Ajrud*, illustration 10; and Suzanne Singer, "Cache of Hebrew and Phoenician Inscriptions Found in the Desert," *BAR* 2/1 (1976), 33.

[102] This is probably the correct use for the bowl even though not explicitly stated as in other dedicatory inscriptions (e.g., the Aramaic Melqart Stele, *KAI* 201). Three late Phoenician examples of interest are a votive inscription of Ba'alshillem, king of Sidon (ca. 400 B.C., *TSSI*, vol. 3, p. 114); a dated dedication from Cyprus (*KAI* 38, 390 B.C. *TSSI*, vol. 3, p. 132); and inscription IX from Umm El-'Amed (late third century B.C., *TSSI*, vol. 3, p. 121) all of which conclude with the request of the deity "May he bless him" יברך. For further comment on Kuntillet 'Ajrud as an "Israelite religious center" (Ze'ev Meshel, "Kuntillet 'Ajrud, An Israelite Religious Center in Northern Sinai," *Expedition* 20 [1978], 50–54) or as an "Israelite caravanserai and attached shrine" (William G. Dever, "Recent Archaeological Confirmation of the Cult of Asherah in Ancient Israel," *HebrSt* 23 [1982], 37–43) see these articles or those already mentioned on the site.

[103] The translation is mine. The RSV fails to include the words "to the Lord" in its translation.

[104] See the discussion of "Substantive Blessing" below in this chapter.

[105] This inscription and Khirbet el-Qôm could well comprise a separate category, "Yahweh with another DN." However, as the identity of "Asherah/asherah" remains uncertain, both inscriptions will be included in this section.

[106] The best photograph (and in color) of this inscription is found in André Lemaire, "Who or What Was Yahweh's Asherah," *BAR* 10/6 (1984), 45. For a depiction of these drawings in relationship to each other (figure 5, p. 9) and a discussion of them, see Pirhiya Beck, "The Drawings from Ḥorvat Teiman (Kuntillet 'Ajrud)," *TA* 9 (1982), 3–68. Beck believes that all of the animal paintings were done by the same person and that that person was aware of "a wide range of decorated objects produced by both Phoenician and North Syrian artisans" (p. 27). For Beck's determination that the two figures represent Bes, see pp. 27–31. She thinks that the Bes figures and lyre player are each by different painters and that they, with the inscriptions, were added to the pithoi after firing, indeed, after they were placed in their final location "the bench room" (pp. 31, 36, 43).

[107] Beck, "Drawings," p. 46.

[108] Ibid., pp. 46–47. Other scholars do not agree. Margalit proposes a new interpretation of the the word אשרתה, *'šrt or *'trt from 'tr "trace" or "footstep" meaning "to follow (in the tracks of)" and signifying "wife" or "consort"—"she (who is) walking behind (her man)" (Some Observations on the Inscription and Drawing from Khirbet el-Qôm," *VT* 39 [1989], 374). In a note, he goes on to point out that on pithos A the two "Bes figures" are drawn in perspective, the second (female) standing behind the first (male) figure (p. 378, n. 18).

[109] For a fuller note on the dating, see the introduction to Kuntillet 'Ajrud 4 in "ברך and שלם" in this chapter.

[110] The reading is Meshel's (*Kuntillet 'Ajrud*, no page number) except for the end of the PN enclosed by brackets in line one which is Hadley's ("Kuntillet 'Ajrud," p. 182).

111 Hadley, "Kuntillet 'Ajrud," p. 182.

112 Meshel, *Kuntillet 'Ajrud*, no page number. His illustration 9 is a photograph of most of the second line, the initial ה (given in the reading on the next page) is cut from the photograph.

113 See the introduction to the discussion of Kuntillet 'Ajrud 4 in "ברך and שלמ" above.

114 Meshel, *Kuntillet 'Ajrud*, no page number. Meshel's description of the inscriptions on plaster are even less complete than those he gives for the inscriptions on the pithoi.

115 In a recent communication Kyle McCarter shared that his reading is אךכ[י] ימם "may he prolonged days." Thus he finds no occurrence of ברך in this inscription though the resulting reading would be understood as a Substantive Blessing.

116 And even in the earlier Aramaic scripts it was written as a full circle with a single diagonal inside of it (Joseph Naveh, *The Development of the Aramaic Scripts*, The Israel Academy of Sciences and Humanities Proceedings, vol. 5, no. 1 [Jerusalem: The Israel Academy of Sciences and Humanities, 1970], p. 8). However, McCarter (personal communication) says that the form of the ט is fine for cursive Phoenician of the period.

117 Since the script of some of the inscriptions is said to have Phoenician affinities (Meshel, "An Israelite Religious Center," p. 53), this difficult letter might be compared to the Phoenician letter form of ל (though, admittedly, the left part of the curve is higher in Phoenician than the right and the reverse is the situation here, cf., Kilamuwa 1). Some form of the verb לבב (KB, p. 471) might be suggested as an alternate reading. KB notes Akkadian *labābu* which means "be enraged" but reports a tenderer meaning in Cant. 4:9 and a wisdom reference in Job 11:12, the two occurrences in BH. Perhaps the Canticles reference is more to the point, the blessing may be an invocation of tender feelings from Yahweh toward the blessed.

118 William G. Dever, "Iron Age Epigraphic Material from the Area of Khirbet el-Kôm," *HUCA* 40/41 (1969–1970), 165. He notes that Cross, in a personal communication, prefers a date of circa 700 (Ibid., n. 53).

119 Reading and translation by Dever (Ibid., p. 159).

120 Ibid., pp. 160–61.

121 Dever freely admits the problems with his reading of the line and even says that perhaps all of line 3 might need "to be divided and understood quite differently" (Ibid., p. 162). Indeed, in his article on the Kuntillet 'Ajrud inscriptions ("Asherah, Consort of Yahweh? New Evidence from Kuntillet 'Ajrud," *BASOR* 255 [1984], 22) he quotes the readings of Lemaire ("Les inscriptions de Khirbet el-Qôm et l'ashérah de Yhwh," *RB* 84 [1977], 599) and Naveh ("Graffiti and Dedications," *BASOR* 235 [1979], 28–29) with approbation.

122 The question regarding the identification of "asherah/Asherah" in this inscription, as in the Kuntillet 'Ajrud inscriptions where it appears, is deferred to others. It is either discussed or bibliography given in each of the articles cited above. For a more recent note to the discussion see Jeffrey H. Tigay, "A Second Temple Parallel to the Blessings from Kuntillet 'Ajrud," *IEJ* 40 (1990), 218. Baruch Margalit has made his own intriguing (and convincing) contribution to the discussion for the view that the reference is actually to the goddess Ashtarte ("Some Observations," pp. 371-78; also see n. 108 above).

123 Interestingly, these leave open the question of why the inscription was found in a burial cave if the request is for future blessing? Is the blessing concerned with afterlife? Is the inscription related to the use of the cave for burial at all? If this inscription does relate to afterlife it is the only blessing inscription in this study that does so. I am inclined to think that the inscription of 'Uriyahu does not have afterlife overtones.

124 Lemaire, "Les inscriptions de Khirbet el-Qôm," p. 599. The italics belong to Lemaire. The transliteration into block letters is mine.

125 Lemaire, "Yahweh's Asherah?," p. 44.

126 Lemaire, "Les inscriptions de Khirbet el-Qôm," pp. 600–01.

Chapter 3: Blessing 83

[127] Patrick Miller Jr., "Psalms and Inscriptions," in *Supplements to Vetus Testamentum, Congress Volume Vienna: 1980*, ed. J. A. Emerton (Leiden: E. J. Brill, 1981), p. 317. The transliteration into block letters is mine.

[128] Ibid., pp. 316–17.

[129] Ibid., pp. 316, n. 13 and 317.

[130] Along these lines see Jeffery Tigay's brief article which proposes a parallel ("A Second Temple Parallel to the Blessings from Kuntillet 'Ajrud." *IEJ* 40 [1990], 218).

[131] Ibid., pp. 317–19. Other parallel uses of these two words are found in Pss. 24:5, 109:26–28, and Zech. 8:13.

[132] Ibid., p. 319, n. 17.

[133] Siegfried Mittmann, "Die Grabinschrift des Sängers Uriahu," *ZDPV* 97 (1981), 142, 144. The transliteration into block letters is mine. Mittmann's readings are very similar to Lemaire's:

1) אריהו.השר.כתבה
2) ברכ.אריהו.ליהוה
3) וממצר ידה לאל שרת הושעלה

[134] Ibid., p. 145.

[135] Ibid., p. 152.

[136] William G. Dever ("Asherah, Consort of Yahweh?," p. 32, n. 6), notes that though the "consonantal rendering may be acceptable, the necessary vocalization seems to make for very awkward syntax." Hadley ("The Khirbet el-Qôm Inscription," *VT* 37 [1987], p. 55) says Mittmann's interpretation, which she translates into English as "the God of his service," is not so much a syntactical problem as it is just "strange."

[137] Karl Jaroš, "Zur Inscrift Nr. 3 von Ḥirbet el-Qôm," *BN* 19 (1982), 36. Jaroš's reading is hard to work with because he remains undecided on so many letters.

1) אריהו.העשר.כתבה
2) ברכ.אריהו.ליהוה
3) ו(מ)מצד[ד]י (ד)(ה)לא[ל]שרתה.הושע לה
4) לאניהו
5) ולאשרתה
6) ו ה

[138] Judith M. Hadley, "Khirbet el-Qôm," p. 51. She notes that if ליהוה and לאשרתה are considered a "compound linguistic stereotype" this line should be read "(and) by his asherah, for from his enemies he has saved him." The transliteration into block letters is mine.

[139] Ibid., pp. 60–62.

[140] "Möglicherweise ist hier ein apotropäischer Fluch literarisch formuliert, den in Chirbet el Qôm die abwehrende Hand blidlich zum Ausdruck bringt." S. Schroer, "Zur Deutung der Hand unter der Grabinschrift von Chirbet el Qôm," *UF* 15 eds. K. Bergerhof, M. Dietrich, and O. Loretz (Neukirchen-Vluyn: Verlag Butzon & Bercker Kevelaer, 1983), p. 199. If this graphic depiction did symbolize a curse then this inscription would be included among those in chapter 5, "Blessing and Curse."

[141] Baruch Margalit, "Some Observations," p. 373. The "tree design" is Margalit's interpretation of the long extraneous marks down the left hand side of the inscription. For his interpretation which ties these to the identity of לאשרתה see pp. 373-75 and 378, n. 18. This also explains his translation of "Asherah" as "consort" in his "lower inscription." He comments briefly on Kuntillet 'Ajrud 3. He argues (quite well) that the drawings on pithos A *are* related to the inscription, contra

Beck ("The Drawings from Ḥorvat Teiman [Kuntillet 'Ajrud]," *TA* 9 [1982], 3–68.). (The inscription which also speaks of "Yahweh and his Asherah.") See n. 108 above. The late thirteenth century B.C. Lachish ewer (decorated with religious symbols and bearing a dedicatory inscription) is compared by Ruth Hestrin to the Kuntillet 'Ajrud 3 text and the drawings which are on pithos A. She makes several valuable comparisons between these two inscribed, decorated vessels but does not suggest a solution to the "Asherah/asherah" problem ("The Lachish Ewer and the 'Asherah," *IEJ* 37 [1987], 220-23).

[142] Joseph Naveh, "Graffiti and Dedications," p. 28. The transliteration into block letters is mine. Miller suggests some serious difficulties with Naveh's reading ("Psalms and Inscriptions," pp. 316–17, n. 14).

[143] Naveh, "Graffiti, and Dedications," p. 27.

[144] Table 1 is a modification of information Naveh ("Graffiti and Dedications") presents on page 28 of his article. The block letter transliteration is mine. "'Ajrud Pithos I" and "'Ajrud Pithos II" are "Kuntillet 'Ajrud 3" and "4" respectively in this study.

[145] Ziony Zevit, "The Khirbet el-Qôm Inscription Mentioning a Goddess," *BASOR* 255 (1984), p. 43. The transliteration into block letters is mine. His reading is:

1) אריהו העשר כתבה
2) ברכת אריהו.ליהוה
3) ומצריה לעשרתה הושע לה
4) לאביהו
5) [ה/ר/באמ/? וללאשרתה
6) [א??רתה

He actually reads all of these letters ר/ומצרריה‎ הלאלשארתהתלההושעלה in line 3, but determines "the sequence of letters conveying lexical information" in this line to be that given above (p. 44). Zevit's translation includes two elements which make it remarkably different from those of others but which are strictly speaking outside the parameters of this study, those being in his interpretation of the form לעשרתה, a) seeing the ending as a "double feminization" (p. 46) and b) regarding the ל as a vocative indicator (p. 46). For a more than adequate rebuttal of both these ideas see Hadley ("Khirbet el-Qôm," pp. 58–59). Zevit's inclusion of several scholars' readings and translations of the text in his article was a good idea but several printing errors have crept in.

[146] Zevit, "Khirbet el-Qôm," p. 44. After a first-hand examination of the inscription, Hadley positively asserts that the "ח" is actually just a couple of incidental scratches in the rock ("Khirbet el-Qôm," p. 61).

[147] Ibid. His references are Deut. 33:11; Judg. 5:2, 9; Pss. 96:2; 100:4; 103:20–22; and Job 2:9 a prose example, but with אלהים rather than יהוה. For contrast he does provide a poetic example (Ps. 135:19–20) which does use את־יהוה after ברכו. Note, however, that each time the imperative is used, various parties are called upon to "bless Yahweh," not another person. Hadley ("Khirbet el-Qôm," p. 54) also makes these objections.

[148] At least Barkay thinks so and restores the suffixes in both of the amulets (*Ketef Hinnom*, pp. 29, 30).

[149] The perfect first person singular forms are relatively rare, occurring only five times and all with Yahweh as subject either followed by the sign of the direct object with a pronominal suffix (Gen. 17:16, 20) or having a pronominal suffix appended to the verb (Gen. 17:16; 26:24; Exod. 20:24). The imperfect first person singular forms are more abundant (eighteen times) but yield the same results with the exception that this form is often directed toward Yahweh (Gen. 24:48; Pss. 16:7, 34:2 with את; Ps. 26:12 without את; Ps. 145:1 with some less direct reference to Yahweh, i.e., "his name").

[150] Michael O'Connor, "The Poetic Inscription from Khirbet el-Qôm," *VT* 37 (1987), 228–29. O'Connor seems to be the only other scholar to accept the "double feminization" and vocative use of ל which Zevit finds in the inscription.

Chapter 3: Blessing 85

[151] William H. Shea, "The Khirbet el-Qôm Tomb Inscription Again," *VT* 40 (1990), 110.

[152] Ibid., pp. 113-14, 116.

[153] Ibid., p. 113.

[154] This is, of course, disregarding the first interpretation of Dever who saw it as a curse to protect the inscription.

[155] See the category "ברכ and שלמ with Substantive Curse" in Chapter 5.

[156] H. Theodore Bossert and U. Bahadir Alkim, *Karatepe: Kadirli and Its Environments, (Second Preliminary Report)*, Publications for the Institute for Research in Ancient Oriental Civilisations, no. 3 (Istanbul: Pulhan Basimevi, 1947), plate 40. The transliteration from Phoenician to block letters is mine, as is the translation.

[157] Scharbert, "ברך," p. 281.

[158] Stanislav Segert, *A Grammar of Phoenician and Punic* (München: Verlag C. H. Beck, 1976), p. 137. The grammar section is 54.252.2. The PN is from text *CIL* 19715 (*Corpus InscriptionumLatinarum*, vol. 8).

[159] Franz Rosenthal, "Azitawadda of Adana," in *ANET*, p. 653.

[160] *KAI*, vol. 2, pp. 36, 38–39.

[161] Zellig S. Harris, *A Grammar of the Phoenician Language*, American Oriental Series, vol. 8 (New Haven, CT: American Oriental Society, 1936), p. 92. Harris wrote too early to include this inscription in his treatment.

[162] Charles F. Jean and Jacob Hoftijzer, *Dictionaire des inscriptions sémitiques de l'ouest* (Leiden: Brill, 1965), p. 44.

[163] François Bron, *Recherches sur les inscriptions phéniciennes de Karatepe*, II Hautes Études Orientales, vol. 11, (Paris: Librairie Champion, 1979), p. 23.

[164] Ibid., p. 32. Bron's discussion of the possibilities is contained in pages 28–32.

[165] *TSSI*, vol. 3, p. 56. He further suggests two more possible Phoenician examples: Pyrgi lines 10–11 (p. 154) הככבמ אל "the stars of El" (which, however, he translates "like the stars above") and Yehaumilk line 4 (p. 94) המזבח נחשת זן "this altar of bronze"; and a tempting biblical example הנחלים ארנונ "the torrents of Arnon" (p. 159).

[166] *TSSI*, vol. 3, p. 56. The Hittite version of this section means something like "man illuminated by the Sun-god" but the translation is not sufficiently dependable to decide the matter. Lipiński, however, thinks that the Hieroglyphic Luwian text can be sufficiently translated to decide the debate in favor of the second option "vizier" ("From Karatepe to Pyrgi, Middle Phoenician Miscellanea," *RSF* 2 [1974], 45). He translates Gen. 41:43, "He mounted him in the chariot of his viceroy and they cried before him: 'The wizier [sic]'" (pp. 46–47). The word המשנה which the RSV translates "second" refering to the Pharaoh's second chariot may also (BDB, p. 1041) refer to the second in rank, cf. 2 Kgs. 23:4. The vizier is the second in rank to pharaoh. The Late Babylonian name of the *abarakku* was *mašennu* corresponding to BH משנה. It should be noted though, that Lipiński's other near parallels for the Genesis passage pertain to pretenders to the throne (2 Sam. 15:1 and 1 Kgs. 1:5) and the BH uses of משנה to mean second in rank are rare and refer only to priests.

[167] W. Randall Garr, *Dialect Geography of Syria-Palestine, 1000–586 B.C.E.*, (Philadelphia: University of Pennsylvania Press, 1985), p. 130.

[168] Meshel, "Did Yahweh Have a Consort?" p. 32.

[169] Meshel, *Kuntillet 'Ajrud*, no page numbers. The reading and translation are Meshel's. Nothing is explicitly said about the relationship of the lines—or even the words—to each other!

[170] Weinfeld, "Kuntillet 'Ajrud," p. 126. Note that he does not translate ברכ in line 2 and his translation is inconsistent, including "the" before day in line 2 but excluding it from the same

context in line 3. As no photograph of this inscription has, to my knowledge, been published, it is not possible for anyone who has not seen the inscription itself to evaluate these readings. Because the information about this inscription is so incomplete, it is hard to reach overall conclusions. Seemingly, the inscription does contain five lines of which three or perhaps four are given. I say "perhaps four" because Weinfeld's reading differs so markedly from Meshel's. If they are the same the spaces between letters have been very inaccurately reported, though, understandably, different readings of letters may occur.

[171] It is used to describe Ba'al's sexual prowess with Anath (*UT* 76:III:9, translated in *UL*, p. 50 or *ARTU*, p. 114), Anath's heart full of laughter at anticipation of slaughter (*UT* 131:7 and 'nt II:25, translated in *UL*, pp. 28 and 18 respectively), and the gold of El's full table (*UT* 51:I:39, translated in *UL*, p. 28).

[172] Two other late inscriptions that provide examples of the passive participle ברי כ with a DN are, first, a funerary stele from an unknown Egyptian site (the Carpentras Stele, *CIS* ii 141) which dates from ca. 390 B.C.:

1) בריכה תבא ברת תחפי תמנחא זי אוסרי אלהא
2) מנדעם באיש לא עבדת וכרצי איש לא אמרת תמה
3) קדם אוסרי בריכה הוי מן קדם אוסרי מין קחי
4) הוי פלחה נמעתי ובין חסיה[י] חיי לעלם]

"Blessed be TB', daughter of THPY, devotee of the god Osiris. Naught of evil did she do, nor calumny against any man did she utter up there. Before Osiris be you blessed; from Osiris receive water. Serve the lord of the Two Justices; and among his favoured ones [live forever]" (*TSSI*, vol. 2, p. 121).

The second inscription is written upon a tablet discovered in a grave in Saqqara, Egypt and dates from ca. 482 B.C. (*CIS* ii 122). The upper portions of the tablet bear an Egyptian funeral scene. (Only the first line and part of the second are relevant.)

1) בריך אבה בר חור ואחתבו ברת עדיה כל [זי] חסתמח קרבתא
2) קדם אוסרי אלהא...

"Blessed be Aba, son of HWR and 'HTBW, daughter of Adiya, both of whom were favoured, faithful. The approach before the god Osiris...." (*TSSI*, vol. 2, p. 119).

[173] Of the sixteen Hebrew inscriptional occurrences, eleven are with Yahweh or in the same context, two have no context, and two have no DN. However, note the pairing of Yahweh with Asherah/asherah in Kuntillet 'Ajrud 3 and Khirbet el-Qôm.

[174] This inscription is included among the target texts from this study because it probably originated in Phoenicia (*TSSI*, vol. 3, p. 71).

[175] Eric Burrows, "Phoenician Inscription from Ur," *JRAS* [no vol.] (1927), 791.

[176] C. Leonard Woolley, "The Excavations at Ur, 1926–7," *The Antiquaries Journal* 7 (1927), 410.

[177] *TSSI*, vol. 3, p. 71.

[178] Ibid. H. L. Ginsberg ("Ugaritico-Phoenicia," *JANES* 5 [1973], 141) and subsequently Gibson (*TSSI*, vol. 3, p. 72) restore a ש rather than a ז in the first line (thus "ivory" rather than "this") for the reasons that זן is a relative pronoun elsewhere only in the Byblian dialect whereas this is Tyro-Sidonian (demonstrated by the feminine suffixes on some of the words) and because the lacuna is larger than needed for a ז and traces of a ש may be visible at the edge of the break.

[179] R. D. Barnett, *A Catalogue of the Nimrud Ivories with Other Examples of Ancient Near Eastern Ivories in the British Museum* (London: British Museum, 1957), p. 226.

[180] René Dussaud, "Une inscription phénicienne découverte à Our en Chaldée," *Syria* 9 (1928), 268. "A cause de dela qu'elle me bénisse pendant les jours de notre maître (un tel) fils (d'un tel)."

[181] Segert, *Grammar*, p. 101, section 51.225.

Chapter 3: Blessing   87

[182] Harris, *Grammar*, p. 91.

[183] Ibid., p. 40; Segert, *Grammar*, p. 140, section 54.345.

[184] *KAI* 29, vol. 2, p. 47. "Möge sie sie in ihren Tagen segnen."

[185] R. Savignac, "Inscription phénicienne d'Ur," *RB* 37 (1928), 259. His translation of the section in question, "Qu'elle la benisse! Aux jours de...."

[186] Birnbaum, "The Sherds," p. 21.

[187] Ibid., p. 20. Though he mentions "seven letters" he only reads six and the photographs only show six. See figure 14. 5–7, 10 (p. 145) for drawings of "high foot bowls."

[188] E. L. Sukenik, "Inscribed Potsherds with Biblical Names from Samaria," *PEQ* 65 (1933), 202. Sukenik is the only commentator to show a seventh letter in his reading. He says (p. 202), that the right part of the ʼ can be "clearly traced" at the broken edge of the inscription.

[189] Ibid., pp. 202–203.

[190] Ibid., p. 203.

[191] Birnbaum, "The Sherds," p. 20.

[192] See Chapter 5.

[193] Lemaire, *Les ostraca*, p. 250.

[194] For example, Larry G. Herr (*The Scripts of Ancient Northwest Semitic Seals*, Harvard Semitic Monographs, vol. 18 [Missoula, MT: Scholars Press, 1978]) gives a seventh century Hebrew example having ל but not בן: לשפטיהו עשיהו (p. 91); an eighth century Hebrew examples without ל or בן: צפן עזריהו (p. 87) and הושע צפן (p. 89).

[195] In addition, perhaps two partially preserved sherds from Beer-Sheba also evidence this absence of בן (Lemaire, *Les ostraca*, pp. 271–73).

[196] Aharoni, *Arad Inscriptions*, p. 118. However, two other, identical vessels from Arad (numbers 102 and 103) perhaps also had sacred purposes and their cryptic inscriptions, interpreted as ק ש, are written on the inside near their rims (pp. 115–17).

[197] Yohanan Aharoni, *Beer-Sheba I: Excavations at Tel Beer-Sheba 1969–1971 Seasons* (Tel Aviv: Tel Aviv University, Institute of Archaeology, 1973), p. 73. The identifying number is the locus/find number. The word קדש is inscribed on the shoulder of a krater.

[198] Yigael Yadin, et al., *Hazor III–IV: An Account of the Third and Fourth Seasons of Excavations, 1957–1958, Plates* (Jerusalem: Magnes Press, 1961), plates CCCLVII:4, 5 and CCCLVIII:4, 5. A single shallow bowl bears the word קדש twice, once the letters are evenly spaced out along the rim and the other time they are written on the outside and preceded by what Yadin calls "an undecipherable" word קדש.יה- the inscription breaks off at some letter before the ʼ and מו is written above the חה of the first word (*Hazor: The Rediscovery of a Great Citadel of the Bible* [London: Weidenfeld and Nicholson, 1975], p. 182).

[199] A distinction might be made though, between votive vessels and sacred vessels.

[200] Or, of course, לבעל. In a late Aramaic example (below), the noun ברכה is the object of the verb שלחת which usually takes שלם as its object (cf. Arad 16, 21, and 40[?], also the Hermopolis papyri i:12; ii:17, etc.). This combination does not occur in one of the target inscriptions. One of the Aramaic inscriptions published by Cowley has a epistolary form which differs interestingly from the formulas seen above, the letter begins

1) אל אמי קויליה ברכה
2) שלחת לכי...

"To my mother Kovelia: a blessing I send to you." (A. Cowley, "Two Aramaic Ostraka," *JRAS* [no vol.] [1929], 108.) Note that David "sent" (שלח) "to greet" (לברך) in 1 Sam. 25:14.

[201] Aharoni, *Arad Inscriptions*, pp. 8, 54.

202 Ibid., p. 54.

203 David Ussishkin, "Excavations at Tel Lachish 1978–1983: Second Preliminary Report," *TA* 10 (1983), 157.

204 Ibid., p. 158.

205 Aharoni, *Arad Inscriptions*, pp. 8–9, 134.

206 Ibid., p. 35. The readings and translation are Aharoni's. His translation does not attempt a literal representation of the Hebrew, e.g., his translation of "God" for יהוה in line 9.

207 Ibid., pp. 35–36.

208 Pardee, *Handbook*, pp. 55–56.

209 Pardee, "Letters from Tel Arad," p. 316.

210 Reading: Albright, "A Prince of Taanach," p. 12; translation, Albright and Mendenhall, "Taanach 1" in *ANET*, p. 490.

211 Reading, Knudtzon, *Di ͞El-Amarna-Tafeln*, vol. 1, pp. 442–43; translation, R. Youngblood, "Canaanite Amarna Letter (EA 96)," p. 24.

212 Dalley, et al., *The Old Babylonian Tablets*, p. 52, text 40, lines 4–5, 12–13. Also see texts 41 and 43 (pp. 52–54).

213 *AP* 30:1–2, pp. 111–13. Also *AP* 31:2; 37:1–2; 38:2; and *BMAP* 13:1 are all similar letters from inferiors to superiors. This greeting is also used in Aramaic letters between equals: *AP* 56:1 אלהיא ישאלו שלמך בכל עדן "may the gods seek after your welfare at all times"; 21:1; 40:1; and 41:1. In Hermopolis i:3 and vi:2, 7–8 (*TSSI*, vol. 2, pp. 129, 140–41), individuals "ask" after the welfare of others. The Aramaic letter of King Adon (*KAI* 266) was written within the time frame of this study (ca. 605–04 B.C.) but is too fragmentary to be treated as a primary text. The restorations of it by both Donner and Röllig and Fitzmyer ("The Aramaic Letter of King Adon to the Egyptian Pharaoh," *Biblica* 46 [1965], 44; and Fitzmyer, *A Wandering Aramean*, pp. 231–42) include the phrase dealt with above. Only the first two-and-a-half lines, as read and reconstructed by Fitzmyer, are included here (quoted from the *Biblica* article, p. 44; note that the transliteration in *A Wandering Aramean*, p. 232, incorrectly reads עברכ for עבדכ in the first line).

1) אל מרא מלכן פרעה עבדך אדן מלך [שלם מראי עשתרת בעלת
2) שמיא וארקא ובעלשמין אלה[א רבא ישאלו בבל עדן וישימו כרסא
3) פרעה כיומי שמין אמין

"To the Lord of Kings, the Pharaoh, your servant 'Adon, the king of [May Astarte, the Mistress] of the heavens and the earth, and Ba'alshamyn, [the great] god, [seek the welfare of my lord at all times and make the throne] of Pharaoh (as) enduring as the days of heaven."

214 Foresti, "Characteristic Literary Expressions," p. 331. He translates ישאל as he does based on the use of the verb in passages like Deut. 14:26.

215 Ibid., p. 333.

216 Torczyner, et al., *The Lachish Letters*, p. 12.

217 Ibid., p. 37. Though the reading and translation are Torczyner's, I have provided marks over questionable letters based on the photographs and drawings provided in the book. (The photographs and drawings in the book were prepared by persons other than Torczyner.)

218 Lemaire, *Les ostraca*, p. 98.

219 Henri Michaud, "Les ostraca de Lakiš conservés a Londres," *Syria* 34 (1957), 41.

220 Pardee, *Handbook*, p. 81.

221 Torczyner, et al., *The Lachish Letters*, p. 123. This inscription is only included because of Torczyner's reconstruction. David Diringer ("Early Hebrew Inscriptions," in *Lachish III: The Iron*

Chapter 3: Blessing

Age, pt. 1 [London: Oxford University Press, 1953], p. 335) reads very few letters of this inscription: only a ש in line 3, only הו in line 4, and only בשלמ at the beginning of line 5.

²²²Torczyner, et al., *The Lachish Letters*, pp. 135, 137. At several points Torczyner's reading of the sherd seems inaccurate. The ב that he includes at the beginning of line 2 is not visible in either the drawing or photograph provided, whereas a clear ו is visible at the end of line 3 of the photograph and appears in the drawing but is not incuded in his reading. Also, the ב he records at the beginning of line 5 is not visible but a study of the photograph indicates the presence of two לs in the line which he does not show.

²²³Ibid., p. 51.

²²⁴Letters 4 and 8 comprise a fourth variation of this formula which, technically speaking, differs significantly from the other letters by their substitution of טב for שלם and therefore are not included in the body of this study. As regards meaning, however, they are quite close and perhaps even synonymous and thus are mentioned here. In a minor way, the two letters also vary from each other: letter 4 interjects the phrase "even now" עת כימ before שמעת טב (and includes ועת after the greeting) unlike 8 where עת כימ serves as the end of the greeting before the substance of the letter. Letter 4:1–2:

1) ישמע.יהוה [את] אדני עת כימ.

2) שמעת טב.

"May Yhwh let hear my lord even now tidings of good" (Ibid., p. 79).
Letter 8:1–2:

1) ישמע יהוה את.אדני ש|למ

2) עת טב עת כימ

"May Yahweh cause my lord to hear good news" (reading from p. 129, my translation). (Note that Diringer's reading (*Lachish III*, p. 335) gives no indication of a ש or כ at the end of line 1 and records the first four letters of line 2 as very questionable.)

²²⁵Ibid., pp. 93, 97. The inscription is very poorly preserved and given here only based on Torczyner's reconstruction. Note, however, that Diringer (*Lachish III*, p. 334), who is usually more cautious in his readings than Torczyner, sees stronger indications of שלמ in line 2 than does Torczyner.

²²⁶W. L. Moran, "A Note on the Treaty Terminology of the Sefîre Stelas," *JNES* 22 (1963), 174.

²²⁷Ibid.

²²⁸Wiseman disagrees with Moran saying *šulummu u ṭubtu* "denotes a state of non-hostility and is used of equals whether or not thay are in covenant bond" ("Is it Peace," p. 313). He also warns against identifying a covenant relationship everytime a certain phrase or inuendo occurs (pp. 312-13).

²²⁹Torczyner, et al., *Lachish I*, pp. 106, 117.

²³⁰Compare the treatment of the other Lachish letters above.

²³¹P. Joüon, "Sur les ostraca hébraïques de Lachish," *RÉS* [no vol.] (1938), 86. He says there must be a ו before the word אתה to justify Torczyner's interpretation; further, he states that עת is always written without final ה. Thus, his word division is את העת הזה. He gives several examples in each case from biblical and inscriptional texts.

²³²Ibid., pp. 86–87.

²³³Ibid., p. 88.

²³⁴William F. Albright, "The Lachish Letters After Five Years," *BASOR* 82 (1941), 22. Hempel's translation is similar, "Es lasse sehen Jahwe meinen Herrn diese Zeit glücklich!" (J. Hempel, "Die Ostraka von Lakiš," *ZAW* 15 [1938], 135). Those of Lemaire (*Les ostraca*, p. 120) "Que Yhwh fasse voir à mon maître ce moment-ci en paix," and *KAI* (vol. 2, pp. 196–97, no.

196) "Möge Jahve sehen lassen meinen Herren diese Zeit (in guter) Gesundheit," are also similar.

[235] H. L. Ginsberg, "Lachish Notes," *BASOR* 71 (1938), 27.

[236] Cyrus Gordon, "Notes on the Lachish Letters," *BASOR* 70 (1938), 17. The jumble of letters on the second line should be divided as "*'ot ha'et hazze*(*h*), this sign of the times." Comparison of this pronoun with the normal form of the feminine pronoun in line 10, shows that the form in line 2 is masculine *ze* (*h*) and actually modifies masculine *'oth*.

[237] Ibid.

[238] There are at least three other inscriptions which may be included as examples. The first example is from a mid-seventh century Aramaic document from Assyria. (As this text is from outside the target area and seems to have originated there, it is not included in the body of this study.) Only the first line is relevant for this study.

(1 [אל א]חי פרור אחוכ בלטר שלמ לש---

"An meinen Bruder Pirʾ-Awurr, dein Bruder Bel-eṭir Gruss!"
(Mark Lidzbarski, *Altaramäische Urkunden aus Assur*, 38 Wissenschaftliche Veröffentlichung der Deutschen Orient-Gesselschaft, [Leipzig: J. C. Hinrichs'sche Buchhandlung, 1921], pp. 8, 14 [an excellent photograph is included as tafel 1]; also *TSSI*, vol. 2, pp. 102–103.)

The second is the seventh of the Hermopolis papyri (ca. 500 B.C.) which, unlike numbers i–iv, has no greeting to a temple in the city of the recipient and, unlike i–vi it has no reference to a deity. (Number viii, the last of the collection, was not accessible for this study but is said to be very fragmentary and without an address [*TSSI*, vol. 2, p. 125].) Perhaps this difference from the others is partially explained by number vii being written by a different hand than all of the others (Ibid., p. 126). Thus it could be said that this scribe did not utilize the same conventional formulas as did the other, or that this scribe's religious beliefs were even different; there is, of course, no way to find out. Note that שלמ occurs several times for "greeting."

(1 אל אמי עתררמרי מנ אחוכי אמי [שלמ וח]ינ שלכת לכי
(2 שלמ אחתי אסורי וזבבו וככי וכענ עליכי מתכל אנה הוי
(3 חזית על ינקיא אלשי שלמ וסרו ושפנית ובניה ופטמונ שלמ
(4 הריוטא ואחתהה לשלמכי שלחת ספרא זנה
Reverse (address)
(5 אל אחתי עתררי מנ א[חוכי א]מי אפי יובל

To my 'mother' 'TRRMRY from your 'brother' 'MY. [Prosperity and long] life I have written wishing you. Greetings to my sister(s) 'SWRY and ZBBW and KKY. Now I am relying on you. Look after these little ones. Greetings to WSRW and to ŠNYT and her children and to PṬMWN. Greetings to HRYWṬ' and his sister. For your welfare I have written this letter. To my "sister" 'TTRY from your ["brother"] 'MY. For delivery to Ofi.

(*TSSI*, vol. 2, pp. 142–43.)

The third inscription is the second of the three fifth century Aramaic letters in the Padua Museum (Italy). The text is designated Pad II (Fitzmyer, "The Padua Aramaic Papyrus Letters," p. 22; or Fitzmyer, *A Wandering Aramean*, p. 227). The text is broken but nevertheless clearly uses שלמ as a greeting four times. (It is quoted here without the address.)

(1 אל אמ[י] יה[ו]י[ש]מע ברכ שלומ בר פטמ[ונ
(2 לכ שלמ גלגל תנה שלמ ינקיה   זכ]
(3 בלא הבה לפחנומ בר נבודלה ויעב[ד
(4 יגרוהי והנ איתי כספ הבי על פמי על[[שלמכ
אמי
(5 שלחת ספרה זנה שלמ מנחמת שלמ]
(6 [א]ל אמי יה[רי]שמע

To my mother, Yahuyishmaʿ; your son, Shallum bar Peṭam[un.] Greetings

Chapter 3: Blessing   91

to you from Galgul here, greetings from the children! [   ] BL' give to Pakhnum bar Nabudalah and let him d[o] they will prosecute him, and if there is money, give (it) according to my instruction. [For your welfare] I send this note. Greetings from Menaḥemet, my mother (?); greetings [ ] Yahuyishma'.

²³⁹Pierre Benoit, J. T. Milik, and Roland de Vaux, *Les grottes de Murabba'ât*, Discoveries in the Judaean Desert, volume II, part 1 (Oxford: Clarendon Press, 1961), p. 7.

²⁴⁰This date is according to Cross ("Epigraphic Notes on Hebrew Documents of the Eighth–Sixth Centuries B.C.: II. The Muraba'ât Papyrus and the Letter Found Near Yabneh-Yam," *BASOR* 165 [1962], 42) and Gibson (*TSSI*, vol. 1, p. 31). Benoit, et al. say it dates from 750 (*Les grottes de Murabba'ât*, p. 95).

²⁴¹Benoit, et al., *Les grottes de Murabba'ât*, p. 96. The readings and translation are theirs. I have only given the first two lines here; the four line list of proper names which was written over the above text is not significant for this study.

²⁴²In iii:5 and vii:1 occurs the phrase "prosperity and life I have sent to you" שלם וחין שלחת לך (Gibson's translation, *TSSI*, vol. 2, pp. 135–36). The phrase "For your welfare I have sent this letter" occurs in a few different forms: in i:12–13 and vi:10 (though broken) it is ספרה זנה לשלמכי שלחת; in ii:17, iv:12–13, and v:9 it is לשלמכן שלחת ספרה זנה (plural); and in iii:13 it is ספרה זנה ל[כן] שלחת לשלמכן (*TSSI*, vol. 2, pp. 129–43).

²⁴³Driver, *AD*. Documents 1, 2, and 3 probably have the formula but are broken, number 5 (pp. 17–18) is fairly well-preserved, lines 1–2 are

1) מן ארשם על ארתהנת שלם ושררת שגיא הושרת ל[כן] וכעת בזנה קדמי שלם
2) אף תמה קדמ[י]ך שלם יהוי

"From 'Aršam to 'Artahant. I send [thee] much (greetings) of peace and prosperity; and now there is peace here with me, (and) there too may be peace with thee!" Document 13 (p. 35), a letter, also provides a fair example.

²⁴⁴See footnote 12 under the discussion of Arad 16. This greeting is found in Akkadian and Ugaritic letters as well as the Tell Fekherye inscription.

²⁴⁵Though it may appear in the damaged Adon papyrus found in Egypt (605–04 B.C.) which may invoke two deities, Astarte and Ba'alshamyn (see footnote 186).

²⁴⁶Cf. also Num 23:9–11: Bala'am does not use ברך but he speaks of the enviable status of Israel as he looks on its host (recognizing the blessing of fertility), and he wishes his own end were righteous like Israel's will surely be (implying the blessing of a full, long life).

²⁴⁷Meshel, "Did Yahweh Have a Consort?" p. 32.

²⁴⁸Hadley, "Kuntillet 'Ajrud," p. 187; and Weinfeld, "Kuntillet 'Ajrud," p. 125. Note that only in this text and in number 5 of the published inscriptions from Kuntillet 'Ajrud is the divine name יהוה written in the abbreviated form יהו (in 5 the preposition ל is appended). It is not clear from Hadley's article how much space occurs between the two segments of the inscription.

²⁴⁹Weinfeld, "Kuntillet 'Ajrud," pp. 125–26. The Proverbs translation is from the RSV, the other translations of biblical material are mine.

²⁵⁰Weinfeld, "Kuntillet 'Ajrud," p. 126. He refers the reader to another article in which he cites the relevant letters (Moshe Weinfeld, "'You Will Find Favor…in the Sight of God and Man' (Proverbs 3:4)—The History of An Idea," in *Eretz Israel*, vol. 16, ed. B. A. Levine and A. Malamat [Jerusalem: Israel Exploration Society, 1982], pp. 93–99 [Hebrew], English summary, p. 255). Those letters are quoted here from their original contexts.

²⁵¹Papyrus Leiden 1 369 in Edward F. Wente, *Late Ramesside Letters*, Studies in Ancient Oriental Civilization, vol. 33 (Chicago: University of Chicago Press, 1967), p. 18. The hieroglyphic text can be found in Jaroslav Cverný, *Late Ramesside Letters*, Bibliotheca Aegyptiaca, vol. 9 (Bruxelles: Édition de la Fondation Égyptologique Reine Élisabeth, 1939), pp. 1–2. Other examples

from the same period include Papyri Bibliothèque Nationale 199, 5–9 + 196, V+198, IV (Wente, p. 21, Cverný, pp. 5–7); and Papyrus Griffith (Wente, p. 32, Cverný, p. 12).

[252] *TSSI*, vol. 2, pp. 64–65.

[253] Ibid., pp. 66–67.

[254] This could be also interpreted as synonymous parallelism, i.e, "prayers heard" || "words accepted."

[255] *TSSI*, vol. 2, pp. 80–81.

[256] Ibid., vol. 3, pp. 94–95.

[257] Maurice Dunand, "Nouvelle inscription phénicienne archaïque," *RB* 39 (1930), 321.

[258] *TSSI*, vol. 3, p. 17.

[259] Pietro Magnanini, *Le iscrizioni fenicie dell'Oriente: testi, traduzioni, glossari* (Rome: Istituto di Studi del Vicino Oriente, 1973), p. 31. The reading are all Magnanini's except the division between מלכ and כ in line 6 where Gibson is followed (*TSSI*, vol. 3, p. 18).

[260] William F. Albright, "The Phoenician Inscriptions of the Tenth Century B.C. from Byblus," *JAOS* 67 (1947), 157.

[261] *TSSI*, vol. 3, p. 17. However, note that in line 9 of his fifth century inscription, Yeḥaumilk gives the names of both his father and grandfather, yet still includes the claim to legitimacy (*TSSI*, vol. 3, pp. 94–95).

[262] This form also occurs three times (lines 2A–B, 7B, and C2) in the late seventh century Phoenician inscription from Cebel Ires Daği (P. G. Mosca and J. Russell, "A Phoenician Inscription from Cebel Ires Daği in Rough Cilicia," *EpigAnat* 9 [1987], 5).

[263] Abou-Assaf, et al., *La statue*, p. 23. Surely the same wish is expressed by לחיי.נבשה in line 7 as well, however the parallel is not as clear as in line 14.

[264] For further treatment of this substantive blessing, see the discussion of Karatepe A.iii.2–A.iv.1 in "ברכ and שלמ with Substantive Curse" in Chapter 5.

[265] Loren F. Fisher, ed., *Ras Shamra Parallels: The Texts from Ugarit and the Hebrew Bible*, vol. 1, Analecta Orientalia, vol. 49 (Rome: Pontificium Institutum Biblicum, 1972), p. 354, no. 549. Biblical parallels are given where the words appear in either order, or simply together but not in parallel. The translation is from *UL*, pp. 77, 79. The transliteration is from *UT*, p. 192.

[266] In these societies while the king lived he reigned so the difference is probably negligible. For the idea that a more specific sense of reign was meant, see the discussion of the Tell Fekherye inscription under "Substantive Blessing and Curse" in Chapter 5.

[267] *TSSI*, vol. 3, pp. 19–20.

[268] Albright, "The Phoenician Inscriptions," p. 157; *KAI* 5; and *TSSI*, vol. 3, p. 20, though Gibson does not restore the name of Yeḥimilk as do the others. René Dussaud ("Les inscriptions phéniciennes du tombeau d'Aḥiram, Roi de Byblos," *Syria* 5 [1924], 146) offers a shorter reconstruction than the others:

(1) ישנ]א אבבעל מלכ ג]בל

(2) ונג]ש גבל במצרמ.לבעלת ולב]על גבל

"Ont offert] Abibaʻal, roi de Ge[bal, Et le Suzerain] de Gebal en Égypte à la Baʻal[t Gebal et à Ba]ʻal Gebal."

[269] Magnanini, *Le iscrizioni*, p. 28.

[270] Albright, "The Phoenician Inscriptions," pp.158–59.

[271] René Dussaud, "Dédicace d'une statue d'Orsokon I par Elibaʻal, Roi de Byblos," *Syria* 6 (1925), 101–17. He says that the request for blessing which concludes the inscription is "une formule liturgique" and its presence on the statue of the pharaoh may have had special power (pp.

Chapter 3: Blessing

110–11).

272 *TSSI*, vol. 3, pp. 21–22.

273 Magnanini, *Le iscrizioni*, p. 30.

274 Albright, "The Phoenician Inscriptions," p. 158.

275 *TSSI*, vol. 3, p. 23.

276 Maurice Dunand, *Byblia Grammata: Documents et recherches sur le développement de l'écriture en Phénicie*, Études et documents d'archaeologie, vol. 2 (Beirut: République Libanaise, 1945), pp. 148–49.

277 Albright, "The Phoenician Inscriptions," p. 158.

278 *TSSI*, vol. 3, pp. 39–40.

279 Ibid., p. 39.

280 Magnanini, *Le iscrizioni*, p. 47.

281 Franz Rosenthal, "Kilamuwa of Y'DY-SAM'AL," in *ANET*, p. 655.

282 So Galling ("The Scepter of Wisdom, A Note on the Gold Sheath of Zendjirli and Ecclesiates 12:11," *BASOR* 119 [1950], 15–16) believes following the lead of the excavator Felix von Luschan and comparing this object with a scepter from Luristan.

283 André Dupont-Sommer, "Une inscription nouvelle du roi Kilamou et le dieu Rekoub-el," *RHR* 133 (1947–48), 29. This option will be discussed below.

284 *TSSI*, vol. 3, p. 40.

285 Ibid., p. 41.

286 Dupont-Sommer, "Kilamou," p. 26.

287 Ibid., p. 27. The nominative masculine plural ends in "-u" and "-i" is the oblique case ending. Since ו is spelled defectively, יה might be expected to be written likewise.

288 Dupont-Sommer, "Kilamou," p. 26. The text is *AP* 30:3.

289 Ibid., pp. 27–29. This DN also appears twice on the early eighth century stele of Ördek-burnu (Mark Lidzbarski, *Ephemeris für semitische Epigraphik*, vol. 3 (Giessen: Verlag von Alfred Töpelmann, 1915), p. 197, photograph tafel 13.) Lidzbarski describes the stele from Asia Minor as essentially non-semitic (p. 206).

290 Dupont-Sommer, "Kilamou," p. 29.

291 Ibid., p. 30. He supports this idea by reference to an inscription of Nabonidus (Louvre catalogue AO 6444) treated by Paul Dhorme ("La fille de Nabonide," *RA* 11 [1914], pp. 113–15, lines 20–22). This is a dedicatory inscription concerning the renovation of the chariot of Lougal-maradda, "...le char, véhicule de sa divinité, insigne de sa bravoure, qui pille le pays de l'ennemi qui est adapté au combat..." *narkabtu rukubu ilutišú semāt qarradutišú šálilāt māt ābi ana taḫazi šúlukāt*. Cf. Isa. 2:6–8 and Cant. 3:9–10, "King Solomon made himself a palanquin from the wood of Lebanon. He made its posts of silver, its back of gold, its seat of purple; it was lovingly wrought within by the daughters of Jerusalem."

292 The reason for the Gezer calendar's inclusion in this study may not be immediately obvious. As will be seen at the beginning of the discussion, most interpreters think it has no religious value. However, a significant minority of commentators think that it does or might; therefore it is included here.

293 R. A. Stewart Macalister, *The Excavation of Gezer 1902–1905 and 1907–1909*, vol. 2 (London: John Murray, 1912), p. 24. Macalister's publication actually post-dated the first scholarly debates on the subject (found in *PEFQS* 41 [1909]) and so he is able to include a summary (pp. 24–28).

[294] William F. Albright, "The Gezer Calendar," *BASOR* 92 (1943), 16–17.

[295] *TSSI*, vol. 1, p. 2. The last word is written vertically on a small protrusion of the tablet in the lower left corner.

[296] Ibid.

[297] Mark Lidzbarski, "An Old Hebrew Calendar-Inscription from Gezer," *PEFQS* 41 (1909), 26.

[298] R. A. Stewart Macalister, "Twenty-first Quarterly Report on the Excavation of Gezer," *PEFQS* 41 (1909), 91.

[299] Sébastien Ronzevalle, "The Gezer Hebrew Inscription," *PEFQS* 41 (1909), 111.

[300] Ibid., p. 112.

[301] Stanley A. Cooke, "The Old Hebrew Alphabet and the Gezer Tablet," *PEFQS* 41 (1909), 308.

[302] Samuel Daiches, "Notes on the Gezer Calendar and Some Babylonian Parallels," *PEFQS* 41 (1909), 117.

[303] Albright, "The Gezer Calendar," p. 21. So also André Lemaire, *Les écoles*, p. 10. The multiple use of the tablet is obvious from the excellent photographs supplied by Macalister (*Gezer*, vol. 3, p. 127).

[304] Albright, "The Gezer Calendar," p. 23. While, admittedly, most scholars read the second letter of the last word as a ב, or combine the ו with the previous word, Albright makes a very good case for his reading. (The multiple use of the text makes several letters difficult to read.)

[305] Ibid., p. 25.

[306] *TSSI*, vol. 3, p. 51. Gibson supplies "of grain" based on a comparison with Gen. 41:29 and Prov. 3:10.

[307] Henri Michaud, *Sur la pierre et l'argile*, (Paris: Delachaux et Niestlé, 1958), pp. 22–23.

[308] Ibid., pp. 27–28.

[309] Wolf Wirgin, "The Calendar Tablet from Gezer" in *Eretz–Israel: Archaeological, Historical and Geographical Studies,* vol. 6, ed. M. Avi-Yonah et al. (Jerusalem: Israel Exploration Society, 1960), p. 11. Gibson also says this interpretation is "attractive." He notes that from a farming point of view, the tablet starts in a strange part of the year, the fruit harvest. He suggests that the Hebrew New Year being in autumn probably had something to do with this (*TSSI*, vol. 1, p. 1).

[310] Ibid., p. 10. He also gives an example from the Mishnah (*Taanith* iii, 8) of "Onias the Circle-maker" reputed to have special abilities.

[311] R. Wünsch, "The Limestone Inscriptions of Tell Sandaḥannah," in *Excavations in Palestine During the Years 1898–1900* (London: Palestine Exploration Fund, 1902), pp. 158–87.

[312] R. A. Stewart Macalister, *A Century of Excavation in Palestine* (London: The Religious Tract Society, 1925), p. 320. These tablets are late third to second century B.C. at the earliest and the vast majority of their contents are Greek, little is in Hebrew. Also, far fewer than 51 are of use in determining their contents. After these caveats, there are several among them that are long enough and complete enough to give an adequate idea of their purpose. For example, number 34 (Wünsch, "Limestone Inscriptions," p. 182, reading and drawing, pp. 173–74; also Macalister, *Century*, p. 320 and photograph facing) is the case of Pankles who has suffered because of Philonides and Xenodikos. They have used imprecations to cause him to suffer headaches and other sickness which have cost him his job. Pankles appeals to the same god (as in their imprecation) to deprive them of speech and the enjoyments of love.

[313] Actually, the tablets vary a good bit in size. While the Gezer tablet is a mere four-and-a-half inches tall by two-and-three-quarters inches wide, the Amman Citadel inscription (which may not be even nearly complete as is) is ten inches tall and eight inches wide.

Chapter 3: Blessing   95

³¹⁴Wünsch, "Limestone," p. 186. He says (p. 182) that he collected more than 400 of these texts. He believes (p. 186) that like writing materials in other cultures that had special powers, "in Palestine, limestone had some superstitious significance, but of what especial [sic] kind we do not know."

³¹⁵Wirgin, "Calendar Tablet," p. 11.

³¹⁶However, Michaud disagrees (*Sur la pierre*, p. 22), noting that, in Judg 8:14, a נער of Succoth captured by Gideon was able to write the names of the 77 chief men of the city. Was the young man a scribe? For a different angle, compare Lachish 3:8–9 (Torczyner, et al., *The Lachish Letters*, p. 51) in which the recipient of a letter, who was a person in authority, claims not to be able to read.

³¹⁷Michaud, *Sur la pierre*, pp. 11–12. Wirgin further points to the tablet's seven lines as significant for their number which, he says, indicate the seven year *shmittah* cycle, observance of which results in good harvests (Deut, 15:1ff. and 31:10ff; "The Calendar Tablet," p. 12*). Wünsch says that there is evidence that these cursing tablets were placed in a public place to be read by passersby ("Limestone Inscriptions." p. 182). One wonders though, if people of the tenth century possessed reading skills comparable to those of people of the third to second centuries.

³¹⁸Victor Sasson, "The 'Amman Citadel Inscription as an Oracle promising Divine Protection: Philological and Literary Comments," *PEQ* 111 (1979), 124–25. For more on his comparison, see the discussion of the Amman Citadel inscriptions under "Substantive Blessing and Curse" in Chapter 5.

³¹⁹According to Fawzi Zayadine and Henry O. Thompson ("The Ammonite Inscription from Tell Siran," *Berytus* 22 [1973], 115) it is the second longest after the Amman Citadel inscription. Cross ("Notes on the Ammonite Inscription from Tell Siran," *BASOR* 212 [1973], 12–14) however, treats the Deir 'Alla inscriptions as Ammonite.

³²⁰Fawzi Zayadine and Henry O. Thompson, "The Ammonite Inscription," pp. 115–16, 127. Kent Jackson (*The Ammonite Language of the Iron Age*, Harvard Semitic Monographs, no. 27 [Chico, CA.: Scholars Press, 1983], p. 37), gives James Sauer's summary description and conclusion about the bottle's function. Sauer says that the bottle is the same size and shape as alabaster, pottery, and glass jars used for spices and precious ointments in the Neo-Babylonian and Persian periods. Its wide mouth, overall shape, and contents (grain) argue against its use as a drinking vessel. The vessel had been closed with a metal lid which was bolted in place by rivets.

³²¹Jackson, *The Ammonite Language*, p. 36.

³²²Readings and translation by Henry O. Thompson and Fawzi Zayadine, "The Tell Siran Inscription," *BASOR* 212 (1973), 9–10. The transliteration to block letters is mine.

O. Loretz ("Die ammonitische Inscrift von Tell Siran," in *UF*, vol. 9, eds. Kurt Bergerhof, Manfried Dietrich, Oswald Loretz, [Neukirchener-Vluyn, W. Germany: Verlag Butzon & Berker Kevelaer, 1977], p. 170); B.E.J.H. Becking ("Zur Interpretation der ammonitische Inschrift vom Tell Siran," *BO* 38 [1981], 275); and J.A. Emerton ("The Meaning of the Ammonite Inscription from Tell Siran," in *Von Kanaan bis Kerala*, eds. W.C. Delsman, et al., [Neukirchener-Vluyn, W. Germany: Verlag Butzon & Berker Kevelaer, 1982], p. 376) all essentially agree with this interpretation.

³²³Zayadine and Thompson, "Ammonite Inscription," p. 136.

³²⁴Ibid., pp. 136–37.

³²⁵Ibid., pp. 138–39.

³²⁶Ibid., p. 139. G. W. Ahlström ("The Tell Siran Bottle Inscription," *PEQ* 116 (1984), 14) agrees that this inscription may have had cultic significance as harvests were usually connected with rituals and religious festivals. He also notes that גיל and שמח were often used in such contexts.

³²⁷Albright, "The Gezer Calendar," p. 23. Also see his translation "The Gezer Calendar," in *ANET* (p. 320) which is only slightly different, "his month is harvest and feasting."

[328] Albright, "The Gezer Calendar," p. 25.

[329] Charles Krahmalkov, "An Ammonite Lyric Poem," *BASOR* 223 (1976), 56.

[330] Ibid., pp. 55–56.

[331] William H. Shea, "The Siran Inscription: Amminadab's Drinking Song," *PEQ* 110 (1978), 108. The possibility of the bottle being a drinking vessel is ruled out by most commentators based on the vessel's enclosure—grain (so Sauer in Jackson, *Ammonite Language*, p. 37, see above; and E. J. Smit, "The Tell Siran Inscription: Linguistic and Historical Implications," *JS* 1 [1989], 112).

[332] Shea, "The Siran Inscription," p. 110.

[333] Robert B. Coote, "The Tell Siran Bottle Inscription," *BASOR* 240 (1980), 93.

[334] M. Baldacci, "The Ammonite Text from Tell Siran and North-West Semitic Philology," *VT* 31 (1981), 364.

[335] Frank M. Cross, "Tell Siran," p. 12. The text is RS 24.252, lines 10–13 published in *Ugaritica V*, p. 553 ḥtkk nmrtk btk ugrt l ymt špš w nʿmt šnt il.

[336] Loren F. Fisher, ed., *Ras Shamra Parallels*, p. 354, no. 549. Biblical parallels are given where the words appear in either order or simply together but not in parallel. The translation is from *UL*, pp. 77, 79. The transliteration is from *UT*, p. 192.

[337] Smit, "The Tell Siran Inscription," p. 113.

[338] Actually, Hebrew showing Phoenician influences. See Meshel's ideas in the discussion of the inscriptions.

[339] The only specific blessing seen in this chapter is long life (which may be interpreted as including these other blessings). Other specifically requested blessings are abundant in the inscriptions which include substantive blessing and curse, see Chapter 5.

Chapter 4
CURSE

Curses without accompanying blessings are found in period inscriptions as warnings on tombs to would be robbers, protection for memorial inscriptions against possible violators,[1] in magical texts, and in mythical texts. In all but the last case, curses appear to be the last resort in situations when conventional means fail to provide needed security: where hidden tombs cannot defeat the cleverness of grave robbers, where respect for the dead does not prevent the living from jealously effacing a predecessor's name from a record of his or her accomplishments, or where sickness can only be prevented by incantations. Curses are also a part of treaties but are usually minimally softened by at least a *pro forma* blessing.[2] In this chapter, inscriptions having forms of ארר with and without formulas, other words for curse, and "Substantive Curse" will all be presented.

ארר Formulas

Formulas with ארר are comparatively rare and come from a relatively narrow time span, the earliest being ca. 700 B.C. and the latest dating from just before the fall of Jerusalem (early sixth century). They are all Judean. The latest of these ארר examples, Lachish 5, will be treated first as it is the only one that may be interpreted as explicitly calling on Yahweh to curse. The Siloam Tomb inscription and the Khirbet Beit Lei inscriptions, none of which include a DN will follow.

With Yahweh

Lachish 5:9–10. This inscription contains the only, possible, explicit curse with Yahweh in the target inscriptions. Introductory remarks on the Lachish inscriptions are included above.[3]

(9) ה.מה לעבדכ.יא
(10) [ר].ביהו.זרע למלכ

Torczyner translates these lines, "What has thy slave that he should curse in (the name of) Yhw seed to the king?"[4] He says the phrase does not mean that a specific individual is to be cursed, but that one's descendants or dynasty are to be cursed. Of the numerous examples that he gives (e.g., 1 Kgs. 21:10; 2 Kgs. 2:24; Lev. 24:11; and Isa. 8:21) only 2 Sam. 16:9 seems a very apt parallel, "Why should this dead dog curse my lord the king?" (in lines 3–4 of this inscription as in several others from Lachish, the writer calls himself a "dog").[5] This verse, like all of the other examples, uses קלל not ארר, yet he says, "this seems to have been a special kind of curse used in ancient Israel, similar to the frequently heard vulgar Arabic curse: *Yikhrib betak* 'May [Allah] destroy thy house.'"[6] While some inscriptions discovered subsequent to the Lachish texts seem to bear out the special use of ארר in Hebrew curses (the Siloam Tomb and Khirbet Beit Lei inscriptions), Torczyner's biblical examples do not show this and it is unlikely that Lachish 5 is best read this way. Likewise, with regard to the reading of יהו in line 10, it seems unlikely that a scribe would use both forms of the divine name in such proximity (lines 7–8 and 10).[7]

Albright's restoration of the two lines in question is מה לעבדכ ייטב או ירע למלכ and his translation is "How can thy servant benefit [o]r injure the king?"[8] The biblical parallels he provides are quite good: 2 Sam. 19:8; 20:6; Num. 11:11; and especially Josh. 24:20. Ginsberg agrees with Albright's reading except that he changes the second י in ייטב for an א but still translates the phrase has having the overall meaning "do anything" and he adds further examples: Lev. 5:4; Isa. 41:23; Jer. 10:5; and Zeph. 1:12.[9] Either of these two readings, and certainly their understanding of the intent of the text, seems the best offered.

Michaud offers yet another reading and translation which varies significantly from those offered above, עבדכ יא טביהו זרע למלכ "Est-ce à ton serviteur qu'apportera Ṭobiyahu de la semence <<lammèlèk>>?"[10] He says that יא at the end of line 9 is a scribal mistake and should read יבא. The last phrase is the question of an inferior to a superior and deals with seed or grain measured according to the royal standard.[11] Lemaire agrees with Michaud.[12]

Without DN

ארר is found without a DN in the rest of its occurrences. All of the occurrences are in Hebrew and from Judean contexts.

Siloam Tomb. This inscription was discovered in 1870 in the village of Siloam (Silwan) just south of Jerusalem. It was cut into the rock of a small chamber. The text is dated to ca. 700 B.C.[13] Very little of the text was deciphered until the work of Avigad in 1953.[14]

(1 זאת [קברת...]יהו אשר על הבית.אין כספ.וזהב
(2 [כי] אמֹ [עצמתו] ועצמת אמתה אתֹה.ארור האדם אשר
(3 יפתֹח את זאת

Avigad translates the inscription, "This is [the sepulchre of...]yahu who is over the house. There is no silver and no gold here, but [his bones] and the bones of his slave-wife with him. Cursed be the man who will open this."[15]

Avigad notes that האדמ is unexpected in the last phrase since biblical Hebrew uses האיש (Deut. 27:15).[16] In the early fifth century Phoenician Eshmunazar

Chapter 4: Curse

inscription, however, two sentences with the same intent begin in similar ways, lines 4 and 20 both prohibit disturbance of the tomb thus

קנמי את כל ממלכת וכל אדמ אל יפתח אית משכב ז

"Whoever you are, be you ruler or be you commoner, let none such open up this resting place."[17] The Eshmunazar inscription likewise includes a claim (line 5) that nothing is in the sarcophagus with the body. The Phoenician Aḥiram inscription (1000 B.C.) contains a similar prohibition against the disturber and includes a substantive curse.[18] The Aramaic Nerab 2 inscription (ca. 700 B.C.) contains both the prohibition against disturbing the tomb and the claim that no valuables "silver or bronze" (line 7) are interred with the body.[19] The fifth to fourth century Phoenician Tabnit inscription contains both these elements as well.[20] Based on his reading of the Siloam tomb inscription, Avigad has produced a very plausible reconstruction of another inscription found on a nearby tomb and containing this same curse-fortified prohibition.[21]

Though this curse does not explicitly call upon Yahweh for its implementation (cf. Josh. 6:26), the broken name of the interred does end in יהו- which, while realizing that one's personal piety is not necessarily indicated by one's name,[22] could be seen to imply a dedication to Yahweh and a dependence on Yahweh for help. This inscription could be accepted then as an implied call for Yahweh to curse the violator.[23]

Khirbet Beit Lei 4. The Khirbet Beit Lei inscriptions are inscribed on the soft limestone walls of a burial cave located about eight kilometers east of Lachish. These inscriptions are dated circa 701 B.C.[24] One wall of the ante-chamber of the tomb contains numerous graffiti, drawings, and inscriptions. The soft limestone of the walls poses the greatest difficulty in the reading of the texts, at points it is difficult to distinguish marks on the wall made prior to the inscriptions from the lines composing the letters and the varying softness of the rock caused some lines to penetrate more deeply than others.[25]

1) א ארר
2) ישרמחר

Naveh suggests an emendation of the inscription to אר[ו]ר ישד מחד[ר] and translates it, "Cursed be he who will rob the chamber." He is not commited to the translation though and even notes that there are no vessels in the tomb to steal.[26]

Bar-Adon suggests that an inscription from En-Gedi may help in the interpretation of this text. He recommends the reading ארר אשר [י]מחה which he translates "Cursed be he who will efface."[27]

Lemaire agrees with Bar-Adon's reading and translation. He suggests that the present position of the letters is a scribal mistake; the extraneous first א being originally intended as the first letter of the second word while the י found there should have been the first letter of the last word.[28]

Miller agrees with Bar-Adon's reading and interpretation. Further, he says that like the inscription found by Bar-Adon at En-Gedi (Naḥal Yishai), this inscription should be seen as standing sentinel over the other inscriptions, guarding them from violation.[29] Whether it should be read as a superscript to the other inscriptions (like En-Gedi) or as a postscript to them (like Tell Fekheryeh, Sefire I.C, or Kilamuwa 1) is not certain.

Khirbet Beit Lei 7. This inscription was also found on the wall of the burial cave. It is inscribed in large letters within the carved outline of what was to be a doorway for another chamber.[30]

1) ארדח

2) רפכ

Naveh does not treat these two lines as belonging to the same inscription, indeed he does not deal with the second line at all. He does suggest that though the ח in ארדח is quite clear, the intention of the scribe may have been to write ארדהו "he is cursed" or "may he be cursed."[31]

Lemaire translates this text, "Maudit soit celui qui t'insulte."[32] He says that the ח at the end of line 1 is the first letter of the word in line 2 and that the last כ is a pronominal suffix on the word חרפ. He points to two biblical passages which use a form of this word with this suffix, Pss. 69:10b(9b) וחרפות חורפיך נפלו עלי "and the insults of those who insult thee have fallen on me," and 79:12 והשב לשכנינו שבעתים אל־חיקם חרפתם אשר חרפוך אדני: "Return sevenfold into the bosom of our neighbors the taunts with which they have taunted thee, O Lord!" He suggests that the pronoun here refers to Yahweh in the same way that the first three inscriptions from this site are prayers to Yahweh[33] and that the extraneous כs are also references or invocations to Yahweh.[34]

Miller agrees with Lemaire's reading and translation. He believes that neither this nor any of the other ארור occurrences in the tomb are curses on someone disturbing it, rather the inscription is a "prayer curse against the common enemies of Yahweh and the petitioner."[35] Miller notes that in Ps. 119:21–22, which pairs forms of these two words, the "accursed ones" (ארורים) are those who cast scorn (חרפה) on the righteous. The scribe of these inscriptions responded against those who taunted him and God by praying that they may be destroyed by Yahweh.[36] Miller cites several convincing examples (Judg. 8:15; 1 Sam. 17:10, 25, 26, 36, 45; 25:39; 2 Kgs. 19:4, 16, 22, 23//Isa. 37:17, 23, 24 and 2 Chr. 32:17; Neh. 3:33–34; 6:13) to show that the taunt consists of a questioning of the power and status of a person and/or his God.[37] He suggests that these inscriptions may be put together with longer inscriptions found in the cave which speak of Yahweh as accepting and redeeming Judah and Jerusalem ("A"), as dwelling in Jerusalem ("B"), and the prayer for salvation ("C") to give an overall picture of an attitude and faith similar to that found in many of the Psalms.[38]

Chapter 4: Curse   101

Khirbet Beit Lei.[39] Naveh reports the occurrence of the root אר/ אור/ארר in isolated words on the northern, southern, and western walls of the burial chamber.[40] He guesses that the numerous instances of forms of אור/ארר might have had "a magical purpose and were designed to lay a curse upon someone, perhaps an enemy."[41] Though technically speaking, these words have no context, in the overall setting of this tomb and the inscriptions found on the wall of the ante-chamber (above) their Judean, Yahwistic context is almost inescapable.[42]

Summary
It is noteworthy that there are no explicit epigraphic examples in which Yahweh is invoked for a curse, if Torczyner's interpretation of Lachish 5 is not accepted. However, the presence of several forms of ארר in the Khirbet Beit Lei burial cave along with prayers to Yahweh may imply that Yahweh was the one invoked to bring about the curse. Likewise, the presence of a name with the theophoric element -יה in the Siloam Tomb inscription curse points in the same direction. The curse in the En Gedi inscription makes the same implication.[43] Biblical examples of ארר with Yahweh are infrequent but not rare.[44] All of the inscriptions having a form of ארר date from a period of just over one hundred years (Siloam Tomb ca. 700 to Lachish ca. 595) and all are not only Hebrew, but Judean as well. Besides Judean Hebrew, this root is only otherwise seen in Akkadian *arāru* where it is used very similarly to the way it is found in the inscriptions and the Bible.[45] Current evidence suggests that ארר came to be used late in the target period and was endemic to Judah. Further, this information lends weight to the theory that the inscriptions show a developmental trend from earlier periods where specific curses were used to later periods where a more general word was used, perhaps with a certain well-known (but not stated) stock of curses implied by that one word.

Other Words for Curse

In addition to the curse with ארר, two other words for curse, אלה and קבב, are found in the inscriptions. Both of these words occur without DNs. אלה is found twice, in the Aramaic Panammu inscription and in the first Arslan Tash incantation (which is of debatable linguistic lineage). קבב occurs only three times and all in the Deir 'Alla texts (whose language is also questioned).[46]

אלה

The first of the two texts which include forms of אלה is the Panammu inscription which appears to describe an undesirable state of affairs resulting from violation of a treaty. The second text, Arslan Tash 1, is an incantation designed to provide protection from harmful spirits. In neither of these is a deity invoked to curse someone or something.

Panammu line 2. This inscription and the statue to Panammu II on which it is written, were made by his son Barrakab. The text, written on the bottom half of the dolerite statue of Panammu II, was found between Zendjirli and Gerçin in 1888, and dates from approximately 730 B.C.[47] Though, strictly speaking, this inscription might be excluded because a curse is not called for but is only mentioned as a past condition, it is included due to the rare occurrence of אלה in line 2.

1) נצב.זנ.שמ.בררכב.לאבה.לפנמו.בר.ברצר.מלכ[.]יאד[י] --.׳-- .-ב-.שנת[.] [מ]פלט[ה.] א[ב]י.פנ[נ]מ[ו.][ב]צ[ד]ק

2) אבה.פלטוה.אלה.יאדי.מנ.שחתה.אלה.הות.בבית.אבוה.וקמ.אלה.הד[ד.ע]מ[ה.] ו[ק----משבה.על[ל.][ו--ו--לאד]וש.---שחת---

This statue has Barrakkab set up for his father Panammu, son of Barṣur, king of [Y'DY...] the year of his deliverance. As for my father Panammu, because of his father's righteousness the gods of Y'DY delivered him from destruction. There was a curse on his father's house, but the god Hadad stood with him, and [...] his throne against [...] he destroyed...[48]

Lidzbarski's transliteration of the word into block letters is אזה[49] but the drawing he provides clearly reads אלה.[50] Cooke notes Lidzbarski's mistake and translates the key phrase "There was a conspiracy in his father's house."[51] Koopmans also disagrees with Lidzbarski's reading and interprets it as confederation or conspiracy ("Eidgenossenschaft, Verschwörung").[52] Donner and Röllig faithfully follow Lidzbarski in his reading (אזה) and translate it as a relative pronoun.[53]

Gibson notices the mistake in the reading like Cooke and Koopmans but he also notices a problem with their translation; they assign the meaning "conspiracy" to אלה but that meaning is unattested elsewhere.[54] Instead, Gibson proposes the more straightforward translation "curse" and compares Deut. 30:7 or Jer. 23:10, "For the land is full of adulterers; because of the curse the land mourns and the pastures of the wilderness are dried up..."[55] Gibson's translation seems more likely, especially in light of stipulations seen in many ancient Near Eastern treaties. Should the parties involved in a treaty (especially the inferior as his were virtually all of the obligations) fail to observe the treaty's terms, dire curses would come upon them. The Sefire steles provide excellent examples: I.A.14–42 and I.C.17–25 are explicit, while I.B.21–45 is only slightly less so in the recurrent use of the phrase "you will have been false to all of the gods of this treaty," and B.1–12 contains a long and formidable list of gods who will "guard" this treaty.[56]

The verb אלה appears seven times in the Bible, four of these occurrences are in the context of an oath (1 Kgs. 8:31//2 Chr. 6:22; Hos. 4:2; 10:4, the last

Chapter 4: Curse 103

two refer to false swearing). The noun occurs over thirty times and is usually translated "oath." The threat included in this "oath" is that failure to fulfill the conditions of the oath brings a curse (cf. Num. 5:23–24 where the woman accused of adultery swears her purity by drinking "the water of bitterness that brings a curse" מי המרים המאררים).[57] It seems likely, then, that Panammu's house was under a curse because he had transgressed treaty conditions but the god Hadad had brought him through those difficulties.[58]

Arslan Tash 1. The seventh century Arslan Tash inscription included here is one of two small limestone plaques purchased at Arslan Tash in Upper Syria in 1933. The plaque is 8.2 cm. tall, 6.7 cm. wide, and 2.2 cm. thick and has a hole in its top apparently used for hanging it. The obverse side bears the images of a sphinx and of a she-wolf in the act of devouring a human figure. The reverse bears the image of a striding warrior or god.[59] This plaque is included because of the occurrence of the word אלה not because it contains an incantation; the second plaque is thus excluded because, though it is an incantation, it does not include a word for "curse" nor is it otherwise an implicit curse.[60] There is a growing view that this inscription is in fact a forgery made in the 1930s. Javier Teixidor says his personal examination of the texts, including handling them, has convinced him of their inauthenticity. He notes that more and more epigraphists and archaeologists are coming to doubt their genuineness.[61]

1) לחשת.לעתא.אלת   (obverse)
2) ססמ.בן פדרש
3) שא אלה
4) ול.חנקת.א
5) מר.בת אבא
6) בל תבא
7) וחצר.אדרכ   (reverse)
8) בל.תדרכנ.כ[נ]
9) רת.לנ.אלת
10) עלמ אשר.כרת
11) לנ.וכל בן אלמ.
12) ורב.דדכל.קדשנ
13) באלת.שממ.וארצ
14) עלמ.באלת.בעל
15) ת--[נ.ארצ.באלת   (bottom edge)
16) א[שת חורנ.קש.דתמ.פי   (left edge)
17) ושבע.צרתי.ושמ   (top edge)
18) נה.אשת.בעלי קש   (right edge)
19) לעפתא.בחדר.חשכ   (on the sphinx)
20) עבד פעמ.פעמ.ללינ
21) [מבות.מ?]חצת.הלכ   (on the she-wolf)
22) סז.זת.לי   23) פתח   (on or near the figure of the god)
24) י.ות   25) וד.ל   26) סז.זת.יצא.שמש

(27 חל.ולד[62]

Incantation: O Flying One, thou goddess, O S-s-m b-n P-d-r-š-š-a, thou god, and O Strangleress of lamb(s), the house I enter enter not, and the court I tread not, for there hath been made with us a bond everlasting! Ashur hath made (it) with us, governor of the beings divine, and burgomaster of all the holy being[s]. By the bonds of heaven and earth, [for] ever (be exorcised), by the bond of Ba'al, [lo]rd of the earth, by the bond of [the Con]sort of Hades, whose spell is pure, and of her sevenfold co-wives, and of the eightfold consorts of Ba'al...O thou that fliest into darkened chamber(s), pass over, right now, right now, O Lilith! Kidnapper, crusher of bones, begone! St! Zt! May her [wom]b be opened and may she give birth(?)! St! Zt! (When) the sun rises, travail and give birth![63]

Du Mesnil, in the *editio princeps*, interprets each occurrence of אלת (lines 1, 9, 13, 14, 15) as a reference to the "déesse" "'Athé" who is the goddess invoked for control over the two night demons.[64] Dupont-Sommer agrees with this interpretation of אלת.[65] Albright believes that the inscription is an incantation to assist women in childbirth[66] and he concurs in the interpretation of אלת.[67]

Gaster agrees with Albright that the plaque was hung up in the room of a woman in childbirth,[68] but differs markedly in his interpretation of אלת. אלת in lines 9, 13, 14, and 15 should be pointed אֱלַת not אֵלַת.[69] He says the vocalization אֱלַת עֹלָם by du Mesnil and Albright (cf. Gen. 21:33) and Albright's emendation of אשר to אשרת in line 10 are incorrect.[70] Likewise, du Mesnil's and Albright's interpretation of כרת "pregnantly," i.e., without ברית yet referring to a covenant, is too awkward.[71] Gaster proposes instead to interpret אלת as the equivalent of Hebrew אלה "bond, covenant, curse, ban" (cf. Akkadian *u'iltu, îltu*) and gives the parallel Deut. 29:11(10) לעברך בברית יהוה אלהיך ובאלתו אשר יהוה אלהיך כרת עמך היום:. He says the phrase in the inscription then "accords exactly" with Hebrew (Isa. 61:8; Jer. 32:40; and Ezek. 37:26).[72] The second occurrence, באלת שממ וארצ "By the bond of heaven and earth" (line 13), can then be seen as an exact counterpart to the Akkadian *nîš šamê u irṣitim* seen in incantations.[73] The third passage באלת בעל [א]דנ ארצ באלת [א]שת חורנ "by the bond of Ba'al, [lo]rd of the earth, by the bond of [the Con]sort of Hades" (lines 14–16, אלת occurs twice) can be interpreted similarly and compared to the Akkadian *nîš ᵈBêl bêl mâtâti*.[74] In this text, אלת combines the ideas of a pact (cf. Deut. 29:11[12]) and of a ban imposed on the demons. The ban idea finds strong parallels in Akkadian texts, e.g. *Šurpu* IV.47 *rakista liprusu, 'ilti lipṭuru, kiṣir lumni liparriru, kasîta liramû* "May they sever the tie, burst the bond, break asunder the shackle of evil, loosen the ban thereof."[75]

Torczyner finds six occurrences of the root אלת in the inscription and

Chapter 4: Curse

treats the text as more generally against night demons.[76] He is the only scholar to interpret אלה in line 1 as "curse" instead of "goddess."[77] His translation of lines 1–5a is unusual enough to give here: "Incantation for the *'Ephata*-demons, the curse of Ssm, son of Pdrsh: Take up the curse and to the she-stranglers say..."[78] The second occurrence of this word he, uniquely, finds in lines 3–4, the first two words of which he points as שָׂא אָלָה. For this he gives the Hebrew parallel נשׂא קינה in Jer. 7:29, "take up a lamentation"; the precise phrase he reads in the inscription does not occur in biblical material.[79] The occurrence of אלה in line 9 he translates as "covenant" in the sense of a covenant confirmed by an oath or curse.[80] The last three occurrences of אלה (lines 13–15) are in the form באלה and are each translated as "with the curse" again with the sense of a treaty curse but this is a curse which binds heaven and earth not only as witnesses but, they are also, "forced by the covenant not to touch those protected by it, and such covenant [sic] included even the spirits descending from heaven or arising out of Sheol to haunt the earth."[81]

Donner and Röllig find four occurrences of אלה in the text (lines 9, 13, 14, 15). They translate the first as "Bund" and the other as "Bann" with the resulting sense of "Fluch," corresponding to Hebrew אלה, should the agreement be transgressed.[82] Rosenthal's interpretation is similar in that he translates each occurrence as "bond," but differs in that he sees אלה in line 1 also as "bond."[83]

Cross and Saley interpret the first use of אלה (line 9) as "covenant" (cf. Torczyner and Donner and Röllig) and the other occurrences as "oath." They also accept the elliptical use of כרת (line 10), without ברית or אלה, which du Mesnil proposed and Gaster rejected.[84] Cross and Saley see a connection between the function of the plaque and the blood of the Passover lamb placed on the door frames to protect the children of the Israelites on the night of the last plague (Ex. 12:22f.).[85] Note that one of the demons in the text is even called the "strangler of lambs." They suggest a further parallel in the contemporary Deuteronomistic practice of inscribing the doorposts of the house with the words of the law (Deut. 6:4–9).[86]

Caquot interprets all of the occurrences of אלה as "pacte" including the use in line 1. He says that the interpretation "malédiction" is only secondary and points to biblical passages where אלה is practically a synonym to ברית (Deut. 29:11, 18–20; Ezek. 16:59; 17:16–19; and Neh. 10:30).[87] The text is then a document of alliance between the god Sasom, mentioned in line 2, and the unnamed author.[88]

Zevit's treatment offers the most comprehensive interpretation of the use of אלה in this text. Basically, his interpretation is similar to Caquot's: he translates each instance of אלה as "covenant."[89] He cites Brichto's well-documented contention that אלה is present "explicitly or implicitly" in every ברית and parallels lines 9–10 אלת עלמ "an eternal covenant" to Jer. 32:40 ברית עולם.[90] The ב on the front of אלה in lines 13–15 is to be understood as a *bet instrumenti* and serves to introduce the object of the verb as the means or instrument by which something is done though the difference in meaning is slight.[91] The plaque belonged to a religious community to which the three parties mentioned—Aššur, all the sons of the gods, and the leader of the council of the gods (lines 10–12)—have granted a covenant. The plaque then testifies to the first non-Israelite covenant theology in the ancient Near East.[92] Three groups of witnesses (lines 13–15) follow the above three guarantors, each introduced by באלה: heaven and earth, Ba'al Lord of the earth, and Horon with his seven wives[93] and the eight wives of Ba'al.[94] Overall, the inscription can be divided into three parts: lines 1–4 are the introduction; lines 5–8 comprise the incantation; and lines 9–18 are the credo.[95] The function of the credo is, Zevit says,

> to explain to the child-stealing demons why they should obey the charge of the incantation. By recording a communal confession of covenant faith, the plaque's owner reminds the demons that he is part of the covenant community. By implication, it may be inferred that one of the covenant stipulations required the gods to protect the community from these malignant figures.[96]

Gibson points out some very significant biblical passages connected to the intent of this inscription.[97]

> Now therefore hear this, you lover of pleasures, who sit securely, who say in your heart, 'I am, and there is no one besides me; I shall not sit as a widow or know the loss of children': these two things shall come to you in a moment, in one day; the loss of children and widowhood shall come upon you in full measure, in spite of your many sorceries and the great power of your enchantments.

This passage, Isa. 47:8–9, is significant because it speaks of loss of children as a curse, because it speaks of acts of sorcery practiced by the people in order to prevent that loss, and because it seems to be nearly contemporary with this inscription. Gibson also notes Deut. 18:10–11, where sorcery and making children "pass through fire" are forbidden.[98] He also points out other texts and inscriptions where heaven and earth are called as witnesses, Deut. 4:26; 30:19; Isa. 1:2; a Hittite treaty between Šuppiluliumas and Kurtiwaza[99]; and the Sefire treaty I.A.11.[100]

קבב

The Deir 'Alla texts contain the only instances of קבב in the target inscriptions.[101] All of the instances occur in broken contexts or contexts which, though whole, are still unclear. In 1967, excavations at Tell Deir 'Alla in Jordan revealed the broken fragments of an inscription written on plaster. The circa 700 B.C. text, written on the plaster of a wall or stele, may have been in a sacred

Chapter 4: Curse   107

precinct. The many fragments bearing letters have been grouped into combinations, the largest of these being identified as "Combination I" and "Combination II." These combinations, and the others (III–XII), may or may not be related to one another. The relationships between the fragments will be discussed as they are deemed relevant to this study.[102]

Deir 'Alla II.17

17) לדעת.ופר.דבר.לענֹה.על.לשׁנ.לכ.נשפט.ומלקב.אמר [103]

A word which is possibly a form of קבב occurs in this line. Van der Kooij proposes a reading of ומלקה without a word divider.[104] Hoftijzer says the absence of the word divider is improbable since this is the last word in red ink.[105] He interprets the word as composed of the conjunction, a מ which is a "neutral interrogative pronoun," ל functioning as a preposition and קב, the infinitive construct of קבב.[106] He translates the line "foolishness and silliness an iniquitous word on (your) tongue. With you we will litigate (?), *it will be impossible for you to curse anymore*."[107] He interprets the words לדעת ופר as "to become completely powerless": the ל is the negation and פר is from the root פרר (cf. Akkadian *parāru* in the G stem).[108] Regarding the interpretation of דבר.לענֹה, he proposes a comparison to Arabic *la'ana* "to curse" and *la'n* or *la'nat* the noun "curse;" Nabatean *l'nh* "curse," verbal root *l'n*;[109] and from the Dead Sea Scrolls דבר לענה.[110] For נשפט, he notes Jer. 25:31 where it is used with ל to mean "litigate with" implying punishment.[111]

The reading of Caquot and Lemaire is תֹ(!)לק(!)י[for Hoftijzer's ומלקב. The word being best interpreted as an imperfect first common plural like the word before it (נשפט) and probably Niph'al. Their translation is "pour connaître la formule de la parole à répondre. Contre la (mauvaise) langue, à ton profit, nous rendrons justice et nous frapperons. Dis/a dit…"[112]

Hackett agrees with van der Kooij that the word is מלקה "punishment, chastisement" from the root לקה (cf. Mishnaic Hebrew and Jewish Aramaic), thus, with the word before it, the pair reads "your judgment and your punishment."[113] This interpretation is in line with what she interprets as the purpose of the text, "to portray what Balaam said orally, and to threaten punishment for the readers of the inscription."[114] Her full translation of the line is "to make known (lit.: 'to know') the account he spoke to his people orally (lit.: 'by tongue'), your judgment and your punishment."[115]

Levine agrees with Hoftijzer in his reading of ומלקב from the root קבה "curse." He sees this as a tie to the biblical Bala'am (Num. 23:7–8ff.).[116] His translation of the line is "To know how to deliver an oracle to his people. You have been condemned for what you have said, and banned from pronouncing words of execration."[117] The person addressed is told that he will no longer be able to interpret oracles to his people.[118] Also in support of Hoftijzer, Greenfield notes that "all of the sure examples of *qbb* in Biblical Hebrew are from the Balaam pericope."[119]

A late seventh century B.C. Phoenician inscription from Cebel Ires Daği in ancient Cilicia provides the only contemporary clear example of the verb (קב) and noun (קבה). In this inscription, the DN בעל קר occurs in close proximity to these forms of the root קבב. The curse is for the protection of a land grant.

(5) ואפ.בעל.כר.ישב.בנ.וקב.מתש.קבת.אדרת
(6) לבל.גזלי.אדמ.שד.אמ.כרמ.בד.שפח.כלש.בכל.אש.יתנ
(7a ל.מתש

"And furthermore, BʻL KR he settled (dwelt) in it, and MTŠ cursed a mighty curse so that no one might wrongfully seize it—field or vineyard—from the possession of the family of KLŠ, of all which MTŠ had given to him."[120]

Deir ʻAlla IX.a.3.

(1) ----מ----------------
(2) ---[--הֹד--].[.מס.כש---כ.-]----
(3) ---].לקב.נקב.כפוה.ו[---]---
(4) ---]אהֹ------ל---.ב---.]---

[121] (traces of lines 5–6 are visible)

Hoftijzer translates line 3, "'he is not cursed' or 'we will not curse' or 'we will not be cursed' (or one of these clauses in question form), his hands (?) ------"[122] He suggests that the proper place for this fragment is just above or below and to the left of line 17 of Combination II.[123] As line 17 of Combination II is another place that the root קבב seems to appear (see the discussion above), this location seems appropriate. Hoftijzer identifies the word לקב as an infinitive absolute from the root קבב with the ל as a negation. The next word he identifies as either a Peʻal or Niphʻal Imperfect first plural or as a Niphʻal Perfect third masculine singular from the same root.[124] Usually, the negation in this construction would stand before the finite verb form rather than the infinitive absolute but this combination of negation + infinitive absolute + verb is not unknown in BH (Gen. 3:4; Amos 9:8; and Ps. 49:8).[125] Examples from other inscriptions are not available. The root קבב does not occur in other Aramaic inscriptions, but is found in a Punic inscription (*CIS* i:4945).[126]

Deir ʻAlla X.a.3.

(1) --------
(2) ------נני----כ.----
(3) ---[.קבת.א-----
(4) ----[-הֹ]---- [127]

Hoftijzer suggests this fragment be placed somewhere between lines 17

Chapter 4: Curse 109

and 26–27 of Combination II but offers no translation. The word קבת could be a Pe'al Perfect third feminine singular, second masculine singular, or first common singular from the root קבב. This word could also be a noun in the singular construct form, perhaps related to rabbinical Hebrew קבה "dome, vault, women's room" (cf. Num. 25:8).[128]

Summary

אלה is used in the Panammu and Arslan Tash 1 inscriptions with treaty overtones. Herbert Brichto says that every oath אלה, implies a curse, the curse is the penalty (implied or explicit) poised to fall on the violator of an agreement sealed with an oath.[129] Though this observation is made of biblical literature, it seems to apply as well to these two instances. The beginning of the Panammu inscription may describe a situation where the oath has already been broken (the specific oath situation is not known), the curse already let loose upon the violator. The Arslan Tash inscription describes the oath itself, an incantation with some of the characteristics of a treaty document. The occurrences of the root קבב in the Deir 'Alla Combinations IX and X are difficult to interpret in such broken contexts and the occurrence in Combination II is dubious.

Substantive Curse

Several "Substantive Curses" are found in the target inscriptions. These are passages in texts which do not use a specific word for curse but whose intent is obviously a curse. These substantive curses do not occur with the name Yahweh,[130] but they do occur both with other DNs and without any appeal to a deity.

With DN

Inscriptions with a substantive curse and an appeal to a deity are Kilamuwa 1 (appeals to Ba'al-Ṣemed, Ba'al-Ḥammon, and Rakkab-El all of whose emblems adorn the stele), Sefire III (threatens with "all the gods of the treaty"), Panammu (calls on Hadad, El, Rakkab-El, Shemesh, and "all the gods of Y'DY"), and Nerab 2 (invokes Sahar, Nikkal, and Nusk).

Kilamuwa 1.13-16. This inscription was found during German excavations of the entrance to the royal palace of Zendjirli in 1902.[131] A relief of the king pointing to four symbols representing three deities (the horned helmet of Ba'al-Ṣemed, the chariot yoke and the winged solar disk together representing Rakkab-El, and the cresent superimposed on a full moon representing Ba'al-Ḥamman) is in the upper left corner and across the top of the stone.[132] A double line divides the two major segments of the inscription (lines 1–8 and 9–16). The inscription dates from 830–825 B.C.

13) ומי.בבנ
14) י אש.ישב.תכתנ.ויזק.בספר ז.משכבמ.אל יכבד.לבעררמ.וברר
15) מ.אל יכבד.למשכבמ ומי.ישחת.הספר ז.ישחת.ראש.בעל.צמד.אש.לגבר
16) וישחת.ראש.בעל חמנ.אש.לבמה.ורכבאל.בעל.בת.[133]

Now, if any of my sons who shall sit in my place does harm to this inscriptionn, may the MŠKBM not honour the B'RRM, nor the B'RRM honour the MŠKBM! And if anyone smashes this inscription, may Baal-Ṣemed who belongs to Gabar smash his head, and may Baal-Hammon who belongs to BMH and Rakkabel, lord of the dynasty, smash his head![134]

In the first eight lines of his inscription, Kilamuwa says that his predecessors accomplished nothing, that the land was surrounded by powerful kings waiting to consume his kingdom, and that he hired the king of Assyria against his enemies (cf. 2 Kgs. 7:6; 16). In lines 8–13, Kilamuwa describes his skill in providing for his subjects ("Him who had never seen the face of a sheep I made owner of a flock," line 6) and deciding disputes between them.[135] The address of the curse (lines 13–16) is two-fold, dealing with two groups who might violate Kilamuwa's inscription.

Lines 13–15. The first group addressed is the king's sons. As has been mentioned above, it was a frequent practice in the ancient world to mar or destroy a previous ruler's written record of achievement.[136] Here the curse for such an offense was that two, apparently important, elements of the population would turn against each other and perhaps cause a civil war. Albright says that the *muškabim* (משכבמ) were the settled population of the land (the group that Kilamuwa takes pride in providing for in lines 10–13) while the *ba'ririm* (בעררמ) were the semi-nomadic Aramaean tribesmen.[137] Most scholars now accept some variation of this theory though, in an early interpretation, Halévy translated משכבמ as "leur tombeau" (cf. Eshmunazar, line 6) and בעררמ as "Ba'al-Ram" (who "semble être un dieu qui administre le monde des morts").[138]

Lagrange says that the verb יזק is from the root נזק "endommager" and here has the nuance of cause since it is followed by the preposition ב.[139] He agrees that the punishment wished on the sucessor is political and is directed against a sucessor who violates the spirit of the stele (i.e., the prosperity that Kilamuwa has brought) by destroying it. He cites the work of Nöldeke who has treated בעררמ as a predecessor of Syriac *b'ryr* "brute, babare."[140] Gevirtz agrees that נזק means "to suffer injury" or in the causative "damage, injury" and notes that it occurs as a noun in Esth. 7:4.[141] He says, contra Lagrange, that if the stem is causative, the presence of the ב on בספר is somewhat difficult and may presuppose an omission, perhaps of שמ "(my) name" so that the inscription would read ויזק [שמ] בספר ז "and shall (e)rase [my name] from this inscription."[142]

Donner and Röllig and Gibson treat the verb as a causitive, Yiph'il, form.¹⁴³ Gibson notes the occurence of the causative (Haph'el) stem in Ezra 4:13, 15, 22 where the verb is used of a loss of revenue to a suzerain.¹⁴⁴ Gevirtz says that the verb כבד (in Pi'el) should be understood as "pay, enrich" (cf. the Qal in Gen. 13:2, "to be rich, wealthy"; Num. 22:17, Balak's promise to Bala'am is also better understood with this meaning). The curse then is that the groups will not repay debts owed to each other.¹⁴⁵

In an early interpretation of the text, offered by Bauer, the curse against the successor is not interpreted as a curse at all, but rather a blessing ("Segenswunsch"). Bauer says that two levels of punishment are elsewhere unknown: one, lesser, level of punishment for damaging the inscription and another, harsher, punishment for destroying the inscription. Instead, he posits a verb root יקן "ausgießen" with the meaning "anointing" with oil. This interpretation also avoids the awkwardness of the ב in בספר which others have tried to explain.¹⁴⁶ The verb כבד should then be interpreted as "sein schwer" or "unangenehm" referring to oppression or antagonism between the two population groups mentioned. The phrase משכבם אל יכבד לבערדם then, is a promise that the successor anointing the stele with oil, thereby honoring it, will continue to enjoy the benefits of goodwill between the two groups.¹⁴⁷ Kraeling agrees that this part of the inscription is a blessing rather than a curse and that the verb יקן should be translated "anoint."¹⁴⁸ The practice of anointing steles is also known from Akkadian texts. In the first instance below, the writer calls for a later renovator to honor the original patron's name:

20) ana EGIR u₄me ina LUGAL.MEŠ [DUMU.MEŠ-ja]
21) ša ᵈŠĀR ù [ᵈ15]
22) ana belut KUR u UN.MEŠ inambu[-ú zikiršú]
23) MU.SAR šiṭir MU-ja l [imurma]
24) Ì.GIŠ lipšúš UDU.SIZ[KUR liqqí
25) itti MU.SAR-e šiṭir [MU-šú liškun]

In later days let one of the kings, [my descendants,] [whose name] Assur and [Ishtar] will nominate to exercise power over land and people, [see] the inscription inscribed with my name [and] let him make a libation of oil, [let him offer] sacrifice. With the inscription inscribed with [his name let him set (it) up].¹⁴⁹

In a second example, the rebuilder states that he has acted with decorum: šiṭir šūm ša ᵐNarām-ᵈSin DUMU ᵐŠarrugina amūrma la únakkír Ì.GIŠ apšúš ⁱᵐᵐᵉʳSIZKUR.SIZKUR aqqi itti musariea aškūnma úteir ašrúšu "The inscription with the name of Naram-Sin, son of Šarru-kin, I found, and I did not displace (it). I anointed it with oil. I offered sacrifices, with my own inscription I set it and I returned it to its place."¹⁵⁰ Bauer's proposed stem יקן has not been found elsewhere with meaning "anoint," though, similar to his suggestion, כבד is found in BH (Qal stem) with the meaning of a battle going hard against someone (Judg. 20:34; 1 Sam. 31:3; 2 Sam. 13:25). An especially good parallel is the use of כבד

in Judg. 1:35, "The Amorites persisted in dwelling in Har-heres, in Aijalon, and in Sha-albim, but the hand of the house of Joseph rested heavily upon them and they became subject to forced labor": ותכבד יד בית־יוסף ויהיו למס.

Lines 15–16. The second group is addressed by ומי with the customary curse against any violator.[151] Lagrange says that Kilamuwa's warning has moved on to others besides his sons since that possibility is "trop invraisemblable."[152] The gods are called upon to deal with the violator (ישחת line 16) as he has dealt with the inscription (line 15 ישחת). Torrey says that ראש should not be read "his head" at all but rather as "an indefinite adverbial accusative" signifying "capital punishment."[153] Gibson says that the verb ישחת may be either Pi'el or Yiph'el meaning, as in BH, "ruin, spoil, pervert, destroy,"[154] though in BH it does not seem to be specifically "smash" as Gibson translates it. He compares the Ugaritic imprecation, "May Ḥoron break, o my son, may Ḥoron break your head, (may) Athtart-name-of-Ba'al (break) your crown! *ytbr ḥrn ybn ytbr ḥrn rišk 'ttrt šm b'l qdqdr*.[155]

Sefire III.[156] The last of the Sefire steles is an extensive list of treaty conditions but does not have the introductory statement naming the parties involved in the treaty, nor does it list the gods who serve as witnesses. Unlike the other two steles from Sefire, III does not contain any sort of blessing and the curses are those implied in general statements which are apparently intended to imply specific curses listed elsewhere. Whether these implied curses are from a missing portion of stele III or whether this stele stood in close proximity to one or both of the other two and assumed their curses, is impossible to say.[157] Because of the broken condition of the text, it can only serve as another witness for the usage of substantive curse, not for the use of curse versus blessing and curse.

Nine times the formula, "you will have been false to all the gods of the treaty which is in this inscription" appears in Sefire III. Four times it occurs in the longer form which includes the reference to the gods (lines 4, 14b, 16b–17a, 23a):

שקרתם לכל אלהי עדיא זי בספרא זנה [158]

Five times the shorter version, "you will have been false to this treaty" (lines 7b, 9b, 19a, 20b, 27b), occurs:

שקרת בעדיא אלן [159]

Interestingly, each time the statement includes the reference to the gods, and only then, it also includes the phrase "which is in this inscription." Though the inscription is incomplete, a sort of pattern of the use of the phrases occurs. The pattern is L, S, S, L, L, S, S, L, S (L = the longer version, S = the shorter version). There does not seem to be an other than stylistic reason for this variation.[160] Dupont-Sommer notes that parallels to both versions of this formula are found in the fourteenth century Hittite vassal treaty between Muršilis II and Duppi-Tešub.[161]

This type of curse may be compared to that used in some of the

Chapter 4: Curse

twentieth century law courts of Europe and the United States. The parties to the treaty oath recognized that the agreement was witnessed by the gods listed as well as the physical elements. Should either party fail in their obligations, the gods were to visit upon them the penalties/curses described in detail (much more detail in Sefire I and II). In modern law courts it is the witnesses who are called upon to swear that they will tell the truth. Indeed, earlier in the century it was the custom to swear honesty by God and/or by the Bible. It is debatable whether the ancient parties to the treaty or the modern witnesses in court were convinced that the gods/God would in fact punish perjury. However, then and now there were/are guaranteed civil punishments should divine retribution prove insufficient motivation.

Panammu lines 21–23.[162]

21) ואמר.במשות.ועל.יבל.אמנ.יסמ[כ].מלכ------.ויבל.יושא.קדמ.קבר.אבי.פנ]מו [
22) וזכר.זנה.הא.פא.הדד.ואל.ורכבאל.בעל.בית.שמש.וכל.אלהי.יאדי] בית[
23) י.קדמ.אלהי.וקדמ.אנש.

Gibson's translation of these lines is "...which this memorial is, then may Hadad and El and Rakkabel, lord of the dynasty, and Shemesh and all the gods of Y'DY...[my house] before the gods and before men."[163]

None of the scholars consulted provides a comprehensive translation of line 21. Cooke says that line 21 seems to be concerned with safeguarding the statue and tomb. Lines 22–23, then, invoke the gods to curse anyone violating the tomb.[164] Gibson says that it "probably concerns the maintenance rites at Panammu's tomb" and suggests a comparison to Hadad lines 15 and following.[165] Both compare it to the ending of the Hadad inscription but each with a different understanding of that inscription: Cooke interprets line 28 of the Hadad inscription as a provision against violators and Gibson says lines 15–24a instruct the heir in his *kispu* obligations and curse him should he fail to fulfill them.

Kraeling does not translate line 21 and finds no blessing or curse formula in lines 22–23. He translates those lines simply as a statement that the inscription is set up as a memorial.[166]

Donner and Röllig say that the line contained a blessing or curse formula or both.[167] Their translation of lines 21–22 is "und er sprach:... *er brachte*...[ ] König [ ] und er brachte...vor dem Grabe meines Vaters Panammuwa [ ] und dieses Denkmal ist es."[168] They propose that במשות is a substantive form from the root נשא "tragen, heben" with the preposition ב attached but are unable to translate it in the context. Also, they suggest that יבל may be a Pa'el Perfect third masculine singular form of the verb יבל "bringen" with אמנ as the direct object.[169]

The verbs נשא and יבל could conceivably be used to speak of means of violating a tomb though they are not the usual terms for violation of a tomb or mutilation of an inscription (cf. the discussions of Karatepe, Zakir, Aḥiram, and Kilamuwa 1). A curse on a potential violator and perhaps a blessing on the king's descendants (cf. Zakir B16–C2) would be expected at the end of a monument like the Panammu inscription and this may be the reason why Cooke and Donner and Röllig look for it here. While there is adequate space for a curse and perhaps a brief blessing in the broken end of line 20 and the gaps of lines 21 and 22, there seems to be little real evidence for them.

Nerab 2. The two Nerab steles were discovered at the village of Nerab seven kilometers SE of Aleppo in 1891 and date from the early seventh century. In this stele, the whole inscription is written above the relief of a seated priest offering a libation before an altar in the company of a fan-wielding attendant. The style of the relief, the name of the priest, and the gods invoked all show a strong Akkadian influence.[170]

1) שאגבר כמר שהר בנרב
2) זנה צלמה בצדקתי קדמוה
3) שמני שם טב והארך יומי
4) ביומ מתת פמי לאתאחז מנ מלנ
5) ובעיני מחזה אנה בני רבע בכונ
6) י והומ אתהמו ולשמו עמי מאנ
7a) כסף ונחש עמ לבשי שמוני

Si'-gabbari, priest of Sahar at Nerab. This is his picture. Because of my righteousness before him, he afforded me a good name, and prolonged my days. On the day I died, my mouth was not closed to words, and with my eyes I was beholding children of the fourth generation; they wept for me, and were greatly distraught. They did not lay with me any vessel of silver or bronze; with my garments (only) they laid me, so that in the future my grave should not be dragged away.[171]

Lines 6–7a. The writer tries to remove any motivation a robber would have for disturbing the grave by claiming no silver or bronze are present in the sarcophagus (cf. the Siloam Tomb as well as the later Tabnit lines 4–5 [ca. 490 B.C.][172] and Eshmunazar line 5 [ca. 480 B.C.][173] inscriptions but contrast that of Batnoam [ca. 500 B.C.] which boasts of the riches enclosed in the sarcophagus[174]).[175] A ninth century Phoenican grave inscription from Cyprus makes a more subtle attempt to dissuade robbers by claiming that no one important is buried inside thus, presumably, the interee would not be accompanied by valuables.[176]

7b) למען
8) לאחרה לתהנס ארצתי מנ את תעשק
9) ותהנסני שהר ונכל ונשכ יהבאשו
10) ממתתה ואחרתה תאבד

Lines 7b–10. Gibson translates these lines, "Whoever you are who do wrong and drag me away, may Sahar and Nikkal and Nusk make his dying

Chapter 4: Curse   115

odious, and may his posterity perish!"[177] Not believing that the disclaimer of treasure will deter thieves, the writer resorts to a curse. לאחרה is usually interpreted as "in the future,"[178] however, Torrey translates line 8, "in order that my coffin may not be plundered (?) by another."[179] The first interpretation lends weight to Gevirtz's translation of ואחרתה as "and his future" in line 10.[180] According to Gevirtz and Gibson, the writer is concerned that his grave (ארצתי) and himself (pronominal suffix on the verb in line 9) will be dragged away (line 8 לתהנס; line 9 ותהנסני).[181] Koopmans prefers the translation "du sollst nicht rauben."[182] This verb is also used in the Zakir stele (B20) where that king is concerned that someone will drag away his stele from its place (ויהנסנה with third masculine singular suffix).

The address of the curse is the inclusive מן את as in Nerab 1:5 and similar to Zakir's ומן in contrast to the more specific introductions of the Aḥiram and Karatepe inscriptions. Is the address stated like this because it is more likely that a "nobody" would rob the grave of a priest than bother the stele of a king, or would a king or ruler be less likely to bother the grave of a priest? Or, perhaps there is no real difference in meaning intended between the forms. The form chosen may simply be a local preference, each form intended to frighten away all would be violators.[183]

The verb תעשק is also found in Sefire III.20 where it means "hinder."[184] In BH it occurs only in Gen. 26:20 and in the Hitpaʻel meaning "contend."[185] The causative of the verb באש (יהבאשו) occurs in line 9. This verb is also seen in BH באש "have a bad smell, stink." Psalm 38:6(5) provides an interesting parallel, "My wounds grow foul and fester" הבאישו נמקו חבורתי. Koopmans notes its appearance in the fifth to fourth century Aramaic Nanasht inscription where the root appears as a noun meaning "böse."[186]

The curses in this inscription and in the first Nerab inscription are very similar; the variations in wording simply show that there was more than one way to express the same substantive curses. The DNs are virtually the same and in the same order, except that Shamash is left out of the second inscription. The two basic curses: 1) that the perpetrator die nastily and 2) that the perpetrator's posterity cease, are included in both and in that order. Nerab 1 adds a reciprocal blessing should the would be thief have a change of heart and guard the tomb and inscription. Nerab 2 adds a disclaimer regarding the value of the grave's contents. Nerab 1 is in the second person. Nerab 2 shifts between second and third person forms.[187]

Line 10. Gevirtz notes that the form מְמוֹתֵי occurs in Jer. 16:4 where it refers to death by disease as in this inscription, ממותי תחלאים ימתו.[188] This form also occurs in Ezek. 28:8 in an oracle against the prince of Tyre לשחת יורדוך ומתה ממותי חלל בלב ימים:. Gevirtz interprets ואחרתה as "and his future" meaning "the rest of his life" though most other commentators, including the original editor Clermont-Ganneau,[189] translate it as "posterity."[190] Gibson points out two important BH parallels in which this word refers to the descendants of the one who is cursed or wicked: Ps. 109:13 "May his posterity be cut off; may his name be blotted out in the second generation!" יְהִי־אַחֲרִיתוֹ לְהַכְרִית בְּדוֹר אַחֵר יִמַּח שְׁמָם; and Ps. 37:37–38 "Mark the blameless man, and behold the upright, for there is posterity for the man of peace. But transgressors shall be altogether destroyed; the posterity of the wicked shall be cut off" שְׁמָר־תָּם וּרְאֵה יָשָׁר כִּי־אַחֲרִית לְאִישׁ שָׁלוֹם: וּפֹשְׁעִים נִשְׁמְדוּ יַחְדָּו אַחֲרִית רְשָׁעִים נִכְרָתָה:[191]

With No DN

In the Aḥiram and Nimrud Hebrew inscriptions and Deir ʿAlla Combinations I and II, substantive curses are used without an appeal to a specified deity. While the Aḥiram inscription seems to be complete and probably never contained an appeal to a deity, the Nimrud Hebrew and Deir ʿAlla texts have substantial broken sections and may well have originally included invocations of a deity or deities to bring about their curses.

Aḥiram . The inscription was discovered during excavations at Byblos in 1923. The sarcophagus, and grave 5 which contained it, were originally used in the thirteenth century. Later study of the sarcophagus has revealed that it originally bore a Pseudo-Hieroglyphic inscription which was partially effaced when the Phoenician text was inscribed (ca. 1000 B.C.). Line 1 of the inscription is written on the upper rim of the sarcophagus and in from the edge in order to avoid the previous inscription. Line 2 is written on the side of the lid.[192]

(1) ארן.ז פעל.[א]תבעל.בן אחרם.מלך גבל.לאחרם.אבה.כשתה.בעלם.

(2) ואל.מלך.במלכם.וסכן.בס>כ<נמ.ותמא.מחנת.עלי.גבל.ויגל.ארן.זנ.
תחתסף.חתר.משפטה.תהתפכ.כסא.מלכה.ונחת.תברח.על.גבל.והא.ימח.
ספרה.לפפ.שבל[193]

Coffin which Ittobaʿal, son of Aḥiram, king of Byblos, made for Aḥiram, his father, when he placed him in 'the house of eternity'. Now, if a king among kings or a governor among governors or a commander of an army should come up against Byblos and uncover this coffin, may the sceptre of his rule be torn away, may the throne of his kingdom be overturned, and may peace flee from Byblos! And as for him, may his inscription be effaced...![194]

Dussaud's translation of the second line is

Et s'(il est) un roi parmi les rois, ou un gouverneur parmi les gouverneurs, qui dresse le camp contre Gebal et qui découvre ce sarcophage sous le dallage, Ḥaṭor (sera) son juge: le trône de son roi se renversera et la destruction fondra sur Gebal tandis que lui (le profanateur) effacera cette inscription à l'entrée

Chapter 4: Curse   117

(?) de l'Hades (?)[195]
Dussaud divides the word תחתסמפ into סמ and תחת, though there is no word divider, and interprets סמ as "flagstone" paricularly those stones used to cover grave pits in the royal necropolis of Byblos.[196] The last two words of the inscription, לפפ שרל, are agreed to be the most difficult by commentators.[197] Dussaud offers two interpretations. First, he suggests translating them as describing the position of the inscription "à l'entree de l'Hades" though he does so with question marks. The phrase "at the mouth of Sheol" (לפי שאול) occurs in Ps. 141:7.[198] Montet favors this interpretation.[199]

In his second interpretation, Dussaud suggests that these words may refer to an implement employed to damage the inscription. He explains לפפ as a "réduplication" of פה and compares Ps. 149:6 (describing a sword) and Isa. 41:15 (describing a threshing instrument), thus the implement of destruction. שרל then, would be a verb "cut" ("trancher"). However, as Dussaud admits, the expected preposition on לפפ would then be ב rather than ל.[200] Donner and Röllig agree with Dussaud in comparing לפפ to BH reduplicated פיפיות (Ps. 149:6; Isa. 41:15). They suggest further parallels (Gen. 34:26; Josh. 6:21; 8:24), but are unable to identify the tool or weapon referred to by שבל, and translate the last phrase, "es soll ausgelöscht werden seine Inschrift mit der Schärfe des (?)."[201]

Vincent recognizes a parallel in the phrases תחתסמ.חתר.משפטה and כסא.מלכה תהתפכ: משפטה ("sa justice") would be the prerogative of governors and מלכה ("pouvoir royal") that of kings, the two classes warned at the beginning of the line.[202] Regarding the phrase ונחת.תברח.על.גבל, Vincent says the ו joins it to the preceding curses but notes that it would be strange to attribute a curse against his own city to a dead king. Therefore, this phrase should be understood in the opposite way: this curse will release supernatural power on the violator preventing him from continuing the siege or attack and peace will hover over Byblos ("et que la paix plane sur Gébal!").[203] Vincent's reading of the end of the inscription differs from that of Dussaud, [ה]ל [ש]שר.לפפ.ז ספר.ימח.והא. The word והא begins another pericope: the first part of the inscription is a warning not to disturb the sarcophagus, this new section is intended to protect the inscription itself (cf. Kilamuwa 1:15 and Sefire I.C.17). The verb ימח is an apocopated imperfect form of the verb מחה (cf. Karatepe A.iii.13).[204] The word לפפ, not elsewhere found in Phoenician, in Aramaic has the sense "entourer, entrelacer, nouer" and corresponds to BH לפת "turn, grasp with a twisting motion" which, in turn, is related to Akkadian *lapātu* which may have the meaning "destroy."[205] The last words should be read שרש לה and refer to the violator's descendants, "his root" (cf. Tabnit 7–8; Eshmunazar 8, 22; and perhaps Nerab 1:11 and 2:10[206]). Vincent says that the greatest fear of the ancient was the end of one's race.[207] The inscription commemorates the dead one perpetually; destroying it will wipe out his memory, his continued existence when his family is no more.

The threat against someone who would do this to Aḥiram is that the same may happen to them.[208]

Torrey differs in his interpretation of the parties addressed from Dussaud and Vincent. He treats ותמא מחנת as "military commander" rather than as a verb construction meaning "to set up camp." He says that תמא is probably the equivalent of Assyrian *tamu-* secondary root of *amū* "speak" corresponding to Arabic *amū* "commander."[209] He points out also that חטר and כסא are masculine in Hebrew and Aramaic but here are feminine as in Akkadian (*kussū* and *ḫaṭṭu*). His translation of the last five words of the inscription is "if he shall destroy this inscription, cover it over or deface it," thus, he says, "The penalty is invoked not only for the destruction of the inscription, but also for any damage to it."[210] He calls attention to the Egyptian and Assyrian practice of erasing the name of an enemy or rival from a record of their accomplishments.[211] The last two words of the inscription are infinitives absolute: לפפ "wrap up, cover over" to prevent those who fearing the curse against the destroyer would obscure the inscription in this way to avoid the curse and שבל a "shaphel" stem of בלל "besmear" to deface the inscription by pouring some pigment or other liquid over it.[212]

Bauer says that ויגל should be interpreted as uncovering the sarcophagus ("die Erde vom Sarkophag wegnehmen") rather than removing it.[213] He says that תברח is a verb of movement and interprets it like Vincent, not as a curse against Byblos, but as the resultant blessing of a lifted siege "und Ruhe möge kommen über Gebal!"[214] Bauer interprets ימח as a Niph'al. Donner and Röllig agree, choosing the passive identification due to the absence of the customary call on a deity to enforce the curse.[215] Gibson also agrees citing the root with the same force and in Niph'al in Exod. 32:32; Pss. 69:29(28); 109:13. The verb is also used in the Karatepe inscription (A.iii.13) for effacing an inscription.[216] The last word of the inscription proper then refers to the violator "er aber soll ausgetilgt werden," the rest (סכרה ז.לפפ.שרל) was added later and simply gives the name and title of the scribe.[217]

Montet notes that interpreters of this text fall into one of two groups: 1) those who wish to correct the text and 2) those who interpret the words in ways uncharacteristic to semitic usage.[218] He says that while it is true that the construction of the curse at the end of the inscription is jerky, it may be inspired by an Egyptian style and was chosen by design to express with more force the supreme menace which will stop the violator. Ittoba'al does not seem worried that his father's tomb will be disturbed by his fellow citizens, but fears violation by an invader. The curse calls for the invader to be punished by the defeat of his enterprise (since peace will return to Byblos) and the loss of his kingdom, but another, more serious punishment awaits: the violator's funerary inscription necessary for his fortunate future will be effaced automatically.[219]

Albright treats עלי as a verb meaning "go up, attack" rather than as an alternate spelling of the preposition על, which also occurs in the line.[220] For the

Chapter 4: Curse   119

first threats in the line תההסכ.תהתפכ.תהתפטה.משפטה.חטר.תהתסכ.כסא.מלכה, Albright gives an Ugaritic parallel from the Ba'al and Mot epic *l ys' 'alt ṯbtk lyhpk ks'a mlkk lyṯbr ht mtpṭk* "he is sure to overturn the chair of your kingship, he is sure to break your judicial sceptre!"[221] A very similar phrase with much the same language also occurs in a biblical passage about a coming day of judgment, Hag. 2:22, I am about to "overthrow the throne of kingdoms; I am about to destroy the strength of the kingdoms of the nations" והפכתי כסא ממלכות והשמדתי חזק ממלכות הגוים. For the word ונחת, Albright relates BH מנחה especially as it is used in 1 Kgs. 8:56, "Blessed is the Lord who has given rest (מנוחה) to his people." He also suggests a helpful Ugaritic parallel *lnht lkht drkt* "on the peaceful throne of authority."[222] Albright's interpretation of the last two words is remarkable, if perhaps too ingenious. He identifies לפפ with Aramaic, Akkadian, and Arabic *lpp* "to wind" in the sense of "wend one's way," שבל is like BH "road, way," so the two words together would be "wayfarer, vagabond." The last curse then is "and as for him, let a vagabond efface his inscription(s)!"[223]

Gevirtz originally disagreed with Albright's interpretation of עלי as referring to invasion saying, "there appears to be no reason for the introduction of a military element" and went on to say the inscription made little sense in light of a military invasion, thus he could see the threats as a warning to a successor and interpret the phrase ונחת.תברח.על.גבל more straightforwardly as "And may peace flee from Byblos!"[224] With his new suggested interpretation of the term תמא he has apparently changed his mind.[225] Gevirtz identifies the word תמא as a noun in construct with the following word מחנת translated "commander of an army" referring to the third group of possible violators, rather than as a verb as previous scholars did.[226] Gevirtz interprets תהחסכ as related to the Aramaic root חסכ "peel" and Akkadian *ḫasāpu* "to pluck out, tear off." He says Albright's translation of the verb as "break" is unwarranted.[227]

Greenfield notes Akkadian parallels to elements in the inscription. The address to possible violators is similar to that found in the Hammurabi Codex *lu šarrum lu belum lu iššakkum u lu awilutum ša šumam nabiat* (XXVIb 40–44).[228] The threat to break the scepter and remove the throne of anyone altering the inscription are also included at the end of the Hammurabi Codex (XXVIb 50–51; XXVIIb 45–46).[229] The curse is also found in the Annals of Sennacherib *ù šumia ^{d}Aššur bêlu rabû abu ilâni^{meš} nakriš lizīssu ^{iṣ}ḫaṭṭa ù ^{iṣ}kussâ likimšúma liskipa palêšu* "May Assur, the great lord, the father of the gods, treat him as an enemy, take away from him the scepter and throne and overthrow his rule."[230] The scepter and throne pair are also found in some neo-Assyrian blessings in which cases the gods are asked to give the king long life (*ūmū arkūti*), eternity (*šanāti darâti*) and "a just scepter and an everlasting throne" (*ḫaṭṭu išartu* and *kussû darû*).[231] The pair described in this way is found in a royal psalm "Your divine throne endures for ever and ever. Your royal scepter is a scepter of equity"

כסאך אלהים עולם ועד שבט מישר שבט מלכותך: (Ps. 45:7[6]; also compare Amos 1:5 and Ps. 89:45).

The list of possible violators of the grave is longer than that seen in other inscriptions (though quite similar to Karatepe A.iii.12ff.) but those other, less specific warnings (like Tell Fekherye lines 10 and 16 מן) may actually be more inclusive. What do the more specific warnings like Ahiram and Karatepe signify? They may indicate that lesser citizens would not dare violate the king's sarcophagus. Thus, they may be warnings against an invader rather than a native successor since simple grave robbers were probably lower socio-economic class people who likely could not read the curse meant to prevent the tomb's desecration. It is ironic that Ahiram's use of the sarcophagus is its second use. This would seem to be the very type of violation that Ahiram's own inscription is intended to prevent.

The verb ויגל may be interpreted in various ways, is it meant to prohibit the opening of the grave or its removal? In BH, the verb גלה may mean "uncover, remove, reveal." It is used of taking people into exile (e.g., Judg. 18:30; 2 Kgs. 17:23; 25:21) of removing one's dwelling (Isa. 38:12, symbolic of the writer's approach to death); and of laying bare enigmatic "secret places" (Isa. 49:10). That objects as well as people were carried off from captive lands is witnessed by the Nimrud Hebrew inscription which seems to have been Assyrian booty and the vessels from Solomon's temple which were carried off by the Babylonians. BH גלה with the meaning "carry into exile" seems to be reserved for passages in which people were taken captive. Biblical references to "secret places" seem to come closest to the meaning in this inscription, the concern seems to be that "the secret places" (or "foundation" Micah 1:6) should not be uncovered rather than that they be carried away (cf. Jer. 49:10; Job 12:22). Note Assurbanipal's action against the defeated Elamites:

> Their secret groves, into which no stranger (ever) penetrates, whose borders he never (over)steps,—into these my soldiers entered, saw their mysteries, and set them on fire. The sepulchers of their earlier kings and later kings, who did not fear Assur and Ishtar, my lords, (and who) had plagued the kings, my fathers, I destroyed, I devastated, I exposed to the sun. Their bones (members) I carried off to Assyria. I laid restlessness upon their shades. I deprived them of food-offerings and libations of water.[232]

The curse "may the scepter of his rule be torn away" (תחתסף.חטר.משפטה) should be viewed in contrast to Hadad 2–3 where the mark of the gods' favor is that they put into his hand "the scepter of prosperity" (חטר חלבבה).[233] The goddess Ištar is called upon specifically to curse (arāru) another's royal rule in Codex Hammurabi (XLIII.103) *ina libbiša aggim ina uzzātiša rabiātim šarrūssu lirūr* and in another place (XLII.52) a god is called upon to break the king's

Chapter 4: Curse

scepter and curse him *ḫaṭṭašu lišbir šimātišu lirūr*.[234] That the scepter would be torn away is then a clear curse since the resulting chaos could hardly bring prosperity to the ruled. Likewise, the curse "may the throne of his kingdom be overturned" (תהתפך.כסא.מלכה) is similar to Sefire I.C.21–24 "so may the gods overturn (יהפכו) th[at m]an and his house and all that (is) in it; and may they make its lower part its upper part! May his scio[n] *inherit* no name!"[235] and it is contrasted by one of the reasons that Hadayisʻi erected his statue in the Tell Fekherye inscription (line 13) "to establish the foundation of his throne" (.לארמורדת כרסאה).[236] In 1 Kgs. 2:45–46, where Solomon's throne (or David's) is secured by the demise of threatening elements, notice that Solomon's blessedness is in tandem to the stability of the throne, והמלך שלמה ברוך וכסא דוד יהיה נכון לפני יהוה עד־עולם: (cf. Isa. 9:3; 14:5; Jer. 48:17; Jer. 1:10).

The next curse does seem illogical if taken at face value, "may peace flee from Byblos" and the comment that a deceased king would hardly curse his city seem true. The parties addressed in the curse could be either foreign invaders or Aḥiram's successors. If the curse is directed against a conqueror already in possession of the city (otherwise, he would hardly have the time to rob graves), then this understanding of the curse may well make sense—"you may well have taken the city, but may it not enjoy peace under your rule!" (cf. the curse of Joshua against the one rebuilding Jericho in Josh. 6:26). The curse could also be directed against a lawful successor who disturbs his predecessor's grave (cf. the king's curse against his unfaithful successors in Hadad 20–24 and the Nimrud Hebrew inscription treated below). In either case the simplest translation of the text could apply.

The last curse of the text is curious if taken as simply an additional curse that the violator's inscription be effaced, since that curse is elsewhere reserved for those who have effaced the inscription in question (e.g. Karatepe A.iii.12–19), while in the Aḥiram inscription the crime is that the sarcophagus has been "uncovered," probably opened and searched for valuables. This curiosity and the fact that effacing the inscription is not listed as a punishable offence may be explained by the circumstance that Aḥiram's use of the sarcophagus itself is a second use and his inscription partially obscures that of the previous resident. Javier Teixidor has recently given a reading of the last words of the inscription based on a new photograph. According to him, the last words of the inscription should be read לפן גבל and the last phrase translated "que son inscription soit effacée à la face de Byblos." This reading and translation seem the best so far offered.[237]

Nimrud Hebrew. This ivory commemorative plaque (officially designated ND 10150) bearing a fragmentary inscription was discovered during the excavation of Fort Shalmaneser in 1961. The plaque probably was meant to accompany a votive offering and bore a curse against anyone tampering with it. It is thought to date from ca. 750 B.C. and have been carried away from Israel after the fall of Samaria in 721 B.C.[238]

(1 ו.ב[   ] יפת [   ]י[
(2]חֹד/רִי.ממלכ.גדל.ח[
(3]א.ומחו[ ]א[ ][239]

Millard's reconstruction of the last two lines is

(2 מאחרי.ממלכ.גדל.ועד.איש.אשר.באו.בע.
(3 ומחו.את.הספר.הזה [240]

     Millard's translation, based on his reconstruction is "(may God curse any) of my successors, from great king to private citizen who may come and destroy this inscription."[241] Millard apparently understands a DN as missing in the curse formula since he begins his translation "(may God curse any) of my successors..." The contrast of the king and the private citizen in the curse is also seen in two Phoenician inscriptions: the eighth century Karatepe inscription A.iii.12ff. and the fifth century Yeḥawmilk inscription (line 11).[242]

     Gibson's translation is "...May (Yahweh) shatter (?)...after me, from great king (to private citizen,...to) come and deface (this inscription)."[243] Gibson reads a ' (with a question mark) as the last legible letter of line 1 and translates "Yahweh" as the avenger apparently because of the Israelite origin he assigns the text.[244] The proposed root of יפת is פתת with which Gibson compares the BH *hapax legomenon* פתת "break up, crumble" used with regard to bread.[245]

     Perhaps יפ in line 1 is better understood as the beginning of יפתח from the root פתח. A comparison with the curse formula guarding the Siloam Tomb (lines 2–3) is perhaps instructive, ארור האדמ אשר (3) יפתֿח את זאת "Cursed be the man who will open this!"[246] Of course, what was not to be opened in the case of the Nimrud Ivory cannot be known since the inscription is broken and was found outside of its original context.[247]

     Deir 'Alla I.[248] This inscription is included among those without DNs because the context is unclear. It could be argued that the council of the gods, a goddess (Shagar-We'ishtar), or a pair thereof (Shgr and Ashtar) do figure in the cursing involved in this text. However, commentators are not clear on what part these deities (or proposed deities) play, nor on the curse content of the inscriptions.

     Hoftijzer summarizes Combination I thus: Bala'am, seer of the gods, had a vision in the night. A goddess (?) gave Bala'am a vision (not recorded) whose content caused him to rise weeping the next morning. His weeping attracted the attention of the people who appointed his uncle to ask him what was wrong. Bala'am responded with an elaborate version of a goddess's message (her mission: to destroy the world with fire) and he appealed for the people to repent. An assembly of the gods met to try to persuade the goddess to change her mind. The modes of destruction are bolts of lightning from heaven, a deluge, the threat of

Chapter 4: Curse

an enveloping darkness, and numerous animals.[249]

McCarter thinks that the two major combinations are parts of the same text with Bala'am serving as a unifying figure tying together several diverse episodes.[250] He believes that the gods called for their message to Bala'am, the inscription itself, to be written down (line 2) so it would be remembered. The doom to come includes the skies being sewn shut to prevent rain and light being replaced by perpetual darkness.[251] He suggests that the goddess thus commissioned may have been She'ol, and the darkness mentioned, the impenetrable darkness of the underworld. The prophet Elijah acting as the herald of Yahweh, announced a three year drought in 1 Kgs. 17:1. There the reason for the punishment is not given; some interpreters believe that is the case as well in Combination I.[252] The punishment of darkness is a parallel to the plague of darkness (חשך) in Egypt which drew distinction between the people of God and their taskmasters (Exod. 10:21–23). Elsewhere in the Bible darkness is also the image of judgment: Isa. 13:10–11; Ezek. 32:7–8; Joel 2:10; 3:3–4; 4:15. McCarter says that the reason for this darkness is not judgment as in the Bible, but because the world order is turned upside-down: the birds, animals, and people mentioned are acting in ways exactly opposed to their characters. "It is cosmic, not moral, evil to which the gods here propose their grim solution."[253] A general parallel to this type of writing is found in Isa. 29:16 "You turn things upside down! Shall the potter be regarded as the clay; that the thing made should say of its maker, 'He did not make me'; or the thing formed say of him who formed it, 'He has no understanding'."

Hackett's summary is very similar to McCarter's. Bala'am, having attended a meeting of the Divine Council, reported back to his people. The gods, upset by unnatural events on earth, asked a goddess to withhold light from earth to punish it: the constant lack of sunlight implies a lack of fertility, one of the basic blessings (cf. Isa. 30:26 where unnatural brightness of the sun and moon are a blessing).[254] Hackett notes that reversal of the natural order is also seen in the Egyptian texts "The Admonitions of Ipuwer," "The Prophecies of Neferti," and "The Complaints of Khakheperre-Sonb."[255] In "The Prophecies of Neferti," the sun god Re withdraws from mankind because of the perversity in nature. Though Hackett agrees that the reversal of the natural order is a sign of catastrophe in other literatures, she says that the syntax of this text demands that this reversal is the reason for the goddess's prouncement of punishment rather than the punishment/curse itself. She holds this interpretation though she acknowledges that two other scholars privately suggested that the series of phrases following כ (line 7) can be interpreted as curses that the gods are pronouncing against the earth.[256] If her interpretation is true, the cause of the gods' anger is not apparent from the extant text. She also acknowledges that this type of curse has strong biblical parallels and cites Isa. 3:4–5, 24.[257]

Levine's interpretation agrees with McCarter's up to a point, but he understands Bala'am as attempting to turn the goddess charged with this judgment duty away from fulfilling the will of the council. Bala'am succeeded in doing this by using execrations and other forms of magic, thus his deeds are recorded in this text, on the wall of the sanctuary.²⁵⁸ Kaufman thinks that Levine's interpretation is correct but goes further and conforms the text to the biblical story saying, "the description of unusual natural phenomena is actually part of a curse that Balaʻam was commanded to utter over the land and its inhabitants."²⁵⁹

Lines 9b through the end of the first combination²⁶⁰ contain material that indicate a reversal of the natural order in which birds, animals, and people do things that are contrary to their normal behavior. Views of commentators vary as to whether these unnatural events are the cause for the announced drought and darkness, or whether they are part of a curse which includes drought and darkness. The substance of these lines is discussed in the following paragraphs.

(7b) כי.ססעגר.חר
(8 פת.נשר.וקל֯.רחמֹן.ענה.ח֯]סד. .[בני.נחצ.וצרֹה.אפרחי.אנפה.דרר.נשרת.
(9a) יוֹנ.וצפר.

For the swift has reproached the eagle, and the voice of vultures resounds. The st[ork has] the young of the NḤṢ-bird and *ripped up* the chicks of the heron. The swallow has belittled the dove, and the sparrow...²⁶¹

Lines 7b–9a. Hoftijzer does an extraordinarily detailed study of each of the birds mentioned in these first lines with regard to their identity, their appearances, and significance in the Bible. He says the context of Isa. 38:14, where the ססעגר is mentioned, suggests that it was a feeble bird. Yet, in these lines, it "reviles" the eagle, the נשר, everywhere regarded as a powerful bird and much admired.²⁶² Hoftijzer thinks this defiance (חרפה²⁶³ cf. 1 Sam. 17:10; 2 Sam. 23:9–10; and Zeph. 2:8) is symbolic of the disobedience of mortal men toward the goddess, a contempt which will now be punished by darkness.²⁶⁴

Caquot and Lemaire agree with Hoftijzer's interpretation and translate רחמֹן later in the line as "vulture," suggesting the translation "chanter" for ענה. They remark that the song of a vulture would be startling and unusual indeed.²⁶⁵ Hackett's addition that the vulture was thought voiceless, augments the unnaturalness of this performance.²⁶⁶

Levine says that Balaʻam describes in these lines what will happen when the darkness comes; it is a scene of alarm with eagles, vultures (symbolizing imminent death), and other birds making shrill sounds that dramatize the coming disaster. The shrieking of the birds at each other would sound like defiance to humans.²⁶⁷

Hoftijzer includes an extensive study of the several birds mentioned in these lines.²⁶⁸ Since the text is broken, the reader cannot know with certainty what these birds are said to do. He suggests that if "the birds of prey from the swamp" (נשרת יונ) are going to inhabit the land of Balaʻam's hearers, the country itself must be about to become a swamp and thus uninhabitable. He gives a biblical parallel in which it is prophesied that Babylon will become a swamp (Isa. 14:23).

Chapter 4: Curse   125

He also cites an Akkadian text of Esarhaddon which describes a destroyed town as overgrown and the abode of birds and fish.²⁶⁹ The birds mentioned are symbols of destruction (because they dwell in a destroyed, forsaken place) and most included here are also found in the biblical list of unclean birds.²⁷⁰

McCarter sees much more specific mayhem expressed in these lines. The בני נחץ are the "young of the NḤṢ bird"; the אפרחי אנפה are "the chicks of the heron"; וצרה means "rip up, tear" (Aramaic צרה and in Syriac it can mean "to rip up" with the claws); for נשרת he compares Akkadian našaru in the D-stem "belittle" and Rabbinic Hebrew and Aramaic נשׁשׁר "tear, mutilate"; יון is a "dove"; and the צפר, though generic for bird in some texts, here refers to something like a "sparrow."²⁷¹ Thus he translates, "The st[ork has] the young of the NḤṢ-bird and *ripped up* the chicks of the heron. The swallow *has belittled* the dove, and the sparrow [ ] and [ ] the staff.²⁷² Hackett agrees, "the peaceful swallow has torn into the dove and the sparrow has done something which we suppose was equally alien to its usual character.²⁷³

(9b) [י.נ.].מטה.באשר.רחלן.ייבל.חטר.ארנבן.אכלו.

"[ ] and [ ] the staff. Instead of ewes the stick is driven along. Hares have eaten [ ]"²⁷⁴

Line 9b. Hoftijzer again understands symbols of destruction. He compares the occurrence of ארנבן "hares" to its use in the Sefire inscription I.A.32 where the hare and other animals are to inhabit ruined Arpad, should it break that treaty.²⁷⁵ The two words for "rod" or "staff" מטה (left dangling from the previous sentence according to Hoftijzer) and חטר are symbols for punishment.²⁷⁶

Caquot and Lemaire say, contra Hoftijzer, that the last word on the line אכלו belongs at the end of the sentence on line 9 rather than at the beginning of the next sentence. Their translation is very similar, however, "à l'endroit où le bâton (=la houlette) menait (paître) des brebis, des lièvres mangent."²⁷⁷

McCarter agrees with Caquot and Lemaire that אכלו belongs with line 9 but disagrees with the syntax of their translation.²⁷⁸ His translation reflects the crazy world pictured by the context, "Instead of ewes the stick is driven along."²⁷⁹ Hackett's translation is virtually the same, "and instead of ewes, it is the staff that is led."²⁸⁰

Levine believes that this line portrays the beating and scattering of flocks and herds according to the wrath of the gods. Likewise, the next line simply says that wild animals will now eat and drink freely in the old pasturing places.²⁸¹

(10 [ ] חד.חפשׁ[ ].שתיו.חמר.וקבען.שמעו.מוסר.גרי.ש
(11a [על. .]לחכמן.

Freemen [ ] have drunk wine, and hyenas have listened to instruction. The whelps of the f[ox]laughs [sic] at wise men.²⁸²

Lines 10–11a. Hoftijzer translates the first part of this broken line, "Eat fear you seekers...drink wrath."²⁸³ This is the fear of those against whom the wrath of the gods has been turned. He compares Job 20:23, "To fill his belly to the full God will send his fierce anger into him, and rain it upon him as his food."²⁸⁴ He translates the end of the line (line 12) "Oh aggrievers (?) listen to the exhortation, Oh adversaries of Sh(gr)(?)"²⁸⁵ The verb שמעו is an imperative and these words are an exhortation to the listeners, thus the described calamities have not yet occurred. The combination שמע...מוסר, Hoftijzer's "listen to exhortation," is also found in biblical wisdom literature (Prov. 1:8; 4:1; 8:33; 19:27; Job 20:3) but is not found in prophetic literature where a parallel might be expected.²⁸⁶

Caquot and Lemaire translate the line "boivent du vin et les hyènes écoutent l'enseignement des petits re[nards (?)."²⁸⁷ A lacuna precedes these words so the drinker is unknown. The larger hyena (and seemingly silly, from the sound it makes) taking instruction from the smaller, yet wily fox is a bizarre picture.

McCarter thinks that the incomplete word חד[ ] was the name of some animal which usually ate hares and that the tables are now turned. McCarter's "freemen[ ] have drunk wine" marks a temporary transition from animal to human subjects. Who "have drunk" are unclear since a gap immediately precedes these words.²⁸⁸ He suggests that perhaps נזר ("Nazarites") might be restored, in keeping with the topsy-turvy character of the text.²⁸⁹ The rest of his translation begins like that of Caquot and Lemaire but ends differently, "and hyenas give heed to chastisement. The whelps of the f[ox]laugh at the wise."²⁹⁰

Levine thinks that a new theme is introduced here and agrees with Hoftijzer's translation of the line. According to him, Balaʿam here starts his attempt to liberate the destroying goddess Shagar-Weʾishtar from the power of the council of the gods.²⁹¹

(11b) יקחכ.ועניה.רקחת.מר[.]וכהנה.
"and the poor woman has mixed myrrh, and the priestess"²⁹²

Line 11b. Hoftijzer translates "] he will not pay heed to the wise ones. She who transmits divine messages, she who makes (holy) perfume with myrrh and the female priest."²⁹³ The verb יקחכ is from the root דחכ "to laugh, not to take seriously"; incorrect conduct toward the wise.²⁹⁴ The rest of the line contains a list of all female, and probably all cultic, personnel. Unfortunately, the text is damaged after this point so the actions of these personnel remain unknown. Hoftijzer suggests that the gender may have to do with the cult of the goddess.²⁹⁵

The translation of Caquot and Lemaire interprets the line as a continuation of the mixed-up world seen in the previous lines, "[un sot?] se moquera des sages et une pauvresse se parfume à la myrrhe et une prêtresse (?)[."²⁹⁶ McCarter concurs with their translation. "The poor woman" (עניה) is understood on the basis of comparison with ענה "humble man" in Zakir A:2.²⁹⁷ Like the others mentioned before her, the poor woman is behaving abnormally.²⁹⁸ Hackett agrees, noting that the cost of myrrh would have put it far out of the reach of the poor. The mention of the priestess, at the end of the line, was probably the beginning of

Chapter 4: Curse 127

another contrast between the rich and the poor.²⁹⁹ Proverbs 30:21–23 provides a strong parallel to this idea of a reversed order, especially as regards human classes and relationships,

> Under three things the world trembles; under four it cannot bear up: a slave when he becomes king, and a fool when he is filled with food; an unloved woman when she gets a husband, and a maid when she succeeds her mistress.

The implication that the world trembles because of these things, not that these things are themselves a curse, is a valuable corroboration of Hackett's interpretation.³⁰⁰

Levine's interpretation of the rest of the text differs markedly. Bala'am took the goddess "Shagar-We'ishtar" to skilled magical practioners—the personnel that Hoftijzer also sees in this line—the oracle, the perfume maker, and the priestess. The rest of the text talks about the magicians employed to free the goddess from her subservience to the coucil of the gods and the task of destruction. Bala'am was successful. The will of the divine council was thwarted: the wild animals were driven out of the grazing areas by domestic animals like the piglet and the other unnatural situations were reversed.³⁰¹

[.לנשא.אזר.קרנ.חשב.חשב.וחשב.ח   ] (12
.שב] (13a

"[ ]to the one whoe [sic] wears a girdle of *threads*. The esteemed esteems and the esteemer is es[teemed.   ]"³⁰²

Lines 12–13a. Hoftijzer treats the first two words (לנשא אזר) as the end of a clause, "for the chief a loincloth."³⁰³ Just as the wise person does not receive becoming respect (line 13 *e.p.*, line 11 others), so here the chieftain is humiliated by having to wear a slave's clothing. Perhaps the broken portion of the text told of some indignity to befall the cultic functionaries listed.³⁰⁴ In Deut. 28:43, the ascendency of the sojourner over the native is a punishment with which the people of Israel are threatened should they be disobedient to their covenant with Yahweh. The humiliation of the mighty by God can be a sign of his power (e.g. Job 12:17–21). It can also be a sign of rebuke on a whole people as in Isa. 24:1–3 where the people of power are brought low but consequently the poorer people dependent on them also suffer (cf. Amos 6:1–8; Lam. 2:2, 10; etc.). Isaiah describes a scene of judgment where the leaders are done away with and boys who become princes are insolent to their elders (3:1–5; cf. Lam. 5:8).³⁰⁵ For the rest of the line, Hoftijzer translates "Oh adversaries, consider, consider, consider, con(sider)."³⁰⁶ The verb חשב is repeated in the imperative three times: the prophet's audience is exhorted to think about the past and be warned.³⁰⁷

McCarter translates this broken line "to the one who wears a girdle of threads. The esteemed esteems and the esteemer is es[teemed   ]."³⁰⁸ The use of קרנ for the threads of a spider's web קורים and metaphorically for human clothing occurs only in Isa. 59:5–6. The context in the inscription is too broken to show its significance.³⁰⁹ The repetition of the root חשב may be seen as a contrasting series of active and passive participles and a continuation of the preceding role reversals.³¹⁰

Hackett follows McCarter in his understanding of the last part of the line but also provides a plausible interpretation of the first part of the line and fine parallels in Middle Kingdom Egyptian literature for the whole. She translates the line "for the prince, a tattered loincloth. The respected one (now) respects (others) and the one who gave respect is (now) respected."³¹¹ In a context bemoaning a chaotic time, the priest Khakheperre-sonb says "He who gave orders takes orders."³¹² The "Prophecy of Neferti" also describes a time of turmoil, in which "One salutes him who saluted."³¹³ And in a remarkable parallel to the first part of the line, Hackett notes another Egyptian text "Lo, [ ] noblewomen, Their bodies suffer in rags," and "See, those who owned robes are in rags, He who did not weave for himself owns fine linen."³¹⁴

[וֹשמעו.חרשׁנ.מנ.].רֹחק. (13b

"]the deaf have heard from far away["³¹⁵

Line 13b. Hoftijzer translates the remaining part of this line "the deaf ones heard from afar."³¹⁶ He compares this to the biblical practice of emphasizing the greatness of an event by saying that those usually considered physically unable to participate yet do so, e.g. Jer. 31:8 the return from exile will be taken part in by the blind, the lame, pregnant women, and even women in labor (cf. Isa. 33:23b; also Isa. 35:5–7). In Ugaritic literature an example is found in the Keret text where unlikely soldiers (a lame man and a newly-wed) join a hyperbolic host and the poorest people (widows and the blind) contribute expansively.³¹⁷ Hoftijzer says it is the distant approach of an army that the deaf hear.³¹⁸ Caquot and Lemaire, McCarter, and Hackett agree with this interpretation.³¹⁹

[וכל.חזו.קקנ.שגר.ועשתר.ל   ] (14

"and everyone has seen those things that decree offspring and young ["³²⁰

Line 14. The extant remnant of this line Hoftijzer translates "and all suffered oppression. Shgr and Ashtar—"³²¹ He says the line likely contained the story of how the two goddesses, "Shgr and Ashtar," rescued those suffering oppression.³²²

Caquot and Lemaire translate "et tous voient restreint le croît des bovine et des ovins."³²³ They also reject Hoftijzer's interpretation of "Shgr and Ashtar" as DNs.³²⁴ McCarter agrees with their translation and notes that the forces governing conception and birth are usually the most hidden in nature.³²⁵ This idea is also seen in the Bible, Job 39:1–4.

Hackett's translation differs though she divides the line like Hoftijzer, "a fool sees visions. The constraint of fertility (lit.: 'offspring')."³²⁶ A lack of fertility is the catastrophe described, not having the animals' mating secrets known. She points to Egyptian parallels from "The Admonitions of Ipuwer,"

Chapter 4: Curse

"Lo, women are barren, none conceive" and "gone are the abundance of children."[327] Though these Egyptian texts refer to human reproduction, the emphasis is the same as in this line.[328] The Sefire treaty curses provide a parallel "[and should seven rams cover] a ewe, may she not conceive" (I.A.21).[329] This curse/punishment is also seen in Hos. 9:11 "Ephraim's glory shall fly away like a bird—no birth, no pregnancy, no conception!"

[לנמר.חניצ.הקרקת.בנ]] (15
[משנ.אזרנ.ועינ] .'] (16a

"]to the leopard. The piglet has chased the young [of] those who are girded, and the eye["[330]

Line 15–16. Hoftijzer translates "as) a panther (makes) a little pig (flee), so she (?) made flee the sons of (?)."[331] He says that lines 15–17 (lines 13–15 others) are recounting past troubles, approaching enemies or oppression, from which the goddess ("she") delivered the people previously. The intent is to show the people the power and faithfulness of the gods in order to push the people to conversion.[332]

Caquot and Lemaire translate the line "la panthère fait fuir le goret."[333] McCarter's translation is substantially the same but he reads a ל on the front of נמר "leopard" and takes into account בנ at the end of the line, "to the leopard. The piglet has chased the young [of]."[334] Hackett agrees.[335]

The rest of the text is too fragmentary for interpretation.

Deir ʻAlla II.[336] Hoftijzer believes that Combination II, or the intelligible parts thereof, is comprised of a series of curses, for which parallels are abundant in the Bible and other ancient Near Eastern literature. These curses include: removal from the tribes of men, maggots, insufficient clothing, death and burial (though he notes that the threat is usually that the dead remain unburied, cf. Nah. 1:14), and that death will take away the child in the womb.[337] The loss of progeny is obviously the opposite of one of the most common biblical blessings; only rarely did a biblical personage consider still-birth to be better than his or her current condition (Jer. 20:17–18; Job 3:11; 10:18). These curses were probably uttered by Balaʻam though his name is not found in this combination.[338] Balaʻam delivered a prophecy which his audience did not like. In response to their rejection of his prophecy, he uttered curses. The audience was displeased and reproached him with stupidity and ignorance saying he "has taken a poisonous word (or a curse word) on his tongue" (line 17). Balaʻam was about to be judged by the people and tensions ran high but peace seems to have returned.[339]

Caquot and Lemaire say that it is impossible from the extant part of the text to recover the overall sense of the second combination. They content themselves with commenting on those phrases where their interpretations do not agree with those of the *editio princeps*.[340] They do think that the overall tone of this combination is brighter than that of the other, speaking of youth and sexual pleasures.[341]

McCarter does not comment in detail about this text, but thinks that the material before and after the rubric, line 17, represents two separate utterances of Bala'am.[342]

Hackett says that Combination II is full of the language of the grave and death. She thinks that the person referred to is a child and that the child is involved in a ritual, perhaps child-sacrifice. The text includes what the parent says to the child and how the parent prepares him or her for death.[343]

Levine says that the text begins with a description of an eternal home built by El, the netherworld. The described netherworld has similarities to that of the Bible and the Mesopotamian myths of She'ol; kings, eternal repose, and the moaning of the dead are mentioned. An unnamed person, probably Bala'am is addressed in the second person and told that his counsel is no longer sought, his oracular powers are defunct. For interferring in divine matters (Combination I), Bala'am is condemned and sentenced to She'ol.[344]

Hoftijzer says that the text consists of curses up through line 11.[345] None of the other scholars compared identified curses herein.

(4 עלמה.רוי.דדנ.כ--]

Line 4. Hoftijzer's translation is, "*a girl* those who were used (?) to be saturated with love—."[346] Though he admits "we do not know what was told about those saturated with the joys of love," he says it is possible that the line contained a curse telling of something unpleasant happening to them.[347]

(5 לה.למ.נקר.ומדר.כל.רטב]

Line 5. Hoftijzer translates the remnant of this line "a blinded one and the whole moistened (?) soil(?)."[348] He suggests that a curse was pronounced against the moistened soil causing it to dry up (Deut. 28:22; Jer. 23:10; Hag. 1:11). He compares Sefire I.A.25 where the kingdom is threatened with becoming "a kingdom of sand" without vegetation.[349]

(6 ירוי.אל.ויעבדאל.בית.עלמנ.בי]

Line 6. "El will saturate and El will make graves in —(or: the house of)."[350] Curses, or announcements of punishment, with root רוי ("saturate") also occur in Jer. 46:10, "The sword shall devour and be sated, and drink its fill of their blood" (cf. Lam. 3:15). The use of בית עלמ for "grave" is previously unknown in older Aramaic but is found in Qoh. 12:5 and a Punic inscription from Malta *CIS* i.124:1.[351] In the Aḥiram inscription, compare the phrase כשתה בעלמ which Gibson translates "the house of eternity" referring, apparently not to the sarcophagus (ארנ) itself, but to the grave which was to hold it.[352] For the imagery of Yahweh making a grave see Nah. 1:14. For an extended biblical text on She'ol and the activities of its inmates see Ezek. 32:18–32.

(7 בית.ליעל.הלכ.וליעל.חתנ.שמ.בית].

Line 7. Hoftijzer translates this line "a traveller will not enter a house neither will enter there a bridegroom a house."[353] He compares this text to Isa. 24:10 where the houses are shut up as a sign of devastation and curse (v. 8). The

Chapter 4: Curse

absence of travellers is also a sign of devastation, cf. Isa. 34:10; Ezek. 33:28 and Zech. 7:14. The absence of the bridegroom is a sign of desolation in Jer. 7:34; 16:9; 25:10; his presence is a sign of restoration in Jer. 33:10–11, "Thus says the Lord: In this place of which you say, 'It is a waste without man or beast,' the cities of Judah and the streets of Jerusalem that are desolate,…there shall be heard again the voice of mirth and the voice of gladness, the voice of the bridegroom and the voice of the bride."[354]

8) ורמה.מנגדש.מנ.פחזי.בני.אש.ומנ.שקי- [

Line 8. "…and vermin from the grave. From the tribes of mankind and from the places (?) of (…*you will be driven away*)" is Hoftijzer's translation.[355] The "vermin" are maggots (cf. Job 21:32) and are paired with the "grave" (She'ol) in Isa. 14:9–11 and Job 17:14. Hoftijzer says the the first words translated in the line are the last of a phrase and the ו on the front of רמה "vermin" signifies that maggots were only one of the unpleasant parties involved.[356] A new clause begins with פחז which he interprets as "something like *tribe, clan*" based on a comparison with Palmyrene and classical Arabic examples.[357] For the second phrase compare Ezek. 25:7, "I have stretched out my hand against you, and will hand you over as spoil to the nations; and I will cut you off from the peoples and I will make you perish out of the countries" (also Ps. 21:10[11] and Job 30:5).

9) --ל־.הלעצה.בכ.ליתעצ.אולמלכה.ליתמלכ.ישב- [

Line 9. "As to counsel one will not ask you for it, and as to advice one will not ask you for it."[358] Hoftijzer says the curse in this line is that someone is excluded from the important position of counselor, a high position. In the Bible the position of counselor is included in lists of people in high positions that God is about to humble (Isa. 3:2–3; Job 12:17–25).[359]

10) נ.מ.---- בנ.תכסנ.לבש.הנ.חד.הנ.תשנאנ.יאנש.הנ.ת[

Line 10. "You will cover with one piece of clothing. Be it that you are hated (or: that you hate) oh men, be it that you…"[360] For the idea that clothing is a vital necessity compare Isa. 3:7 (also Gen. 28:20; Isa. 58:7; Ezek. 18:7, 16) and that God provides the stranger, the widow, and the orphan with food and clothing (Deut. 10:18). Also note the obligation of the creditor to return the garment taken in pledge to its owner (Exod. 22:26; Deut. 24:10–13, and the Meṣad Ḥashavyahu ostracon).[361] The curse is probably of the type in which a needed object is available only in very limited supply (cf. Isa. 4:1; Amos 4:8; Lev. 26:26).[362] In Deut. 28:48 and Amos 2:16, nakedness is part of the punishment of God. This type of curse is the reverse of that seen in Sefire I.A.21–24 (and II.A.1–3) "and should seven mares suckle a colt, may it not be sa[ted]" (1.A.22)[363] and Tell Fekherye lines 19–22 "Though 100 ewes suckle one lamb, let it not be satisfied" (line 20) where the abundance available is yet unable to satisfy the need.[364] The last part of the line may have contained a curse similar to that found in Lev. 26:17, "those who you hate will rule over you." The opposite of this curse appears as a blessing in Gen. 24:60.

(11 אשׁמ̇--- .תֿחת.ראשׁכ.תשׁכֿב.משׁכבי.עלמיכ.לחלק.ל]ֿ

Line 11. "Under (or: she will smash) your head. You will sleep the sleep of death as a portion for (or: to perdition..."[365] Concerning the first phrase, Hoftijzer hesitates to use the translation in parenthesis because of the broken context, but sees no curse in the extra-parenthetical translation except a vague parallel to Ezek. 32:27 where the mighty men gone down to She'ol lie with their swords under their heads.[366] Concerning the second phrase, the Bible also uses sleep as a metaphor for death, in Ps. 13:3(4) the psalmist seeks deliverance from his enemies lest he "sleep the sleep of death." Jeremiah 51:39, 57 is more specific, וישנו שנת־עולם ולא יקיצו "and they shall sleep a perpetual sleep and they will not awake."

(12 אד̇--כ̇.-כ̇--ט̇ר̇-----כל--ה.בלבב.מנֿ.נאנח.נקר.בלבבֿה.נאנח.-].

Line 12. "] (he/they will say) in (his/their) heart: who is sighing, is a blinded one sighing in his heart?["[367] Hoftijzer says the נקר is not only a blinded one (Peʿal passive participle), but one who has had his eyes pierced, i.e. put out, as a sign of humiliation, cf. Num. 16:14; Judg. 16:21; 1 Sam. 11:2; Prov. 30:17.[368] Blindness is a punishment in 2 Kgs. 25:7 and Sefire I.A.39,[369] and a symbol of punishment in Deut. 28:28 and Zech. 12:4. Blindness is also used as a symbol in announcements of doom, Deut. 28:29; Isa. 59:10; Zeph. 1:17; and Lam. 4:14. Sighing in reaction to calamities is found in Isa. 24:7; Joel 1:18; and Lam. 1:4, 8, 11, 21. For sighing in one's heart, cf. Lev. 26:36 and the Ugaritic Legend of Aqht 1:34–35 *tbky.pǵt.bm.lb tdm'.bm.kbd* [370] "Pǵt weeps in her heart, She sheds tears in the liver."[371]

(13 בת.שׁמה.מלכנ.יחזוֿ.----.תֿ.ליש.במיקח.מות.על.רחמ.ועל]ֿ.-[.

Line 13. "]of horror. Kings will see [ ] of the lion(?). Why (?) Death takes away the child still in (?) the womb and the child..."[372] One of Esarhaddon's vasssal-treaties includes the curse, "May the Lady of the gods, the mistress of creation, cut off birth from your land; may she make rare the cries of little children in the streets and squares."[373] Death is viewed as an individual in parts of the Bible, e.g., Jer. 9:21 "For death has come up into our windows, it has entered our palaces, cutting off the children from the streets and the young men from the squares" (cf. Isa. 28:15).[374] At points, Jeremiah (20:17f.) and Job (3:11; 10:18) despised their lives so much that they preferred even what was considered an accursed fate.[375]

Line 17 has been treated above.[376] The rest of this combination is too fragmentary for substantive commentary.[377]

Summary

Of the inscriptions which invoke a deity for substantive curse, Kilamuwa 1 is a curse designed to protect an inscription and apparently warns two groups, an heir who shall suffer civil unrest and any other who is threatened with having one of the gods named "smash his head." Sefire III invokes a group of gods not

Chapter 4: Curse

specified to enforce a treaty. The only real substantive curse is the undefined threat "you will be have been false to all the gods which are in this treaty."[378] Panammu lines 21–23 are damaged but appear to be a curse intended to protect a memorial inscription. Four deities are asked to take some action now missing in a lacuna. The Nerab 2 inscription contains a curse to deter grave violators. Three deities are requested to kill the grave robber unpleasantly and do away with his family.

The inscriptions without DNs are Aḥiram, Nimrud Hebrew, and Deir 'Alla Combinations I and II. The Aḥiram inscription contains a three-fold curse against the ruler (king, governor, or army commander) who violates his tomb. The grave disturber is threatened with having his rule disrupted: his scepter torn away, his throne overturned, and peace fleeing from the captured land. This curse is unique among the target inscriptions in that, though the inscription appears complete (unlike the others in this section), no deity is explicitly invoked to bring about the curse. Further, it is the only Phoenician target inscription without a DN. The Nimrud Hebrew inscription is heavily damaged and may have originally included a DN. It is interpreted by its publisher as a curse designed to protect the inscription of which it is a part. Deir 'Alla Combination I does refer to a council of the gods early in the text but it is debatable whether a specific goddess (goddesses) is (are) mentioned and the relationship between the council and the substantive curses is disputed. Deir 'Alla Combination I is interpreted by some scholars as largely composed of substantive curses. Chaos reigns. All types of animals and people are described as doing things contrary to their natures.[379] Biblical parallels are abundant in which such conditions indicate a state of curse. Deir 'Alla Combination II is much more variously interpreted, only the original editor seems to think that substantive curse is included and that much like Combination I.

Conclusions

Though there are no explicit instances in the inscriptions where a deity is called upon to curse with one of the general words used for curse,[380] several forms of ארר in the Khirbet Beit Lei burial cave along with prayers to Yahweh, the fragmentary En Gedi inscription, and the presence of a name with the theophoric element יה- in the Siloam Tomb inscription imply that Yahweh was the one invoked to bring about the curse in each of these instances. Biblical examples of ארר with Yahweh are rare but present.[381] All of the inscriptions having a form of ארר are Judean Hebrew and date from shortly after the fall of Samaria (Siloam Tomb ca. 700) to just before the fall of Jerusalem. Present evidence indicates that ארר was used only late in the target period and only in Judah. The only other language in which the root is found is Akkadian (including the contemporary Neo-Assyrian) and there it is used in much the same way. This information indicates a general developmental trend in the inscriptions from earlier periods when specific curses were used exclusively to later periods where a more general word was used as an alternative for specific curses.³⁸² The occurrence of קבב in the Cebel Ires Daği inscription is very similar to ארר in the Hebrew inscriptions. In it, the verb קב and the noun קבת occur in close proximity to the DN בעל קר. The actual curse in unspecified. In this later period, the general word may have represented and implied a comprehensive stock of curses.

The Panammu inscription and Arslan Tash inscription which use אלה use it in an oath or treaty context with a legal air. The occurrences of קבב in the Deir 'Alla inscriptions are problematic. The texts containing substantive curses are from outside the territory of ancient Israel and vary widely in the type of inscription they are a part of: from memorial inscription (Kilamuwa 1, Panammu) to funerary text (Aḥiram, Nerab 2) to treaty document (Sefire III) to what may be a votive text (Nimrud Hebrew).³⁸³ All of these curses are intended to protect the inscription of which they are a part.³⁸⁴ Deir 'Alla Combinations I and II are from a much rarer type, a religious and mythological text.

---

¹Gevirtz says that since the curses are levelled against potential future alterers of the inscription or violators of a grave on which the inscription occurs, they serve as prohibitions "protecting the monument by threatening divine punishment for malefeasance" ("West-Semitic Curses and the Problem of the Origins of Hebrew Law," *VT* 11 [1961], p. 140). This consideration gives the curses a legal air.

²These blessing and curse combinations will be treated in Chapter 5.

³See "שלמ Alone" in Chapter 3.

⁴Harry Torczyner, et al., *Lachish I (Tell ed Duweir): The Lachish Letters* (London: Oxford University Press, 1938), p. 97. The readings and translation are his. The above reading represents his "reconstruction." After a comparison of Torczyner's reading to the drawings and photograph provided, I have marked the א in line 9 and the י in line 10 as questionable which he sees as fairly certain. The ב (line 10), which he thinks dubious, appears clear.

⁵מי עבדכ כלב. However, in this biblical passage, note that the speaker is not calling himself a "dog" while that is the case in the Lachish letters. Thoug hhe could have added 2 Kgs. 8:13 where the speaker *does* refer to himself that way.

⁶Torczyner, et al., *Lachish Letters*, p. 97. Perhaps "house" equals "dynasty," cf. Zakir B.16–C2 in "Substantive Blessing and Curse" Chapter 5. However, it could simply refer to the building, cf. Ezra 6:11. The Arabic is probably cognate to Akkadian *karābu* through Old South Arabic (see the treatment of Akkadian *karābu* in Chapter 2). If so, this would appear to be an euphemistic use of the root for curse.

⁷Cyrus H. Gordon ("Notes on the Lachish Letters," *BASOR* 70 [1938], 18) disallows the shorter form in letter 9, line 8. Albright agrees ("The Oldest Hebrew Letters: The Lachish Ostraca," *BASOR* 70 [1938], 15) and says this applies to letter 5 as well. However, Albright's statement, true at the time, that the form יהו is not extant in preexilic texts, must be revised in light of Kuntillet 'Ajrud 5 (see "ברכ Alone" in Chapter 3).

⁸Albright, "The Oldest Hebrew Letters," p. 15. The transliteration to block letters is mine. Even a comparison with the drawings and photograph provided in *The Lachish Letters* (pp. 90–91)

Chapter 4: Curse   135

shows that the first relatively clear letter of the last line is a ט. J. Hempel ("Die Ostraka von Lakiš," *ZAW* 15 [1938], 134) agrees with Albright, "was ist's mit deinem Knecht, daß er Gutes tue...Böses dem König." So does Roland de Vaux ("Les ostraka de Lachis," *RB* 48 [1939], 196), "Que peut faire ton serviteur en bien ou en mal pour le roi," with the sense, "Ton serviteur ne peut rien pour le roi." René Dussaud's translation ("Le prophète Jérémie et les lettres de Lakish," *Syria* 19 [1938], 260) agrees in substance as well, though his reading differs, "Voici! Qu'importe à ton serviteur (que) Tobiyah soit de race royale?" הנ מה לעבדכ זה? טביהו זרע למלכ.

[9] H. L. Ginsberg, "Lachish Notes," *BASOR* 71 (1938), 26. The modified reading actually comes from a private communication he had with Cyrus Gordon (n. 5).

[10] Henri Michaud, "Ostraca de Lakiš conservés a Londres," *Syria* 34 (1957), 47.

[11] Ibid., pp. 48–49. Gibson (*TSSI*, vol. 1, pp. 44–45) prefers Albright's reading of line 9 to Michaud's (though he says the last letter of the line is definitely an א) but says Michaud's reading of line 10 is preferable. He interprets the last two lines as separate phrases, "What has your servant to do with it?...Tobiah...seed for the king." He believes that the first phrase may have reference to a conspiracy but is unable to give a cogent interpretation of the last phrase, which he, nevertheless, calls "clear" in light of that understanding.

[12] André Lemaire, *Inscriptions hébraïques, Tome I: Les ostraca* (Paris: Les Éditions du Cerf, 1977), p. 117.

[13] Nahman Avigad, "The Epitaph of a Royal Steward from Siloam Village," *IEJ* 3 (1953), 150. His dating is determined by paleography. Based on the presence in the inscription of the phrase אשר על הבית, Avigad says the *terminus ante quem* must be the 586 B.C. destruction of the temple.

[14] *TSSI*, vol. 1, p. 23.

[15] Avigad, "The Epitaph of a Royal Steward," p. 143. The reading and translation are his. It is, of course, tempting to restore the name Shebanyahu in the first line in light of Isa. 22:15–25 where שבנא אשר על־הבית is denounced for the richness of his tomb. The form שבניה is found in Neh. 9:4.

[16] Ibid., p. 146.

[17] *TSSI*, vol. 3, pp. 106–108. Actually line 20 substitutes עלתי "what is over me" for משכב in line 4.

[18] Ibid., p. 14. See the treatment of this inscription in the "Substantive Curse" section of this chapter.

[19] Ibid., vol. 2, p. 97. See the treatment of this inscription in the "Substantive Curse" section of this chapter.

[20] Ibid., vol. 3, p. 103.

[21] In 1946, A. Reifenberg ("A Newly Discovered Hebrew Inscription of the Pre-exilic Period," *JPOS* 21 [1948], 134–37) found a tomb inscription in the vicinity of the two Siloam Tomb inscriptions discovered by Clermont-Ganneau in 1870. The inscription was very fragmentary. Reifenberg was able to make out ten letters but did not hazard any comprehensive translation. Avigad ("The Epitaph of a Royal Steward," p. 148) has proposed a reading based on his translation of the Clermont-Ganneau Siloam inscriptions ("inscriptions" because two were discovered but the second of the two is very short, fragmentary, and not relevant for this study. That inscription is treated in Nahman Avigad, "The Second Tomb-Inscription of the Royal Steward," *IEJ* 5 [1955], 163–66).

(1 קברת
(2 אשר יפ[תח]

Avigad's translation ("The Epitaph of a Royal Steward," p. 148) is "sepulchre of...whoever shall open..." This inscription would very probably have included a curse against the one opening the tomb similar if not identical to the Siloam Tomb inscription discussed above.

136   Chapter 4: Curse

Likewise, a fragmentary Phoenician grave inscription from Cyprus apparently uses the same verb in what appears to be a curse-warning:

(1)ש.פ.ש[
(2 [מ.מלכ.הא.אמ].
(3 יפ[תח.הקבר.].
(4]ו.כאיא

The inscription (ca. 675 B.C.) is designated as Chytroi *RÉS* 922 and was published by A.M. Honeyman ("The Phoenician Inscriptions of the Cyprus Museum," *Iraq* 6 [1939], 106) but without translation. Masson and Sznycer translate it, "]qui [qu'il puisse être (?)...o]u bien (qu'il soit) un roi, ou bien (qu'il soit)[un homme quel-conque (?)[...qui'il n'ou]vre [pas cette] tombe [...] car il n'y a pas [" (Olivier Masson and Maurice Sznycer, *Recherches sur les Phéniciens à Chypre*, Hautes Études Orientales II, vol. 3 [Paris: Libraire Droz, 1972], p. 106.) Compare the warning addresses of the Aḥiram inscription above.

[22] Saul Olyan, *Asherah and the Cult of Yahweh in Israel*, SBL Monograph Series, no. 34 (Atlanta: Scholars Press, 1988), p. 36. Also see the Chapter 1, note 3 above.

[23] Why Yahweh is not specifically called upon then is not known. This understanding depends on the assumption that at least by this time (700 B.C.) any magical ideas of curse had been superseded by the more orthodox idea that Yahweh was the arbiter of blessing and curse, rather than that special individuals had that power independent of Yahweh. See the remarks on the views of Johannes Pedersen and Sigmund Mowinckel in the "History of Interpretation" section of Chapter 2.

[24] This is actually the earliest date given for the inscriptions, which are all dated to approximately the same period. See the next footnote for more information.

[25] Joseph Naveh, "Old Hebrew Inscriptions in a Burial Cave," *IEJ* 13 (1963), 74–76. The readings of the inscriptions are Naveh's unless otherwise specified, but the numbers of the inscriptions are according to Lemaire ("Prières en temps de crise: les inscriptions de Khirbet Beit Lei," *RB* 83 [1976], 558–68). I am treating the inscriptions from Khirbet Beit Lei with the other pre-exilic inscriptions though opinion is divided as to their actual date: scholars like Naveh and Lemaire consider them to date around 700 B.C., while Cross who treats only the first three inscriptions thinks that the inscriptions can date from no earlier than the sixth century B.C. (perhaps 587; Frank M. Cross, Jr., "The Cave Inscriptions from Khirbet Beit Lei," in *Near Eastern Archaeology in the Twentieth Century* [Garden City, NY: Doubleday & Company, Inc., 1970], pp. 299–306, see below, note 38). (Actually, in his last footnote, Cross says the curse formulas may be older than the ones he treats.) Gibson (*TSSI*, vol. 1, p. 57) and Miller ("Psalms," p. 326) agree with Cross. Viewing these inscriptions as the prayers of someone fleeing some sort of enemy and probably an invader provides a very plausible background against which to interpret them and gives them strong ties to the Psalms, whatever historical situation gave birth to them.

[26] Naveh, "Old Hebrew Inscriptions," pp. 78–79. He says that the skeletal remains were undisturbed leading him to think that nothing had been taken from the tomb.

[27] P. Bar-Adon, "An Early Hebrew Inscription in a Judean Desert Cave," *IEJ* 25 (1975), 231. See the discussion of the En-Gedi inscription in Chapter 5.

[28] Lemaire, "Prières en temps de crises," pp. 561–62. The last letter is certainly questionable and looks much more like a ר than a ה.

[29] Miller, "Psalms and Inscriptions," p. 332.

[30] Naveh, "Old Hebrew Inscriptions," p. 80.

[31] Ibid. He does mention that around the inscription are several single letters including כ and ר. In fairness to Naveh, Lemaire's "second line" is very poorly inscribed and two other extraneous כs are in close proximity, one before each line.

[32] Lemaire, "Prières en temps de crises," p. 562.

Chapter 4: Curse

[33] See note 38, just below, for the text of those larger inscriptions.

[34] Lemaire, "Prières en temps de crises," pp. 562–63.

[35] Miller, "Psalms and Inscriptions," pp. 323–24.

[36] Ibid., p. 324.

[37] Ibid., pp. 324–27.

[38] Ibid., pp. 327–28. The texts of those three inscriptions is given below. The reading of "B" is particularly difficult and may be read as a prayer for deliverance. Cross ("Cave Inscriptions," p. 304) had previously suggested, and rejected, the idea that the three main inscriptions were not funerary inscriptions at all but were prayers for deliverance written by people fleeing an invading force (which he thinks was the Babylonian invasion of 587). Cross thinks, however, that the curse inscriptions were from an earlier time and did have funerary functions (n. 25).

Inscription A

1) ני יהוה אלהיכה ארצה
2) ערי יהדה וגאלתי ירשלם

"I am Yahweh thy God: I will accept the cities of Judah, and will redeem Jerusalem."

Inscription B

נקה יה אל חנן נקה יה יהוה

"Absolve (us) O merciful God! Absolve (us) O Yahweh!"

Inscription C

הושע [י]הוה

"Deliver (us) O Lord [sic]."

These readings and translations are those of Cross ("Cave Inscriptions," pp. 301–02; other readings are available in Naveh ("Old Hebrew Inscriptions," pp. 84–87), Miller ("Psalms and Inscriptions," pp. 321–23, 328), and Lemaire ("Prières en temps de crise," pp. 559–61). The designations "A, B, C" are Naveh's.

[39] These inscriptions received no independent designations.

[40] He does not specify the precise forms found nor on which wall. He interprets all of these as forms of ארר.

[41] Naveh, "Old Hebrew Inscriptions," p. 81.

[42] The idea that the inscriptions all date from the same period seems generally accepted (e.g., Naveh, "Old Hebrew Inscriptions," p. 92; and Miller, "Psalms and Inscriptions," p. 332).

[43] See "ברכ" and "ארר" in Chapter 5.

[44] For example, Josh 6:26 "Joshua laid an oath upon them at that time saying, 'Cursed before the Lord be the man (ארור האיש לפני יהוה) that rises up and rebuilds this city, Jericho. At the cost of his first-born shall he lay its foundation, at the cost of his youngest son shall he set up its gates.'" Note that the oath is laid on them by Joshua, but it is attributed to Yahweh and thus brought to pass by him 1 Kgs. 16:34 (cf. Gen. 3:14, 17; 1 Sam. 26:19; Jer. 11:3).

[45] See the treatment of *arāru* in the "Semantic Survey" of Chapter 2.

[46] Within the target area inscriptions. However, the late seventh century, Phoenician Cebel Ires Daği inscription also includes noun and verb forms. See the discussion of Deir 'Alla II.17 in this chapter.

[47] *KAI*, vol. 2, p. 223.

[48] *TSSI*, vol. 2, pp. 78–79.

⁴⁹Mark Lidzbarski, *Handbuch der nordsemitischen Epigraphik*, (1962; rpt. Hildesheim: Georg Olms, 1898), vol. 1, p. 442.

⁵⁰Ibid., vol. 2, table 23. Earlier in the line, אלה is translated as "gods" (notice the full spelling of the plural construct in line 22 אלהי). The letters are just as clear, if not clearer, in the second occurrence.

⁵¹*NSI*, pp. 173, 175. Cooke cites articles by Müller ("Altsemitischen Inschriften von Sendschirli," *WZKM* 7 [1893], 33–70, 113–40) and Nöldeke ("Bemerkungen zu dem aramäischen Inschriften von Senschirli," *ZDMG* 47 [1893], 96–105), neither of which were available to me. Kraeling (*Aram and Israel or the Aramaeans in Syria and Mesopotamia*, Columbia University Oriental Studies, vol. 13 [New York: Columbia University, 1918], p. 125) agrees with this interpretation and gives the details of this conspiracy in some detail (!). Theodor Gaster ("A Canaanite Magical Text," *Orientalia* 11 [1942], 65, n. 2) cites the meaning "bond, conspiracy" and this text as proof in interpreting אלה in the Arslan Tash inscription.

⁵²Koopmans, *Aramäische Chrestomathie*, vol. 1, pp. 70–71 (translation of the text and discussion), vol. 2, pp. 16–17 (transcription of the text).

⁵³*KAI*, vol. 2, p. 225. Their translation is "*Auf Grund der [Gerechtigkei]t (2) seines Vaters haben ihn die Götter von Ja'udi aus seinem Verderben errettet, das im Hause seines Vaters entstanden war*" (p. 223).

⁵⁴*TSSI*, vol. 2, p. 82. K. Lawson Younger, Jr. ("Panammuwa and Bar-Rakib: Two Structural Analyses," *JANES* 18 [1986], p. 92, n. 9) says that אלה may mean either "curse" (citing *DISO*, p. 14) or "conspiracy" (citing Eduard Y. Kutscher, "Aramaic," in *Linguistics in South West Asia and North Africa*, Current Trends in Linquistics, vol. 6, ed. Thomas A Sebeok [The Hague: Mouton, 1970]). Kutscher, in turn, cites Cooke (*NSI*, p. 352).

⁵⁵*TSSI*, vol. 2, p. 82.

⁵⁶Joseph A Fitzmyer, *The Aramaic Inscriptions of Sefîre*, Biblica et Orientalia, no. 19 (Rome: Biblical Institute Press, 1967), pp. 12–21.

⁵⁷Brichto (*The Problem of Curse*, p. 24ff) has argued convincingly for the inclusion of the implied curse in oath, especially in the use of the word אלה.

⁵⁸Compare Sennacherib's succession treaty (*VAT* 11449, *PKTA* 31 in Parpola and Watanabe, *Neo-Assyrian Treaties and Loyalty Oaths*, State Archives of Assyria, vol. 2 [Helsinki: Helsinki University Press, 1988], p. 18) obverse lines 7–12 and reverse lines 2–6, where numerous gods are invoked in an "indissoluble, grievous curse" *ar[-rat la napšur mar[u-uštu]* (obv. line 11)

⁵⁹*TSSI*, vol. 3, p. 78. For drawings see p. 186.

⁶⁰The language of the text has been vigorously debated. Some scholars treat it as Aramaic, some as Phoenician, some as Hebrew, and some are content merely to recognize it as Canaanite (see the articles referred to below for the specific arguments).

⁶¹Javier Teixidor, "Les tablettes d'Arslan Tash au Musée d'Alep," *AulaOrientalis* 1 (1983), 105. Also Teixidor, "Book Review Article: J. C. L. Gibson's *Textbook of Syrian Semitic Inscriptions*, vol. 3," *JBL* 103 (1984), 454. He marks the whimsical nature of the language and the very physical lightness of the plaques among his arguments against their authenticity. Pierre Amiet of the Louvre concurs, based on art history reasons, with Teixidor's doubts about the texts ("Observations sur les 'Tablettes magiques' d'Arslan Tash," *AulaOrientalis* 1 [1983], 109).

⁶²Comte du Mesnil du Buisson, "Une tablette magique de la région du moyen Euphrate," in *Mélanges Syriens offerts a Monsieur René Dussaud* (Paris: Librairie Orientaliste Paul Geuthner, 1939), pp. 422, 424. This is the *editio princeps*. The line numbers are not from the original but are the ones regularly used by scholars dealing with the inscription. For the sake of completeness and clarity, the following last lines according to *TSSI* (vol. 3, p. 82) are given: (27) סמם (28) חלפ.ולדר עפ (29). The text is quite difficult to read and many variations have been offered. Variants will be given here as they are significant for the parts of the text under discussion.

Chapter 4: Curse 139

[63] Gaster, "A Canaanite Magical Text," p. 44.

[64] Du Mesnil, "Une tablette magique," p. 424. This includes lines 9b–11a which he translates as "la déesse de l'univers ('Athé), qui a fait alliance avec nous." "Alliance" is not an interpretation of אלה, but is implied from כרת. (This understanding of his interpretation, not clear from his article, is due to Gaster ["A Canaanite Magical Text," p. 65].)

[65] André Dupont-Sommer, "L'inscription de l'amulette d'Arslan-Tash," *RHR* 120 (1939), 134–35.

[66] William F. Albright, "An Aramaean Magical Text in Hebrew from the Seventh Century B.C." *BASOR* 76 (1939), 5. This interpretation is also suggested by du Mesnil at the end of the *editio princeps* ("Une tablette magique," p. 434).

[67] Albright, "An Aramaean Magical Text," pp. 8–9.

[68] Gaster's abundant examples of similar magical texts and ideas are convincing (pp. 41, 43, 45ff.). His attempt to approach the text "as a monument of folklore and not only of Semitic archaeology" (p. 43) is very helpful. An interesting example from Egyptian literature is translated by John A. Wilson, "Magical Protection for a Child," Berlin Papyrus 3027 (*ANET*, p. 328). Jeremiah 9:21 provides a good parallel, "For death has come up into our windows, it has entered our palaces, cutting off the children from the streets and the young men from the squares."

[69] Gaster, "A Canaanite Magical Text," p. 43, n. 1a.

[70] Ibid., p. 65.

[71] Ibid., p. 65, n. 1; he does allow one such use, 2 Chr. 7:88.

[72] Ibid. He cites many other examples as well but the parallel is not strong unless one accepts אלה as a synonym for ברית.

[73] Gaster ("A Canaanite Magical Text," pp. 44, 65) gives specific examples from the *Lamashtu* plaque I.i.13; I.ii.30; II.i.56; II.iii.11; etc. (Louvre AO 8184, treated in F. Thureau-Dangin, "Rituel et amulettes contre Labartu," *RA* [no vol.] [1921], 195–98).

[74] Gaster, "A Canaanite Magical Text," pp. 44, 54. The text is *Utukkē limnûti* V.iii.30, 54. The English translation is from p. 44.

[75] Ibid., p. 66. He also gives an example from later in the same text VII.86 *ina šiptika elliti ša balâti 'iltašu puṭurma* "by thy pure life-giving spell, burst thou his bond!"

[76] Harry Torczyner, "A Hebrew Incantation Against Night Demons from Biblical Times," *JNES* 6 (1947), 18.

[77] Ibid., pp. 19–21. Though other scholars do interpret it as other than "goddess" (see below). Torczyner says that the use of both forms—אלה (line 3–4) and אלת (lines 1, 9, 13, 14, 15)—shows that a morphological distinction is made between the construct and absolute states (p. 19). He also says that the form אלת is not known from Hebrew or Phoenician (though it does occur in Ugaritic).

[78] Ibid., p. 28.

[79] Ibid., p. 21.

[80] Ibid., p. 22. Several of Torczyner's translations vary significantly from other com-mentators since he interprets the text as "pure biblical Hebrew." The significance of this understanding is seen in his interpretation of אשר (line 10) as "which" instead of as the name of the deity Aššur (cf. Gaster, "A Canaanite Magical Text," p. 58) whom other scholars see as the chief guarantor of the treaty.

[81] Torczyner, "A Hebrew Incantation," pp. 23–24. His reading of line 15 באלת תחת ארצ ends with an awkward, final, dangling use of אלה which he compensates for by translating (lines 14–15) "with the curse by Ba'al below the earth" and by seeing the last באלת as either a scribal mistake (a misplaced word) or as a characteristic of a mass-produced incantation (poor quality).

[82] *KAI*, vol. 2, pp. 44–45.

[83] Franz Rosenthal, "Incantations: The Amulette from Arslan Tash," in *ANET*, p. 658.

[84] Frank M. Cross, Jr. and Richard J. Saley, "Phoenican Incantations on a Plaque of the Seventh Century B.C. from Arslan Tash," *BASOR* 197 (1970), 45 and note 18.

[85] Ibid., p. 48.

[86] Ibid., pp. 48–49. Compare also *AP* 81:30 תפלה זי כסף and the discussion above of the Ketef Hinnom silver scrolls.

[87] André Caquot, "Observations sur la première tablette magique d'Arslan Tash," *JANES* 5 (1973), 46.

[88] Ibid., pp. 46–47.

[89] Ziony Zevit, "A Phoenician Inscription and Biblical Covenant Theology," *IEJ* 27 (1977), 117–18. Every occurrence is translated "covenant" except the one in line 1 which he translates as "goddess."

[90] Ibid., p. 114, referring to Brichto, *The Problem of Curse*, p. 70.

[91] Ibid., p. 115. He cites Gesenius §119 o, p, q for this usage.

[92] Ibid., p. 116. Note that Gaster sees each of these three parties as different titles of Aššur "governor of the beings divine, and burgomaster of all the holy beings" ("A Canaanite Magical Text," p. 44).

[93] Or, just "the wives of Horon" (Albright, "An Aramaean Magical Text," p. 9). Isaiah 28:15 (and 18), is a striking parallel, נסתרנו: כי אמרתם כרתנו ברית את־מות ועם־שאול עשינו חזה שיט שוטף כי־עבר לא יבואנו כי שמנו כזב מחסנו ובשקר "Because you have said, 'We have made a covenant with death, and with Sheol we have an agreement; when the overwhelming scourge passes through it will not come to us; for we have made lies our refuge, and in falsehood we have taken shelter." The use of אלה ("curse") as synecdoche of the part for the whole שבועה ("oath") has been well-argued by Brichto (*The Problem of Curse*, p. 24); by extension, since every covenant is sealed by an oath and every oath contains a curse against the violator, אלה and ברית are also closely tied (cf. Gen. 26:28–31 and Deut. 29:9–20) (Brichto, *The Problem of Curse*, pp. 26–31). So the word ברית in the above passage actually functions as a synonym of אלה in the Arslan Tash text. Albright and Gaster both comment that the god Horon is a god closely connected with the nether-world; Albright says that "Hauron" seems to be "a manifestation of Ba'al connected particularly with the netherworld," ("An Aramaean Magical Text," p. 9, n. 25) and Gaster includes extensive notes on the identity of Horon in this and other literatures ("A Canaanite Magical Text," pp. 61–62) The "wife of Horon" would therefore be the counterpart of the consort of the god of the underworld (Ibid., p. 62). The biblical passage speaks of a covenant made with death, with Mot (literally מָוֶת) and with the grave—She'ol—easily a parallel with Arslan Tash and the sort of magic condemned by the prophets (cf. Nicholas Tromp, *Primitve Conceptions of Death and the Nether World in the Old Testament*, Biblica et Orientalia, no. 21 [Rome: Pontifical Biblical Institute, 1969], pp. 99–106).

[94] Zevit, "Biblical Covenant Theology," pp. 112, 117.

[95] Ibid., p. 118.

[96] Ibid.

[97] *TSSI*, vol. 3, p. 78. One of Gibson's interesting suggestions about the text is that the language was an intentional mix of several different languages in order to "impart a 'magical' flavour to their texts and thus increase the potency of the incantations" (pp. 79–80).

[98] Ibid. Was this a sacrifice of the child as most interpreters think, or was this a ritual act which dedicated the child to a deity and then received him or her back into the community? For an examination of the possibilities and other sources see Robert P. Carroll, *The Book of Jeremiah: A Commentary*, Old Testament Library (Philadelphia: Westminster, 1986), pp. 222–23.

[99] John A. Wilson, "Egyptian and Hittite Treaties," in *ANET*, p. 206.

Chapter 4: Curse   141

¹⁰⁰Fitzmyer, *Sefîre*, p. 13, also see his note on p. 38. The Zakir stele B.25–26 is similar "the gods of heaven [and the gods] of earth" (*TSSI*, vol. 2, pp. 12–13). See Fitzmyer's comments (p. 38) on the Sefîre treaty where he notes the possibility that these entities were considered to be deities.

¹⁰¹Slightly outside the target area but within the period is the Phoenician Cebel Ires Daği inscription. See Chapter 2, note 38.

¹⁰²Jacob Hoftijzer ("The Prophet Balaam in a 6th Century Aramaic Inscription," *BA* 39 [1976], 11; the date in the title was a typographical error) said the two large combinations may or may not have been related, although Combinations III–IX do relate to one or the other; but P. Kyle McCarter, Jr. ("The Balaam Texts from Deir 'Allā: The First Combination," *BASOR* 239 [1980], 49) has provided strong evidence that the fragments are related. He pointed out that the two larger fragments were found at the extreme ends of a three-and-a-half by one-and-a-half meter area and all of the smaller fragments were found within the intervening space. Thus the two are from the same inscription: Combination I from the beginning and Combination II from near the end.

¹⁰³Jacob Hoftijzer and G. van der Kooij, eds., *Aramaic Texts from Deir 'Alla* (Leiden: E. J. Brill, 1976), p. 177, hereafter *ATDA* This is the *editio princeps*. See the discussion of the Deir 'Alla combinations under "Substantive Curse" in this chapter for the complete text of the inscription.

¹⁰⁴Ibid., p. 136.

¹⁰⁵A few words and scattered lines in the combinations are set off from the rest of the text by the use of red ink. The intention of this designation is debated. For some of the opinions, see the treatments of Combinations I and II in the "Substantive Curse" section of this chapter.

¹⁰⁶*ATDA*, p. 247.

¹⁰⁷Ibid., p. 181. The italics are his and indicate the more certain translations.

¹⁰⁸Ibid., p. 246.

¹⁰⁹Ibid. For the noun see *CIS* ii:211:8; 217:8; for the verb see *CIS* ii:197:5; 198:3; 199:4; etc.

¹¹⁰*ATDA*, pp. 244–45. DJD i:28b:I:1; cf. דבר ברכי 28b:III:22. He notes that לענה also designates a poisonous herb in BH: Jer. 9:14; 23:15; Lam. 3:15. Especially interesting is Amos 5:7, where Israel is said to have turned "justice" (משפת) into "wormwood" (לענה).

¹¹¹*ATDA*, pp. 244–45.

¹¹²André Caquot and André Lemaire, "Les textes araméens de Deir 'Alla," *Syria* 54 (1977), 207. The transliteration to block letters is mine. They apparently misunderstand the *e.p.* at this point because they say Hoftijzer treats this line as a reponse of the gods to Bala'am. However, Hoftijzer clearly says, both in the *e.p.* (*ATDA*, p. 245) and in his article ("The Prophet Balaam," p. 13), that these are the words of his hearers, i.e. his people.

¹¹³Jo Ann Hackett, *The Balaam Text from Deir 'Allā*, Harvard Semitic Monographs, vol. 31 (Chico, CA.: Scholars Press, 1984), p. 73; hereafter designated *BTDA*.

¹¹⁴Ibid.

¹¹⁵Ibid., p. 30.

¹¹⁶Baruch A. Levine, "The Deir 'Alla Plaster Inscriptions," *JAOS* 101 (1981), 202. He notes (n. 25) that this cluster of prepositions occurs in early Phoenician and recurs in late Hebrew, but he gives no examples, referring only to his unpublished Ph.D. dissertation, "Survivals of Ancient Canaanite in the Mishnah," (Brandeis University, 1962), pp. 4 and 63, n. 6.

¹¹⁷Ibid., p. 201.

¹¹⁸Ibid. The word דבר is to be taken as a technical term for "oracle," cf. Jer. 18:18.

¹¹⁹Jonas C. Greenfield, "Some Phoenician Words," *Semitica* 38 (1990), 155-58.

¹²⁰Paul G. Mosca and James Russell, "A Phoenician Inscription from Cebel Ires Daği in

Rough Cilicia," *EpigAnat*9 (1987), 5-6. The transliteration to block letters is mine. This inscription is not included as one of the target texts because its location is too far outside the target area. It is, however, a very significant example of a general word for curse in a Phoenician text of the target period. As is noted in the conclusions, the extant inscriptions show a trend toward use of general words for curse, away from specific curses, nearer the end of the target period.

Mosca and Russell favor the interpretation of בעל קר as a DN perhaps "Lord of the Furnace" in view of the curse that follows. They seem to assume that this deity is being invoked to bring about the curse. They do allow that בעל קר may be interpreted as referring to humans, thus "lords of the crucible" designating metal workers (p. 14). They say that by the curse "MTŠ underscores his full surrender of any claim to the property in question and paves the way for the transfer of ownership to MSNZMŠ" (p. 15).

[121] *ATDA*, p. 174. The transliteration into block letters is mine.

[122] Ibid., p. 182. Only line 3 is complete enough for translation.

[123] Ibid., p. 261. The words of this line in Combination II are in red, as are words in some other parts of this text. Apparently, these words had some special importance. Opinions about their significance differ. McCarter suggests that red was used for passages that refer, in some way, to the inscription itself ("The Balaam Texts from Deir 'Allā," p. 49).

[124] *ATDA*, p. 262.

[125] Ibid. Note, however, that in each of his biblical examples the form is actually לא; the negation is not attached to the infinitive absolute.

[126] The text of this inscription is quoted with translation in the first note to Chapter 5.

[127] *ATDA*, p. 177.

[128] Ibid., p. 263.

[129] Brichto, *The Problem of "Curse,"* pp. 24ff.

[130] Substantive curses do occur frequently with the name Yahweh in the Bible. A sterling example of this is Ps. 109:6–20, especially vv. 14–15. This psalm contains a comprehensive and creative catalogue of curses against the psalmist's adversary. These curses include the usual shortness of life, lack/loss of progeny, and hunger; as well as the wish that the adversary would be under the supervision of a wicked (רָשָׁע) man, that the adversary's prayer become sin (וּתְפִלָּתוֹ תִּהְיֶה לַחֲטָאָה v. 7), and that his sin be remembered by Yahweh.

[131] *TSSI*, vol. 3, p. 30.

[132] This is pictured and described in an article by Yigael Yadin ("Symbols of Deities at Zinjirli, Carthage, and Hazor," in *Near Eastern Archaeology in the Twentieth Century*, ed. James A Sanders [New York: Doubleday & Company, 1970], pp. 201–04).

[133] Pietro Magnanini, *Le iscrizioni fenicie dell'Oriente: testi, traduzioni, glossari*, (Rome: Istituto di Studi del Vicino Oriente, 1973), pp. 45–46. Only the last, and relevant, lines of the inscription are included here. The rest of the text may be found in Magnanini or Gibson (*TSSI*, vol. 3, pp. 34–35).

[134] *TSSI*, vol. 3, pp. 34–35.

[135] O'Connor says that the first part of the inscription (lines 1–8) is made up of two sections: 1) an "ideological history lesson" (the failure of his predecessors to integrate the kingdom into the area's power structure) and 2) a description of the kingdom's woes and how he solved them. The second part of the inscription (lines 9–16) is also two-part: 1) a description of the kingdom's prosperity during his reign and 2) the curse directed at his successors (Michael O'Connor, "The Rhetoric of the Kilamuwa Inscription," *BASOR* 226 [1977], 24). O'Connor does not give much attention to the structure or contents of the curse, saying only that an expected "blessing section" is missing and that the curse section refers to Kilamuwa's successors (p. 24).

[136] See the discussion of the Zakir stele in the "Substantive Blessing and Curse" section of

Chapter 5.

¹³⁷William F. Albright, "Notes on Early Hebrew and Aramaic Epigraphy," *JPOS* 6 (1926), 85. Albright is building on the ideas of Lidzbarski (*Ephemeris für semitische Epigraphik* [Giessen: Verlag von Alfred Töpelmann, 1915], vol. 3, pp. 235–36).

¹³⁸Jacob Halévy, "Les inscriptions du Roi Kalumu," *RS* 20 (1912), 29. Halévy lists three other early translations (pp. 20–23). Kraeling agrees with Lidzbarski (*Ephemeris*, vol. 3, p. 233) that the משכבם are inhabitants of a lower order; he says that Halévy's interpretation "sepulchers" "is impossible" (*Aram and Israel*, p. 91, n. 1).

¹³⁹M. J. Lagrange, "La nouvelle inscription de Sendjirly," *RB* 21 (new series no. 9) (1912), 257.

¹⁴⁰Ibid. He does not credit the work in which Nöldeke presents this study. Lagrange does not seem certain of the political consequences nor of the relationship between the two groups mentioned.

¹⁴¹Stanley Gevirtz, "West-Semitic Curses and the Problem of the Origins of Hebrew Law," *VT* 11 (1961), 141, n. 4. He cites B. Landsberger (*Sam'al: Studien zur Entdeckung der Ruinenstaette Karatepe*, Veröffentlichungen der Türkischen Historischen Gesellschaft VII, series no. 16 [Ankara: Türkischen Historischen Gesellschaft, 1948], pp. 52–53) for the discussion of the simple and causative possibilities of the verb.

¹⁴²Gevirtz, "West-Semitic Curses," p. 141. Michael O'Connor says that Lidzbarski (*Ephemeris*, vol. 3, p. 235) and Gevirtz ("West-Semitic Curses," p. 141) "unnecessarily regard the *b*(*spr*) as anomalous," though he provides neither examples nor reasoning for his contrary view ("Rhetoric," p. 23).

¹⁴³*KAI*, vol. 2, p. 34; *TSSI*, vol. 3, pp. 35, 39.

¹⁴⁴*TSSI*, vol. 3, p. 39.

¹⁴⁵Gevirtz, "West-Semitic Curses," pp. 141–42, n. 5. He also recommends an Akkadian example from Ugarit (text 16.251) *ù šawittena 1 meat ḫurâṣa šarra bêlšu uktabbid* "Et Šawittenu, de 100 (sicles d')or, le roi, son maître, a 'honoré'" (Jean Nougayrol, *Le Palais Royal d'Ugarit*, vol. 3, part 1, Mission de Ras Shamra, vol. 6, ed. Claude F. A. Schaeffer [Paris: Imprimerie Nationale, 1955], pp. 108–09).

¹⁴⁶He says that Lidzbarski explains it as a scribal error for הספר (Hans Bauer, "Die כלמו-Inschrift aus Sendschirli," *ZDMG* 67 [1913], 689). Bauer does not cite the work in which Lidzbarski states this view.

¹⁴⁷Bauer, "Die כלמו-Inschrift," p. 689. Though other scholars have noted, with some mild surprize, the absence of a blessing formula (Bauer mentions Littmann and Nöldeke, more recently O'Connor has made a similar observation "Rhetoric," p. 24), no one, writing recently, appears to share Bauer's explanation.

¹⁴⁸Kraeling, *Aram and Israel*, p. 91.

¹⁴⁹Ebbe E. Knudsen, "Fragments of Historical Texts from Nimrud—II," *Iraq* 29 (1967), 60–62, fragment 4378 B, col. iii, lines 20–25. These lines are preceded by Ashurbanipal's account of his pious renovation/rebuilding of the Temple of Nabu in Calah and this quote is followed by a curse against anyone who destroys the inscription. Other examples are found in the inscriptions of Sennacherib (Daniel D. Luckenbill, *The Annals of Sennacherib*, University of Chicago Oriental Institute Publications, vol. 2 [Chicago: The University of Chicago Press, 1924], pp. 98, 101, 116) and Sîn-šarra-iškun (A. K. Grayson, "Cylinder C of Sîn-šarra-iškun, a New Text from Baghdad," in *Studies on the Ancient Palestinian World*, ed. J. W. Wevers and D. B. Redford [Toronto: University of Toronto Press, 1971], pp. 165–66). Further references to the practice of anointing inscriptions and images of deities may be found in *AHw*, vol. 2, pp. 843–44.

¹⁵⁰*VA* 2536, col. 3, lines 8–10, the transliteration and translation are the work of Professor Thomas Smothers (The Southern Baptist Theological Seminary, Louisville, Kentucky). Another example may be found in col. 2, lines 42–45.

[151] Lidzbarski, *Ephemeris*, vol. 3, p. 235, "die üblichen Flüche gegen etwaige Schänder."

[152] Lagrange, "La nouvelle inscription," p. 257.

[153] Torrey, "The Zakar and Kalamu Inscriptions," p. 369.

[154] *TSSI*, vol. 3, p. 39. He also notes its ocurrence in Hadad lines 27ff, especially the Haph'el verb form in line 29 (Ibid., vol. 2, 68–69, 75).

[155] Ibid., vol. 3, p. 39. This quotation is from *CML*, p. 102. Gibson notes that *qdqdr* should be emended to *qdqdk* (n. 57).

[156] For information concerning the origins of and relationship between the Sefire steles, see the beginning of the treatment of Sefire I in the section "Substantive Blessing and Curse" in Chapter 5.

[157] Nothing is known of the actual discovery of the stele.

[158] Fitzmyer, *Sefire*, pp. 96, 98, 100.

[159] Ibid. Note that three times, lines 14, 19, 27, there is also the difference, for no apparent reason, that the verb is singular rather than plural as elsewhere (cf. I.B.23, 27–28, 33, 38 which have the same phrase). Line 23 has the plural form and line 38 has the singular form. Interestingly, Fitzmyer has restored the middle two occurrences with the singular form (*Sefire*, p. 18). He does note the changes from singular to plural and says they mark references to the king alone (singular) as opposed to the king and his people (plural). He first thought that the final מ was enclitic (p. 107). Hillers points out that the change of pronoun and verb form in the Sefire formula is also seen in the Mati'ilu treaty and Deuteronomy (he rejects the idea that the switch in Deuteronomy is attributable to editorial expansion). With regard to Deut. 28, he says the shift merely means "the author momentarily abandoned the rather artificial habit of addressing the nation as one individual" (*Treaty-Curses*, pp. 32–33); certainly this is the case with the Sefire and Mati'ilu treaties as well.

[160] Likewise, it is interesting to note that two forms of the verb שקר appear, שקרת and שקרתם, though there is no consistent correspondence between the form used and the length of the statement, the longer statement uses the plural form of the verb three times and the singular form once; while the short form of the statement uses the singular form of the verb four times and the plural form once. All of the verbs address Mati'el as king of his land and so there would seem to be no reason to change forms. Elsewhere in the Sefire steles this alternation also occurs: in I.B the formula group occurs five times, twice mentioning the gods, three times not, two of the five times the formula is complete, once referring to the gods (longer version with plural verb form), and once not (shorter version with singular verb form), the other three occurrences are not complete but are all reconstructed by Fitzmyer (one referring to the gods, two not, and all with singular verb form) with the longer version of the formula. In II.B the formula group occurs twice, once with shorter version of the formula with no reference to the gods and having the plural form of the verb (broken, not verifiable), once referring to the gods having the longer formula and the singular verb form. The formula occurs so few times in II.B that nothing can be said definitively about the style. In I.B the formula occurs five times but is partially obscured three times so little can be decided about the style. Based on a comparison to III, however, where the formula referring to the gods always takes the longer version and usually uses the plural form of the verb (3 to 1) and the formula not referring to the gods takes the shorter form and usually uses the singular form of the verb (4 to 1), if the writers were the same or used the same styles (alternating between forms) then Fitzmyer's reconstructions in I.B are not all correct.

[161] Dupont-Sommer, "Une inscription araméenne inédite de Sfiré," *BMB* 13 (1956), 37.

[162] For general information on this inscription, see the beginning of the discussion of line 2 above "Other Words for Curse."

[163] *TSSI*, vol. 2, pp. 80–81. The reading of the text and translation of the inscription are Gibson's though he provides no translation of line 21 terming it "untranslatable." The gaps within brackets at the ends of the lines, as presented here, are not intended to accurately represent the size of lacunae in the text.

[164] *NSI*, p. 180. His translation does not significantly enforce this interpretation though, "and

Chapter 4: Curse   145

said ? and concerning ? surety (?)[ ] king [ ] and ? [ ]before (?) the sepulchre of my father Pa[nammu] and a memorial is this (?). Also may Hadad and El and Rekub-el, lord of the house, and Shamash, and all the gods of Ya'di [ ]before the gods and before men" (pp. 174–75).

[165] *TSSI*, vol. 2, p. 85. For a commentary on Hadad lines 15–24a, see the discussion of "Substantive Blessing and Curse" in Chapter 5.

[166] Kraeling, *Aram and Israel*, p. 127.

[167] *KAI*, vol. 2, p. 229.

[168] Ibid., p. 224. The italics are theirs.

[169] Ibid., p. 229. The verb יבל also occurs in lines 6 and 14 of the inscription. Elsewhere it is found only in later works (e.g., *AP* 2:9, 13; Aḥiqar 48, 52; etc.).

[170] *TSSI*, vol. 2, pp. 93–94. The Nerab 1 inscription is discussed in Chapter 5 in the section "Substantive Blessing and Curse."

[171] Ibid., p. 97. The readings and translation are Gibson's.

[172] Ibid., vol. 3, p. 103.

[173] Ibid., pp. 106–07.

[174] Ibid., p. 100.

[175] Clermont-Ganneau thinks that the mention of silver and bronze but not gold as in other inscriptions from Zendjirli (Barrakab 1:11 and Panammu lines 10–11; as well as Tabnit lines 4–5 and Batnoam) is significant. He says that it shows this community's relative poverty ("Les stèles araméennes de Neîrab," *Études d'archéologie orientale* 2 [1897], 204).

[176] The following inscription is significant though it is from outside the target area of this study. This funerary text of unknown provenance was found in the Cyprus museum in Nicosia in the late 1930s. Portions of the final seven lines are preserved though the top, bearing information about the deceased, is missing. The script and the presence of word dividers date the inscription from the early to mid-ninth century B.C. (*TSSI*, vol. 3, p. 28). Honeyman notes that though "the stone is not typical of the well-known Phoenician sites" its reddish sandstone is found in south-east Cyprus (Honeyman, "The Phoenician Inscriptions of the Cyprus Museum," p. 106). Masson and Sznycer disagree, saying that their examination of the stone reveals it to be a yellowish limestone (*Recherches sur les Phéniciens a Chypre*, p. 13, n. 2).

(1) הא.אי.מפת.וראש.אש
(2) ש[מ.לקבר.זא.שעל.הגבר.זא.
(3) י[נש]י.ויאבד.ה[מט]מ[מא.אית.הא]רנ
(4) זא[ ] בנ.יד.בעל.ובנ.יד.אדמ.ובנ[
(5) יד.ח[בר.אלמ]   ].[[עני.ל|
(6)   ]ש[
(7)   ]לינ.|

This reading is Honeyman's (Ibid., p. 107). His translation is "This is no magistrate or ruler that is [pla]ced in this tomb which is over this man. He who [de]files [this] sar[cophagus will be for]gotten and will perish whether by the hand of Ba'al or by the hand of man or by [the hand of the as]sembly of the gods[." He says that the inscription was probably built into the entrance of a tomb as a tomb curse.

The translation of Masson and Sznycer is "]Et l'homme qui [ ]vers (ou: en) ce tombeau-ci, car sur cet homme-ci[ ] Et que fasse périr[ ]-ci l'hom[me] entre les mains de Ba'al et entre les mains de 'DM et ent[re les mains de] -'R dieux[ ] le [ ]-nom (? ?)[" (*Recherches sur les Phéniciens a Chypre*, p. 15).

[177] *TSSI*, vol. 2, p. 97. The readings and translation are his.

[178] Koopmans, *Aramäische Chrestomathie*, vol. 1, p. 94, "für die Zukunft, in Zukunft."

¹⁷⁹Charles C. Torrey, "New Notes on Some Old Inscriptions," *ZA* 26 (1912), 90. Cooke notes that the לs on לתהנס and לאחרה may both be treated as negatives repeated for emphasis and translated "in order that thou—other one—shouldest not plunder" but says that this is "almost intolerably harsh." The ל on לאחרה may be a preposition and the word thus translated "for another" in the phrase "so that for another (?) thou shouldest not plunder my couch" (*NSI*, pp. 190–91; cf. Sefire II.C where someone is bribed to damage an inscription). Halévy interprets לאחרה as אחרה לא, a negation followed by the first person singular imperfect form of the verb חרה and translates the phrase, with a question mark, "On m'a placé avec mon vêtement afin que je ne me fâche pas" (Jacob Halévy, "Les deux stéles de Nerab," *RS* 4 [1896], 283–84). This interpretation seems extremely unlikely: this root occurs only in Hadad, line 23, and there is a noun.

¹⁸⁰Gevirtz, "West-Semitic Curses," pp. 147–48, n. 6. See the discussion below.

¹⁸¹Ibid., p. 147; *TSSI*, vol. 2, p. 97. Notice that in Eshmunazar 5–6 (יעמסן), as here in line 9 (ותהנסני), the 'interee' asks that they not remove "me" as well as the box (Eshmunazar חלת), or grave (Nerab 2:9, ארצתי).

¹⁸²Koopmans, *Aramäische Chrestomathie*, vol. 1, p. 94.

¹⁸³This would seem likely since Zakir's stele includes the less specific formula.

¹⁸⁴Fitzmyer, *Sefire*, p. 117. There it is not in a curse formula but describes treaty conditions regarding the return of runaway slaves.

¹⁸⁵In the same verse it is used as a GN as well.

¹⁸⁶Koopmans, *AramäischeChrestomathie*, vol. 1, p. 94. Also, *TSSI*, vol. 2, p. 153, where it is translated "damage." Koopmans notes Akkadian cognates *ba'āšu* and *bîšu*. *CAD* interprets *ba'āšu* as "to smell bad, be of bad quality" (also a form *bu'ušu* "cause to smell bad, besmirch, cast aspersions") and *bîšu* as an adjective "malodorous, bad quality" ("*ba'āšu*" and "*bîšu*" in *CAD* B, vol. 2, pp. 4–5 and 270 respectively).

¹⁸⁷Nerab 1:10–11 ומות לחה יכטלוך ויהאבדו זרעכ

Nerab 2:10–11 יהבאשו ממתתה ואחרתה תאבד Cooke observes the inscriptions' correspondence as well, saying, in his treatment of the meaning of the phrase ומות לחה יכתלוכ in Nerab 1:10, that though its meaning is uncertain, "The general sense must be the same as ממתתה יהבאשו" (*NSI*, p. 189).

¹⁸⁸Gevirtz, "West-Semitic Curses," p. 147, n. 5. He also notes its appearance in Targumic Aramaic ממותא.

¹⁸⁹Clermont-Ganneau, "Les stèles," p. 194.

¹⁹⁰Gevirtz, "West-Semitic Curses," pp. 147–48, n. 6. He notes that the usual words referring to descendants are זרע or שם, though he admits that his own translation is also ambiguous. He suggests the same parallel to Ugaritic *'uḥryt* that he recommends for אשרכ in Nerab 1:10. Hoffman's translation reflects this interpretation as well, "und sein Ausgang möge verloren gehn" ("Aramaische Inschriften aus Nêrab bei Aleppo, neue und alte Götter," *ZA* 11 [1896], 222). See the discussion of Nerab 1 under "Substantive Curse." Gibson (*TSSI*, vol. 2, p. 97), Jean and Hoftijzer (*DISO*, p. 25, "אחר" VI), Donner and Röllig (*KAI*, vol. 2, p. 276), and Cooke (*NSI*, p. 191) are among those that interpret it as "posterity."

¹⁹¹*TSSI*, vol. 2, p. 98. The first passage is a clear example of this meaning; the second less certain. He also notes Dan. 2:28, where the word occurs with a meaning more akin to that which Gevirtz favors, "but there is a God in heaven who reveals mysteries, and he has made known to king Nebuchadnezzar what will be in the latter days (באחרית יומיא)." Note, however, that the judgement context of Ps. 109:13 compares better to the Nerab usage than does the Daniel passage. The reference to the deceased's "children of the fourth generation" (line 5) may give some credence to the interpretation that ואחרתה in the last line refers to the perpetrator's descendants.

¹⁹²*TSSI*, vol. 3, pp. 12–13; and L. H. Vincent, "Les Fouilles de Byblos," *RB* 34 (1925), 183. Martin makes a convincing arguement that the sarcophagus was used twice and originally

Chapter 4: Curse   147

bore an inscription which was 300 years older. The side of the sarcophagus used for the tenth century Phoenician inscription is the side facing the door and, interestingly, in light of the curse of the inscription, these second users apparently tried not to damage the earlier inscription, though they did not quite succeed ("A Preliminary Report After Re-examination of the Byblian Inscriptions," Orientalia 30 [1961], 72–75).

[193] Magnanini, Le iscrizioni, p. 29.

[194] TSSI, vol. 3, p. 14.

[195] René Dussaud, "Les inscriptions phéniciennes du tombeau d'Aḥiram, roi de Byblos," Syria 5 (1924), 136.

[196] Ibid., p. 140.

[197] His reading of שׂרל for the last word disagrees with (שׁבל) accepted by most scholars (Ibid., p. 141).

[198] Ibid. How he arrives at his understanding of שׂרל as שׁאל is unclear. He does compare Akkadian arallu, a poetic word for the nether world, sometimes as the abode of the dead (CAD A, vol. 1, pt. 2, pp. 226–27).

[199] Pierre Montet, Byblos et l'Egypte, quatre campagnes de fouilles a Gebeil 1921–1922–1923–1924, Bibliothèque archaéologique et historique, vol. 11 (Paris: Librairie Orientaliste Paul Geuthner, 1928), p. 238.

[200] Dussaud, "Les inscriptions phéniciennes," p. 141.

[201] KAI, vol. 2, p. 4.

[202] Vincent, "Les fouilles de Byblos," p. 187.

[203] Ibid.

[204] Ibid.

[205] Ibid., p. 188.

[206] The Phoenician Tabnit and Eshmunazar inscriptions date from the fifth century B.C. and so are not treated among the target inscriptions in this study. They may be found in TSSI, vol. 3, pp. 102–03 and 106–09 respectively. The early seventh century Aramaic inscriptions from Nerab are treated in this study (Nerab 1 under "Substantive Blessing and Curse" in Chapter 5 and Nerab 2 under "Substantive Curse" in this chapter) but the interpretation of the phrase in the curse portion thought to refer to the descendants of the violator is subject to different interpretations.

[207] Cf. Hadad lines 17–22 (TSSI, vol. 2, pp. 66–69).

[208] Vincent, "Les fouilles de Byblos," pp. 190–91.

[209] Charles C. Torrey, "The Aḥiram Inscription of Byblos," JAOS 45 (1925), 273.

[210] Ibid., p. 274.

[211] Ibid. He points specifically to the inscriptions of Adad-nirari I of the fourteenth century B.C., Tiglath-pileser I of the twelfth century B.C. and Ašur-naṣirpal I of the ninth century B.C. which mention these acts of vandalism (see Eberhard Schrader, ed., Keilinschriftliche Bibliothek, Sammlung von assyrichen und babylonischen Texten, vol. 1 [Berlin: H. Reuther's Verlagsbuchhandlung, 1889], pp. 6–8, 46–47, 120–22).

[212] Torrey, "The Aḥiram Inscription," p. 274. He refers to an inscription of Ašur-naṣirpal for an example of this prohibition in an Assyrian text ša...ina piššatí ikatamušu (Schrader, Keilinschriftliche Bibliothek, vol. 1, p. 120, line 58).

[213] H. Bauer, "Eine phönikische Inschrift aus dem 13. Jahrh.," OLZ 28 (1925), 131.

[214] Ibid., p. 132.

[215] KAI, vol. 2, p. 4.

[216] *TSSI*, vol. 3, p. 16.

[217] Bauer, "Eine phönikische Inschrift," pp. 133–34. Rosenthal (*ANET*, p. 661) also interprets the end of the text this way.

[218] Montet, *Byblos et l'Egypte*, p. 238. Interestingly, he treats the two lines as seperate inscriptions (pp. 236–37).

[219] Ibid.

[220] Albright, "The Phoenician Inscriptions," p. 155 and n. 20. He also restores על after עלי which he says dropped out due to haplography.

[221] Ibid., p. 156, n. 25. The text is *UT* 49:VI:27–29, the transliteration is from Gibson (*CML*, pp. 74–75) and the translations is from *ARTU*, p. 97. The phrase also occurs elsewhere in Baʻal and Mot III.iii.18 (*ARTU*, p. 37).

[222] Albright, "The Phoenician Inscriptions," p. 156, n. 26. The transliterated Ugaritic text may be found, with another translation, in *CML*, p. 50.

[223] Albright, "The Phoenician Inscriptions," p. 156 and n. 27.

[224] Gevirtz, "West-Semitic Curses," p. 147.

[225] The originally article was "West-Semitic Curses" (p. 147, n. 1), the subsequent article ("A Spindle Whorl with Phoenician Inscription," *JNES* 26 [1967], 14) simply included a note about this new suggestion, not a re-examination of the entire inscription.

[226] Gevirtz, "A Spindle Whorl," p. 14. He says that it is related to Greek ταμιας "overseer, manager" though the Greek references date only to the fifth-fourth centuries and usually designated a military paymaster. In the Iliad (IV.84), Zeus is called ταμιας πολεμοιο τετυκια "overseer of armed conflicts." Accordingly, Gevirtz rejects Torrrey's earlier suggestion that the word is connected to Akkadian *tamû*. "The form *tamû* with meaning 'commander' is unattested in Akkadian and highly improbable" ("West-Semitic Curses," pp. 146–47, n. 8).

[227] Gevirtz, "West-Semitic Curses," p. 146, n. 2. See *CAD*, Ḫ vol. 6, p. 122.

[228] Jonas C. Greenfield, "Script and Inscription," in *Near Eastern Studies in Honor of William Foxwell Albright*, ed. Hans Goedicke (Baltimore: Johns Hopkins Press, 1971), p. 254.

[229] Ibid., p. 255.

[230] Luckenbill, *The Annals of Sennacherib*, p. 130,. col. VI, lines 81–83. The reading and translation are Luckenbill's, except that the text has been normalized as elsewhere in the study.

[231] Greenfield, "Scripture and Inscription," p. 255.

[232] Luckenbill, *Ancient Records*, vol. 2, p. 310, §810.

[233] *TSSI*, vol. 2, p. 64.

[234] "*arāru*" in *CAD* A, vol. 1, pt. 2, p. 235. Compare the late Neo-Assyrian foundation text of Sin-šarra-iškun whose protective curse ends in part, may the gods "overturn his throne, curse his reign, take from him the sceptre, bind his arms, (and) make him sit in bonds at the feet of his enemy" (Grayson, "Cylinder C of Sin-šarra-iškun," p. 166).

[235] Fitzmyer, *Sefire*, pp. 20–21.

[236] Abou-Assaf, et al., *La statue*, p. 23. The transliteration to block letters and the translation are mine.

[237] Javier Teixidor, "L'inscription d'Aḥiram à nouveau," *Syria* 64 (1987), 139–40.
Another, very brief inscription, crudely written, was found on the wall of the Aḥiram tomb shaft about half-way down. Montet speculates that the sealed mouth of the tomb was not enough to prevent robbers from entering and the warning was placed where, having broken in, the thieves could not miss it (*Byblos et l'Egypte*, p. 217, drawing p. 216). Dussaud reads it as

Chapter 4: Curse   149

1) לדעת.
2) הני בדלכ.
3) תחת זנ

and translates it, "Avis! Voici! Ta perte (est) ci-dessous" (Dussaud, "Les inscriptions phéniciennes," p. 143). He treats לדעת as an infinitive from the root ידע meaning "notice, warning." בדל means "séperrer, couper, détacher" in ritual language "exclure, expulser" for the reason of impurity or violations of the law (cf. Isa. 56:3). The warning was insufficient to stop the robbers: the grave was violated, apparently, in the eighth century B.C. (Dussaud, "Les inscriptions phéniciennes, p. 143). If this interpretation is correct then the text contains a substantive, though unspecified, curse. All commentators are not agreed, however, Bauer interprets the inscription as a mere epitaph. Believing the correct division of the second line to be ב.דלכ, he translates it "Siehe, ich liege im Staube hier unten" (Bauer, "Eine phönikische Inschrift," p. 136). Other scholars, including the more recent commentators, read the second line as הניפד (Vincent, "Les fouilles de Byblos," p. 189, n. 1); Albright redivided the words הנ יפד.לכ ("The Phoenician Inscriptions," p. 156, n. 30); as do Donner and Röllig (*KAI* 2, vol. 2, pp. 4–5); Magnanini (*Le iscrizioni fenici*, p. 29); and Gibson (*TSSI*, vol. 3, p. 17). Vincent's hypothetical יפ׳ is, Albright says, unattested linguistically and should simply be פד related to BH פיד "misfortune, ruin" ("The Phoenician Inscriptions," p. 156, n. 30). The later commentators agree with Albright's interpretation, seeing the inscription as a preliminary warning to the full curse formula found on the sarcophagus.

[238] *TSSI*, vol. 1, p. 19. The inscription is treated in this study because of its supposed Israelite provenance, however, M. Heltzer ("Eighth Century B.C. Inscription from Kalakh [Nimrud]," *PEQ* 110 [1978], 6) says the letter types of this text are closer to those of the Mesha inscription and proposes a provenance of Ammon or Moab.

[239] Alan R. Millard, "Alphabetic Inscriptions on Ivories from Nimrud," *Iraq* 24 (1962), 45.

[240] Ibid., p. 47.

[241] Ibid.

[242] Ibid. The Karatepe inscription is treated in the section "ברכ and שלמ with Substantive Curse" in Chapter 5.

[243] *TSSI*, vol. 1, p. 19.

[244] Ibid., p. 20.

[245] Ibid. See BDB, p. 837. The noun פת for a "fragment" or "morsel" of bread occurs several times.

[246] Avigad, "The Epitaph of a Royal Steward," p. 143. This inscription is treated above in the section "ארר."

[247] It should also be noted that BDB (p. 836) also gives a second root פתח which means "engrave." Perhaps the first line of the Nimrud Ivory once gave the name of the person who made the inscription.

[248] For information on the discovery of the texts see the discussion of Deir 'Alla II.17 above in this chapter.

[249] Hoftijzer, "The Prophet Balaam," p. 13.

[250] McCarter, "The Balaam Texts from Deir 'Allā," p. 49.

[251] Ibid., p. 51.

[252] This absence of reason might not be so unusual. Alexander Heidel, in his discussion of the Babylonian flood story of the Gilgamesh Epic, notes that unlike the biblical flood account, the reason for the Babylonian flood seems to be divine caprice—only as an afterthought is human sin said to have necessitated it (*The Gilgamesh Epic and Old Testament Parallels*, 2nd ed. [Chicago: The University of Chicago Press, 1949], p. 225).

[253] Ibid., pp. 58–59.

254 *BTDA*, p. 76.

255 Ibid., p. 75. All of these Egyptian texts are treated by Miriam Lichtheim (*AncientEgyptian Literature*[Berkeley, CA.: University of California Press, 1973], vol. 1, pp. 139–63). Translations of the first two texts may also be found in *ANET*, pp. 441–44 and 444–46 respectively. Michael O'Connor also notes this contrast in his study of Kilamuwa 1 though the one causing the change from lack to plenty in that context is the king ("Rhetoric," pp. 24–25).

256 *BTDA*, p. 47, n. 37.

257 Ibid.

258 Levine, "The Deir 'Alla Plaster Inscriptions," pp. 195–96.

259 Stephen A. Kaufman, "Review Article: The Aramaic Texts from Deir 'Alla," *BASOR* 239 (1980), 72. (Kaufman had access to Levine's article before it was published, thus he "agrees" with Levine though the publication date of Levine's article is later.)

260 These line numbers ("9b through the end") are according to the *editio princeps* and pre-date two breakthroughs in the study of the texts: 1) the determination that two of the other combinations may be fitted into Combination I and 2) that two of the first parts of Combination I, first thought separate, should be pushed together. This being the case, the numbers of the *editio princeps* now differ from those generally accepted by other scholars. For a more exact summary of these newer readings see note 261. The line numbers referring to the first Combination used below will be according to the new numbering scheme. In a recent personal communication McCarter said that Hoftijzer told him he now accepts the new numbering system.

261 McCarter, "The Balaam Texts from Deir 'Allā," p. 51, readings and translation. The block letter transliteration is mine throughout. Caquot and Lemaire ("Les textes araméens de Deir 'Alla," pp. 189–208), pull down fragments I a and b to align them with I c and d, also they hypothesized a connection of fragments VIII d and XII c with each other and the first combination (p. 193). McCarter placed these last two fragments in the proper place in the first combination (p. 51). For a more precise explanation of this alignment see Hackett (*BTDA*, p. 21) who called the later part of this alignment to McCarter's attention.

McCarter's reading is thus used here instead of the *editio princeps* because it contains these now accepted readings. In comparing this reading to the *editio princeps*, the reader should be aware that the line numbers will not agree; line 1 in McCarter's version corresponds generally to lines 1 and 3 of the *editio princeps*, line 2 to lines 2 and 4 of the *editio princeps*, and the other line numbers here are then two off from the *editio princeps*, e.g., line 3 corresponds to line 5 of the *editio princeps*, line 4 to line 6, etc. The block letter transliteration is mine.

262 *ATDA*, pp. 200–01. In Deut. 32:11–14; Isa. 40:31; etc. it was admired as a powerful bird, also Akkadian texts (Ibid., nn. 41–42).

263 Also compare Lemaire's interpretation of Khirbet Beit Lei 7 above where he sees an occurrence of this root ("Prières en temps de crise," pp. 562-63).

264 Ibid., p. 201.

265 Caquot and Lemaire, "Les textes araméens de Deir 'Alla," p. 199. Their interpretation does not differ from Hoftijzer's at this point but they do draw more attention to the situation's strangeness.

266 *BTDA*, p. 46.

267 Levine, "The Deir 'Alla Plaster Inscriptions," p. 199.

268 *ATDA*, pp. 203–04. His conclusions are not certain though. For example, the צפר may be a small bird (Prov. 26:2 and Ps. 84:4) or a bird of prey (Ezek. 39:4). It can serve as the name of a woman צפרה, the wife of Moses, or of a man צפור, the father of the king of Moab (Num. 22:2). It would seem to be a general word for bird; would a king's father be named after a small bird or would a woman be named after a bird of prey?

269 Ibid., p. 204. The Akkadian text is in Borger (*Die Inscriften Asarhaddons Königs von*

Assyrien, Archiv für Orientforschung, supplement 9 [Graz: Ernst Weidner, 1956], p. 14): URU_{mes} [ú]šabšima gan (GI) apparâte^{mes} ù gišṣarbati ina qerbišú magal iširma úṣarrišá papallu iṣṣurât šamê nûnê^{mes} apsî ša la nibi ina qerbišu ibbašuma "Stadt liess er sein. Sumpfrohr und Weiden wuschen dort üppig und trieben Schösslinge. Vögel des Himmels und Fische der Wassertiefe gab es dort ohne Zahl."

[270] *ATDA*, p. 206. Cf. Isa. 13:21; 34:11, 13, 15; Jer. 50:39; and Mic. 1:8.

[271] McCarter, "The Balaam Texts from Deir 'Allā," p. 55.

[272] Ibid., p. 51.

[273] *BTDA*, p. 49.

[274] McCarter, "The Balaam Texts from Deir 'Allā," pp. 51–52.

[275] *ATDA*, p. 205, and n. 62.

[276] Ibid., p. 205. In biblical texts מטה is used as an image of punishment in Prov. 26:3 and Exod. 21:20. The word חטר best translated "sceptre" is also found in the Hadad inscription lines 3, 9, 20, et al. Cf. Isa. 28:27 where מטה and שבט are used in parallel.

[277] Caquot and Lemaire, "Les textes araméen de Deir 'Alla," p. 199.

[278] McCarter, "The Balaam Texts from Deir 'Allā," p. 55.

[279] Ibid.

[280] *BTDA*, p. 29.

[281] Levine, "The Deir 'Alla Plaster Inscriptions," p. 199.

[282] McCarter, "The Balaam Texts from Deir 'Allā," pp. 51–52; readings and translation.

[283] *ATDA*, p. 179.

[284] The text used by the RSV for this passage is emended by its editors.

[285] *ATDA*, p. 179.

[286] Ibid., p. 209. The warning aspect of this phrase is strongest in Prov. 19:27 of the biblical passages.

[287] Caquot and Lemaire, "Les textes araméens de Deir 'Alla," pp. 199–200.

[288] McCarter, "The Balaam Text from Deir 'Allā," p. 55. Hackett suggests that the "hares" might now make their meal on "wolves" (*BTDA*, p. 50).

[289] McCarter, "The Balaam Text from Deir 'Allā," p. 58.

[290] Ibid., p. 56.

[291] Levine, "The Deir 'Alla Plaster Inscriptions," p. 199.

[292] McCarter, "The Balaam Texts from Deir 'Allā," pp. 51–52; readings and translation.

[293] *ATDA*, p. 180.

[294] Ibid., p. 211.

[295] Ibid., pp. 212–14. He includes very in-depth research on the nature and function of these personnel complete with Akkadian parallels from Mari.

[296] Caquot and Lemaire, "Les textes araméens de Deir 'Alla, " p. 200.

[297] McCarter, "The Balaam Text from Deir 'Allā," p. 56. Note, however, that scholarly opinion is divided over the phrase אש ענה אנה in the Zakir/Zakkur stele. Recently, Alan Millard has reiterated and strengthened the argument that the phrase refers to Zakkur's place of origin rather than being a formulaic phrase indicative of religious humility ("The homeland of Zakkur," Semitica 39 [1990], 47-52).

[298] Ibid., p. 58.

[299] *BTDA*, pp. 51–52.

[300] Why a fool נבל being filled with food would rank with these other reverses might seem unclear; perhaps a fool would not work and thus starve. This understanding seems to be supported by comparison to a Middle Kingdom Egyptian text "He who begged dregs has overflowing bowls" (Lichtheim, "The Admonitions of Ipuwer," *Ancient Egyptian Literature*, vol. 1, p. 156).

[301] Levine, "The Deir 'Alla Plaster Inscriptions," pp. 199–200. The details of this interpretation are not relevant for this study.

[302] McCarter, "The Balaam Texts from Deir 'Allā," pp. 51–52; readings and translations.

[303] *ATDA*, p. 180. Cf. BH נשׂיא Num. 7:2; Josh. 22:14; etc. and for אזר Job 12:18.

[304] Ibid., p. 214.

[305] Ibid., p. 215.

[306] Ibid., p. 180.

[307] Ibid., p. 216.

[308] McCarter, "The Balaam Text from Deir 'Allā," p. 56.

[309] Ibid.

[310] Ibid., pp. 56, 58.

[311] *BTDA*, p. 29.

[312] Lichtheim, *Ancient Egyptian Literature*, vol. 1, p. 148.

[313] Ibid., p. 143.

[314] These quotes are from the "Admonitions of Ipuwer" (Lichtheim, *Ancient EgyptianLiterature*, vol. 1, pp. 152 and 156 respectively).

[315] McCarter, "The Balaam Texts from Deir 'Allā," pp. 51–52; readings and translations.

[316] *ATDA*, p. 180.

[317] Ibid., p. 217. The text is Keret lines 96–104 in *UL*, p. 69 (*ARTU*, pp. 195–96, Keret I.ii:32–51). Isaiah 29:18 is similar though the emphasis there is probably that those with handicaps or weaknesses will have those conditions mended by Yahweh.

[318] Ibid.

[319] Caquot and Lemaire, "Les textes araméens de Deir 'Alla," p. 201; McCarter, "The Balaam Text from Deir 'Allā," p. 58; *BTDA*, p. 53.

[320] McCarter, "The Balaam Texts from Deir 'Allā," pp. 51–52; readings and translation.

[321] *ADTA*, p. 180.

[322] Ibid., pp. 218–19.

[323] Caquot and Lemaire, "Les textes araméens de Deir 'Alla," p. 201.

[324] Ibid. They reject his argument and most of his proposed biblical parallels. For his argument, see *ATDA*, p. 273.

[325] McCarter, "The Balaam Text from Deir 'Allā," p. 58.

[326] *BTDA*, p. 29.

[327] Ibid., p. 55, quoting from Lichtheim (*AncientEgyptianLiterature*, vol. 1, pp. 151 and 154 respectively).

[328] *BTDA*, p. 55, n. 50. This footnote reference is the only indication Hackett gives to show

Chapter 4: Curse 153

that her translation does not refer to human reproduction.

[329] Fitzmyer, *Sefire*, p. 15.

[330] McCarter, "The Balaam Texts from Deir 'Allā," pp. 51–52; readings and translation.

[331] *ATDA*, p. 180.

[332] Ibid., pp. 220, 279.

[333] Caquot and Lemaire, "Les textes araméens de Deir 'Alla, " p. 202.

[334] McCarter, "The Balaam Text from Deir 'Allā," p. 58.

[335] *BTDA*, p. 29.

[336] See the treatment of Combinations II:17; IX:a:3; and X:a:3 above for comments concerning the relationships of the combinations to one another and concerning the texts' discovery.

[337] Hoftijzer, "The Prophet Balaam," p. 16.

[338] Ibid., p. 13. Hoftijzer does argue that a fragment bearing the interjection "O, Balaam!" may well be placed in this combination or just above it. However, the only two such fragments listed in the *editio princeps*, VIIId and XIIc (*ATDA*, pp. 175–78, 181–82), have been convincingly placed by Caquot and Lemaire in Combination I ("Les textes araméens de Deir 'Alla," p. 193).

[339] Hoftijzer, "The Prophet Balaam," p. 11. Also see *ADTA*, p. 280.

[340] Caquot and Lemaire, "Les textes araméens de Deir 'Alla," p. 202.

[341] Ibid., pp. 202–03.

[342] McCarter, "The Balaam Texts from Deir 'Allā," p. 59, n. 5.

[343] *BTDA*, pp. 84–85. Hackett's translation is quite close to that of the *editio princeps* at most points, but as she does not interpret substantive curse in Combination II, her interpretation will not be included in this study.

[344] Levine, "The Deir 'Alla Plaster Inscriptions," p. 196.

[345] *ADTA*, p. 280. Lines 1–3 are completely unintelligible.

[346] Ibid., pp. 174, 180–81. The text and translation used below are those of the *editio princeps* (*ATDA*). The italics are Hoftijzer's and indicate translations and insertions that are "completely uncertain" (p. 179). Hoftijzer actually includes a total of 37 lines in his reading of Combination II; lines 21–29 are "untranslatable" and lines 30–37 contain very few readable words so that his translation is extremely sketchy (p. 181). The other scholars consulted do not comment on these lines at all.

[347] Ibid., p. 221. It seems likely that Hoftijzer's tendency to see curses in these fragmentary lines is due to his identification of the Bala'am of this text (Combination I, at least) with biblical Bala'am primarily known for proficiency in pronouncing blessings and curses (*BTDA*, p. 77).

[348] *ATDA*, pp. 174 (text) and 180 (translation).

[349] Ibid., p. 223 and n. 109. Fitzmyer thinks this refers rather to the potential destruction of the land by Assyria (*Sefire*, p. 45).

[350] *ATDA*, pp. 174 (text) and 180 (translation).

[351] Ibid., p. 224.

[352] *TSSI*, vol. 3, p. 14. For further discussion of this idea, see André Parrot, *Maledictions et Violations de Tombes* (Paris: Librairie Orientaliste Paul Geuthner, 1939), pp. 174–75.

[353] *ATDA*, pp. 174 (text) and 180 (translation).

[354] Ibid., pp. 225–26.

[355] Ibid., pp. 174 (text) and 180 (translation).

356 Ibid., pp. 226–27.

357 Ibid., p. 227. The italics are his. He says "an exact definition is impossible."

358 Ibid., pp. 174 (text) and 180 (translation).

359 Ibid., p. 230.

360 Ibid., p. 180.

361 Ibid., p. 231. For the Meṣad Ḥashavyahu (Yavneh Yam) inscription, a discussion of this concern, and further bibliography, see *TSSI*, vol. 1, pp. 26–30.

362 *ATDA*, p. 232.

363 Fitzmyer, *Sefîre*, p. 15.

364 Abou-Assaf, et al., *La statue*, p. 20. The translation is mine.

365 *ATDA*, pp. 174 (text) and 180 (translation).

366 Ibid., p. 235.

367 Ibid., p. 180.

368 Ibid., p. 237.

369 Fitzmyer, *Sefîre*, p. 15.

370 *UT*, p. 245. Pointed out by Hoftijzer, *ATDA*, p. 237.

371 *UL*, p. 94.

372 Ibid., p. 180.

373 Donald J. Wiseman, *The Vassal-Treaties of Esarhaddon* (London: The British School of Archaeology in Iraq, 1958), pp. 61–62, lines 437–39.

374 Tromp, *Primitive Conceptions of Death*, pp. 160–62. And, of course, in Ugaritic literature, e.g. *UT* 67:I:6–8 and 51:VIII:15–20.

375 *ATDA*, p. 239.

376 See "Other Words for Curse" in this chapter.

377 Hoftijzer's readings and translation through line 18 are given below. For his readings and translation of lines 19–37, see *ATDA* (pp. 174, 180–81).

(14) -עֲלָ[.|.]----ר.--.ת.שׁמה.כב-----ח.יֹלֹבֹ.לבב.נקר.שׁהה.כי.אתה.ל-|

(15) לקצה.שׁ---ח----וזל.-גֹדרטש----שאלת.מלכ.ססה.וש|א|לֹ|ת

(16) ה.--ו---חזנ.רחק--מכ.שאלתכ.למ.-|

(17) לדאת.ופר.דבר.לעֲנֹהֹ.על.לשנ.לכ.נשפט.ומלקֹבֹ.אמר|

(18) ולנ.שתי.למלכ-----ת.ח--.---עשק.כת.|

------ horror --------will the heart of an *impotent* blinded one be *firm*, when comes the ---( ) ----- *a plastered wall*. What a king asks for is a horse, what ( ) a(sk)s for ( ) ----------far --- what you ask for is ---(*they said to Balaam: you did take* in) foolishness and silliness an iniquitous word on (your) tongue. With you we will litigate (?), *it will be impossible* (*for you*)*to curse anymore*; -- say ( ) and we have beverage (?) --------( )

In August of 1989, a symposium on the Balaʿam texts was held at the University of Leiden. The proceedings of that symposium have recently been published under the editorship of the original editors of the *editio princeps*. Several aspects of the study of the texts are covered in the work, among them: archaeological context, palaeography, general interpretation, language, details of the texts, and the relation of the texts to the Old Testament (*The Balaam Text from Deir ʿAlla Re-evaluated: Proceedings of the Symposium held at Leiden, 21–24 August 1989* [Leiden: Brill, 1991]). Unfortunately, I was able to gain access to the work only as my own study was sent to the

Chapter 4: Curse   155

publisher.

³⁷⁸The gods are not mentioned by name though they may be the same ones listed in Sefire II and III or another list may have been in a portion of the text now missing. Likewise, a portion of the text specifying the curses may have originally been included. The other Sefire treaties contained both specific lists of gods, specific curses, and the general threat of condemnation found in the threat "you will have been false to all the gods which are in this treaty."

³⁷⁹Though, as was seen above, whether the conditions described are punishments/curses or the things being punished is unclear.

³⁸⁰Except the Phoenician Cebel Ires Daği inscription (see the discussion of Deir 'Alla II.17). Notice that the following conclusion regarding general words for curse (e.g., ארר, קבב) is not negated by this late seventh century inscription.

³⁸¹For example, Josh. 6:26 "Joshua laid an oath upon them at that time saying, 'Cursed before the Lord be the man (ארור האיש לפני יהוה) that rises up and rebuilds this city, Jericho. At the cost of his first-born shall he lay its foundation, at the cost of his youngest son shall he set up its gates.'" Note that the oath is laid on them by Joshua, but it is attributed to Yahweh and thus brought to pass by him 1 Kgs. 16:34.

³⁸²As stated elsewhere, it must be recognized that though the evidence is consistent, it is far from abundant.

³⁸³This inscription is considered by its editor to be of Israelite (Northern Kingdom) provenance.

³⁸⁴Many more examples of this type of curse will be treated in the next chapter. In those texts a blessing is included in the text.

Chapter 5

Blessing and Curse

There are several combinations of the elements of blessing and curse. The various possiblilies include the use of the specific words ברכ and ארר, ברכ and שלמ with a Substantive Curse, or a Substantive Blessing and Curse. As noted before, "Substantive Blessing" and/or "Substantive Curse" occur when a deity is called upon explicitly (DN mentioned) or implicitly (DN not specified) to benefit or harm the addressee when the addressee either 1) does something the speaker wishes of them (thus a request to "bless") or 2) does not do something the speaker wishes or does something the speaker wishes them not to do (thus a request to "curse").

ארר and ברכ

The combination ברכ and ארר is found only in the En Gedi inscription and it occurs here with what may be the DN Yahweh.[1]

The En Gedi inscription was found in a cave overlooking the Dead Sea one-half kilometer south of Naḥal Yishai. The cave is divided into small upper and lower chambers, both of which show signs of numerous short-term habitations from the Chalcolithic through Byzantine periods. The inscription is dated circa 700 B.C. Parts of nine lines of writing are found written in black ink on a large stalactite in the upper chamber.[2]

1) ארר.אשר.ימחה
2) ----[נה]---
3) --[יה]----
4) ברכ.יהו[--
5) ---[ב]ׄ---
6) ברכ.בנ[י]----] מלכ
7) ברכ אדנ[י]----
8) ------------------
9) ------------------

Bar-Adon translates this inscription, "Cursed be he who will efface[ ] Blessed be YHW[ ]Blessed be BGY[ ]king Blessed be 'DNY["[3]

If this reading is correct with ארר in line 1 and "Yahweh" in line 4, it is the only time ברכ and ארר occur with Yahweh in an extant inscription.[4] However, it should be said that even then the idea that Yahweh is to bring about the curse of line 1 is no more than an implication at best. There are no epigraphic examples as clear as Jer. 17:5–8 where the man (הגבר) is cursed (ארור) who trusts in man, but blessed (ברוך) who trusts in the Lord. ברך and ארר are juxtaposed several times in BH, including Gen. 12:3; 27:29; Num. 22:12; 23:7; and 24:7. Also, apart from Kuntillet 'Ajrud 5 where ברכ הא ליהו occurs, this is the only occurence of ברכ יהוה.[5] The grammatical forms of ברכ and ארר seem to be the passive participle written defectively, in light of the contemporary Siloam Tomb inscription where the full spelling occurs.[6]

Mitchell says the major difference between the biblical use of ברך and the extra-biblical use of ברכ in NW Semitic is that the root is used for praise of divinities only in Palmyrene dedication formulas from the second to fourth centuries A.D. (aside from post-biblical Jewish literature) and most authors have concluded that the Palmyrene use of ברכ is due to Jewish (biblical) influence. The use of ברך for the praise of God, then, is an entirely inner-biblical development.[7] However, since line 4 of the above seems to be an accurate reading and this inscription dates from the eighth to seventh century B.C., Mitchell is incorrect: here is a very early use of ברכ with Yahweh as object rather than subject.

After the preliminary standard warning to any would-be grave disturbers in lines 1–2, two letters of a name (perhaps with the theophoric element יה-) occur in the third line. Perhaps this is the name of the deceased and is followed by a statement about his character. Line 4 may begin a doxology (cf. BH ברוך אתה יהוה). The two letters in line 5 are not enough to propose a specific restoration, but may have been some substantive praise. The second word on line 6 may be the beginning of some divine attribute, cf. Ps. 72:18–19 where ברוך יהוה is paralleled to שם ברוך. The space between בני and מלכ is only large enough for a few (3–4) letters; Bar-Adon's drawing shows indications of another letter following מלכ, "king of..."(?). In line 7, אדני would then be a designation for Yahweh. Though it would be unusual to have both "Yahweh" and "Adonai" referring to God in the same context, this is not unknown. Psalm 130 uses both יהוה and אדני throughout.[8] The phrase ברוך אדני occurs only in Ps. 68:20. The two phrases ברוך יהוה and ברוך אדני occuring together are unknown in the Bible. Alternatively, "Adonai" may be interpreted as "my master" (cf. Arad 18:1; 21:3, 4, etc.), and if so, was probably followed by a PN (no letters are visible in the photographs but Bar-Adon's drawing indicates others).[9] Lines 8–9 are shown as blank in Bar-Adon's transliteration but his drawing shows traces of letters that could be the word ארר on one of those lines.[10] Miller compares this inscription with those from Khirbet Beit Lei and notes that the warning not to disturb the text comes after that inscription, while at En Gedi it serves as a "kind of superscription to an extended series of blessings."[11]

Lines 2–9 could also have originally held a series of PNs, perhaps names of a group of people (or family?) fleeing an invasion. In this case the [יה]

Chapter 5: Blessing and Curse   159

in line 3 and יהו] in line 4 could both be theophoric elements of PNs. Likewise, biblical PNs are available as parallels to the other letter combinations in the text.

Summary

As was noted in the summary to the first section of Chapter 4, ארר is found only in Judean inscriptions near the end of the target period. It is not found in any other language nor in closer association to any god other than Yahweh. It does not occur in a precise parallel to biblical ארור האיש לפני יהוה (Josh. 6:26) anywhere in the inscriptions. However, the possibility that line 4 of the En Gedi inscription should be read ברכ.יהוה may indicate that Yahweh is also being invoked to protect by the curse in line 1.

ברכ and שלמ with Substantive Curse

The Karatepe inscriptions discussed below form a mediating category between inscriptions which have ברכ and שלמ and those which include a substantive curse. While they include forms of both ברכ and שלמ, these elements are parts of larger inscriptions that also include substantive curses.

The Karatepe inscriptions included in this study are the Phoenician portions of a Phoenician-Hieroglyphic Hittite bilingual inscription found in 1945, east of modern Kadirli near the Ceyhan River in Turkey.[12] Various epigraphic features of the inscriptions make it possible to date them from the mid-eighth century to the early seventh century[13]; Donner and Röllig date them circa 720 B.C.[14] Three exemplars of the Karatepe inscription exist in varying degrees of completeness. The copy now known as "A" was the last found and is the most complete.[15] Copy A and copy B (most poorly preserved) are both gate inscriptions. Because they are so nearly identical, only A will be treated of the two. "A" is written on four separate orthostats (one of the columns is written across two of the orthostats, one of which is very narrow) and finishes on a stone lion.[16] Exemplar "C" is written on four sides of a statue of Ba'al and was probably kept in a temple. All three inscriptions are virtually identical excepting that the curses protecting C (the statue) are slightly different from those protecting A and B (the gate inscriptions).

Karatepe A

(2b) וברכ בעל כר[נ]
(3) תריש אית אזתוד חים ושלמ
(4) ועז אדר על כל מלכ לתתי בעל כרנתריש
(5) וכל אלנ קרת לאזתוד ארכ יממ ורב
(6) שנת ורשאת נעמת ועז אדר על כל מל

And may Baal KRNTRYŠ bless Azitiwada with life and health

and powerful strength above every king! May Baal KRNTRYŠ
and all the gods of the city give to Azitiwada length of days,
and many years, and a pleasant old age, and powerful strength
above every king!"[17]

The request to be blessed with long life (A.iii.2–4, 4–6) is similar to that found in the Tell Fekherye inscription lines 7 and 14 and Kilamuwa 2, as well as the Yeḥimilk, Elibaʻal, Abibaʻal, Shipiṭbaʻal, and Nerab 2 inscriptions among others.[18] An Old Babylonian letter greeting formula was ^{d}Šamaš liballiṭka lu šalmāta lū balṭāta "May Shamash grant you life; may you be well, may you live."[19] A. M. Honeyman translates the beginning of the blessing as "And Baʻal KRNTRYŠ blessed 'ZTWD with life and prosperity..." thus making the text a retrospective statement rather than a request for blessing.[20] This is a request not for ברכ and שלמ but that Azitiwada may be blessed with שלמ. In this context, שלמ may mean safety from enemies as it seems to in the Lachish letters (Lachish 2:1–3; 3:2–3; 5:1–2[?]; 7:1–2[?]; 9:1–2) or it may, more probably, mean welfare as is the case, for example, in the Tell Fekherye text (lines 8–9, see the discussion under "Substantive Blessing and Curse"). Strength asked in conjunction with life is also found in biblical blessings like the blessing of Asher in Deut. 33:25b, "...and as your days, so shall your strength be" וכימיך דבאך (cf. Deut. 34:7). In each of the versions (A.iii.2–3 and C.iii.16–17) the sign of the direct object אית precedes the name of Azitiwada at the beginning of the blessing. This is true also in the fifth century Yeḥaumilk inscription following a form of the verb ברכ.[21]

First Kings 3:14 contains an interesting parallel: if Solomon will be faithful to the commandments of Yahweh, as was his father, Yahweh promises, "I will lengthen your days" והארכתי את־ימיך. Important points of comparison include: 1) the biblical passage is part of an address by Yahweh in response to extraordinary sacrifices and a prayer, while Azitiwada's inscriptions should certainly be understood as his attempt to show his worthiness for the blessings he asks. 2) Azitiwada bases his worthiness for this blessing on his benevolent administration of foodstuffs, his provisions to make travel safe, and rebuilding cities in obedience to the gods; whereas Solomon asks for the wisdom to administer his people (probably hoping for the same success that Azitiwada claims) and is given the blessings that he did not ask for. 3) Solomon is promised (v. 13) superiority over every other king "so that no one will compare with you all your days" אשר לא־היה כמוך איש במלכים כל־ימיך which is substantially what Azitiwada asks for "And may Baʻal KRNTRYŠ bless Azitiwada with life and health and powerful strength above every king!" (lines 2–4). Malachi 2:5 is also significant at this point, "my covenant was with him for life and peace" בריתי היתה אתו החיים והשלום.

Michael Barré says that lines 2b–6 divide into two parallel sections "like the two panels of a diptych."[22] Each part begins with a verb followed by a divine name, then the recipient (royal name), and the substance of the blessing. וברכ in

Chapter 5: Blessing and Curse

the first panel is paralleled by יתן in the second, (cf. Ps. 29:11 יהוה עז לעמו יתן
יהוה יברך את־עמו בשלום). חים in the first panel is parallel to ארכ ימם in the
second. The first section has three blessings of one word each; the second section
has three blessings of two words each plus the repetition of the last blessing of
section one (see Table 3[23]). The two word blessings of part two are significant
because each is composed of the same types of elements: ארכ ימם quantity and
time, שנת רב quantity and time, רשאת נעמת time and quantity.[24] The repetition
of the last blessing (ועז) brings the total number of blessings to seven and gives
symmetry to the two parts.[25] Compare also the fifth century Yeḥaumilk inscription which has strong similarities: גבל אית יחומלכ (9) מלכ גבל ותחוו והארכ
ימו ושנתו על גבל (8) תברכ בעלת "May the Mistress of Byblos bless Yeḥaumilk
king of Byblos and give life to him and prolong his days and his years over
Byblos."[26]

Table 3

Karatepe Royal Blessing (A.iii.2–6) Structure

	lines 2–4a	lines 4b–6
verbs	וברכ	לתתי
DN	בעל כר[נ]תריש	בעל כתריש וכל אלנ קרת לאזתוד
RN	אית אזתוד	לאזתוד
blessings	חים	ארכ ימם
		ורב שנת
	ושלמ	ורשאת נעמת
	ועז	ועז
	אדר על כל מלכ	אדר על כל מלכ

Barré further insists that שלמ used in parallelism with חימ has the
specific meanings "health, well-being" rather than the more general "health,
well-being, peace, good relations," etc., and that עז used with שלמ denotes the
special strength or power which enables one to live a long healthy life.[27]

(7) כ וכנ הקרת ז בעלת שבע ותרש ועמ
(8) ז אש ישב בנ יכנ בעל אלפמ ובע
(9) ל צאנ ובעל שבע ותרש וברבמ ילד
(10) וברבמ יאדר וברבמ יעבד לאז
(11) תוד ולבת מפש בעבר בעל ואלמ

And may this city be owner of plenty (of grain) and new wine;
and may they bear many (children), and as they grow many
become powerful, and as they grow many serve Azitiwada and

the house of Mopsos, by the grace of Baal and the gods!²⁸

Lines iii.7–11 contain a request for blessings of plenty on the city. In an Ugaritic incantation text apparently recited in a New Year's festival banquet, the king of Ugarit implores Ba'al to bless the city in the coming year, zk.dmrk.[ln]k (?).ḥtkk.nmrtk.btk.ugrt.l.ymt.špš.w yrḫ w n'mt.šnt.il ²⁹ "Let your strength, your protection, your power, your patronage, your beneficial might be in the middle of Ugarit throughout the days of the Sun and Moon and the happiest of Ilu's years."³⁰ The blessings asked by Azitiwada are among those most basic and most desired of blessings: numerous offspring for both humans and animals and plenty of food (cf. Gen. 12:2; Deut. 1:11; 14:28–29).

(12) ואמ מלכ במלכמ ורזנ ברזנמ אמ א
(13) דמ אש אדמ שמ אש ימח שמ אזתו
(14) ד בשער ז ושת שמ אמ אפ יחמד אי
(15) ת הקרת ז ויסע השער ז אש פעל א
(16) זתוד ויפעל לשער זר ושת שמ עלי
(17) אמ בחמדת יסע אמ בשנאת וברע יסע
(18) השער ז ומח בעל שממ ואל קנ ארצ
(19) ושמש עלמ וכל דר בנ אלמ אית הממלכת הא ואית המלכ הא ויאת
(IV) 1) אדמ הא אש אדמ שמ אפס

Now, if a king among kings, or a prince among princes, or any man who is a man of renown, effaces the name of Azitiwada from this gate and puts up his own name, or more than that, covets this city and pulls down this gate which Azitiwada made, and makes another gate for it and puts his own name on it, whether it is out of covetousness or whether it is out of hatred and malice that he pulls down this gate—then let Baalshamem and El-Creator-of Earth and the eternal Sun and the whole generation of the sons of the gods efface that kingdom and that king and (iv.1a) that man who is a man of renown!³¹

In contrast to the blessing, lines A.iii.12–A.iv.1 contain a substantive curse. Azitiwada calls on the gods Ba'alshamem, El-Creator-of-Earth, the eternal sun, and "the whole generation of the sons of the gods" (lines iii.18–19) to avenge him should anyone damage the gate and its inscription. The attempt to include any possible vandal in the curse, anyone with the temerity to disturb a king's inscription, is also found in other inscriptions: the Aḥiram tomb inscription includes kings, governors, and army commanders in its warning. ואל.מלכ.במלכמ.וסכנ.בס[כ]נמ. ותמא.מחנת,³² and the Nimrud Ivory bearing a Hebrew inscription includes great kings and private citizens ממלכ.גדל.ועד.איש.³³ Other inscriptions are simpler though still inclusive: Siloam tomb ארור האדם אשר יפתח את זאת (lines 2–3)³⁴; Tell Fekherye text וזי.ילד.שמי (line 11) and מנ.ילד.שמי (line 16)³⁵; Zakir 1.B.16

Chapter 5: Blessing and Curse

[וכל.]מנ[36]; Kilamuwa 1:15 ומי[37]; Sefire I.C.17–18 ומנ...ויאמר[38]; and Nerab 1 and 2 מנ את.[39]

The curse is actually against anyone who would attack the city, thus destroying the gate (an enemy), or someone who would efface Azitiwada's name from this gate (a jealous successor). The same verb (מח in lines 13 and 18) is used both for the forbidden activity against the gate and the punishment asked of the gods against the enemy. If the person "effaces" (ימח) the name of Azitiwada from the memorials describing his accomplishments, the vandal himself and his own land shall be "effaced" (ומח) by the gods.[40] This verb is found in other inscriptions with the same intent: Aḥiram line 2; Zakir B.15–16; Sefire I.A.42; and En Gedi as well as Karatepe C.iv.15. The biblical root מחה can be used with the same force, to "wipe out, blot out, or obliterate" a name or the remembrance of a person (e.g., Exod. 32:32–33, Moses from the book of Yahweh; Deut. 29:19–20[20–21], in the context of a curse) or a people (Exod. 17:14; Deut. 9:14; 2 Kgs. 14:27; Ps. 9:6) from memory.[41] The kings who wrote these inscriptions were objecting to becoming anonymous. Genesis 12:2 shows the Hebrew concern that the name be carried on to future generations. Second Samuel 18:18 is especially valuable for comparison: Absalom set up a monument to himself because he had no son. The concern with being remembered is closely tied with the blessing of offspring, one of the most basic blessings in the ancient world.

1b) אפס
2) שם אזתוד יכן לעלם כם שם
3) שמש וירח[42]

Only may the name of Azitiwada last for ever like the name of the sun and the moon![43]

The final lines of A (iv.1–3) contain a request for enduring honor for the name of Azitiwada which will last as long as the sun and the moon. The bilingual inscription of Samsu-Iluna contains the wish that the gods give him life and health that will last as long as Sin and Šamaš *šūlmam ù balaṭam šakima* *ᵈEN.ZU ù ᵈUTU daríum a[n]a qî[š]tim liqíšušum ana šeriktim lišrukušum.*[44] Psalm 72:5 contains a prayer that the king may live as long as the sun and moon endure,[45] and Ps. 89:37 asks that the reign of the Davidic house, like the moon, be "established forever." A late Phoenician inscription from Pyrgi (Italy) uses a similar phrase to call for the endurance of a deity's statue ושנת למאש אלם בבתי שנת כם הככבם אל "So (may) the years granted to the statue of the deity in her temple (be) years like the stars above."[46]

Karatepe B. Karatepe examplar B is a gate inscription like A. It so closely parallels A that the various treatments of the Karatepe texts do not even give the text in full but merely note some minor differences. In the points of interest for this study it so nearly resembles A that the text will not be given but will be understood as another example of the types of blessing and curse found in

A (except that it does not include the last blessing on Azitiwada, the part which would parallel A.iv).⁴⁷

Karatepe C

(16b) וברך
(17) בעל כרנתריש אית אזתוד בח
(18) ימ ובשלמ ובעז אדר על כל מלכ
(19) לתתי בעל כרנתריש לאזתוד
(20) ארכ יממ ורב שנת ורשאת נעמת
(IV 1) ועז אדר על כל מלכ
lines concerning sacrifice
(6b) וכנ [ה]קרת ז בעל[ת]
(7) שבע ותרש ועמ ז א[ש]
(8) ישב בנ יכנ בעל [א]לפמ ו
(9) בעל צאנ ובעל שב[ע ו]תר[ש]
(10) וברבמ ילד וברבמ [י]אדר
(11) ובר[ב]מ יעבד לאזתוד ול
(12) בת מ[ו]פ[ש] בעבר בעל ובעבר אלמ
(13) וא[מ] מלכ במלכמ ורזנ ברזנמ
(14) [מ] [א]מ אדמ אש אדמ שמ אש יא
(15) מ[ר] למחת שמ אזתוד בסמל
(16) א[ל]מ ז ושת שמ אמ אפ יחמד
(17) א[י]ת הקרת ז ויאמר אפעל
(18) סמל זר ושת שמי עלי ואי
(19) ת סמל האלמ אש פעל אזתוד
(20) בעל כרנתריש אש בבא מלח ב
(21) ננחל אמ אט-ל....אל]...
(V 1) [ ] רנב--ר [ ]
(2–4) [ ]
(5) מ[ ]ר [אפס שמ]
(6) אזתוד[י] יכנ לעלמ כמ שמ
(7) שמש וירח ⁴⁸

So may Baal KRNTRYŠ bless Azitiwada with life and with health, and with powerful strength above every king! May Baal KRNTRYŠ give to Azitiwada length of days, and many years, and a pleasant old age, and powerful strength above every king...And may this city be owner of plenty (of grain) and new wine; and may this people who dwell in her be owners of oxen, and owners of sheep, and owners of plenty (of grain) [and] new wine; and may they bear many (children), and as they grow many become powerful, and as they grow many serve Azitiwada and the house of Mopsos, by the grace of Baal and by the grace of the gods! (iv.6–12) And may this city be owner of plenty (of grain) and new wine; and may this people who dwell in her be owners of oxen, and owners of sheep, and owners of plenty (of grain) [and] new wine; and may they bear many (children), and as they grow many grow powerful, and as they grow many serve Azitiwada and the house of Mopsos, by the grace of Baal and by the grace of the gods! Now, if a king among kings, or a princes among princes, or any man who is a man of renown, gives orders for the name of Azitiwada to be effaced from the statue of this god, and puts up his own name, or more than that, if he covets this city and says, I will make another statue and put my own name on it, and the statue of the god which Azitiwada made, (that of) Baal KRNTRYŠ, which is at the king's entrance in(to) [Only may the name] of Azitiwada last for ever like the name of the sun and the moon.[49]

This inscription, on a statue of Ba'al, (C.iii.16–C.iv.1 and C.iv.6–v.7) is almost exactly the same as the "A" section.[50] The blessings asked for (חים, שלם, and עז [A.iii.3–4]), are prefixed by a ב (C.iv.17–18) marking each of them as direct objects of the verb. Another noteworthy difference is that the "C" version removes the action of effacing the inscription one step from the king, prince, or man of renown: should such a one "give orders for the name of Azitiwada to be effaced" יאמ[ר] למחת שם אזתוד (lines 14–15) or "if he says, I will make another statue and put my name on it" ויאמר אפעל סמל זר ושת שמי עלי (lines 17–18) that one will be just as liable to the curse as the person in "A" doing those things. Apparently, the person who wishes the inscription effaced thinks to avoid the curse by not doing the actual damage himself. The inscriber disallows this loophole, cf. Sefire II.C.1–10 where the provision is written more fully.[51] The phrase וכל קרת אלן which appears in A.iii.5 is not found in C.iv.19. Honeyman thinks this is a tactful omission since this copy of the inscription is upon the statue which he takes to represent Ba'al himself.[52] As the relevant part at the end of this copy of the inscription is missing, one may not know whether Ba'al alone or all the above mentioned gods were called upon to avenge damage done to the inscription.

Summary

The mid-eighth century Karatepe inscriptions combine elements found in both earlier Byblian Phoenician inscriptions requesting the blessing of long life (Yeḥimilk, etc.) and the later Hebrew and Edomite more general requests for blessing which use ברכ (Kuntillet 'Ajrud, Arad, Lachish, Ḥorvat 'Uza, etc.). To date, the combination of ברכ with שלמ and substantive curse have only been found in this inscription and with this DN. ברך in earlier inscriptions (before the late sixth to fifth centuries) appears most often in Phoenician, Hebrew, and the "Edomite" Ḥorvat ʿUza inscriptions, but not in Aramaic. The target Aramaic inscriptions follow the earlier pattern of requesting a specific blessing, rather than using a general ברך formula.

Substantive Blessing and Curse

Just as there are instances in the inscriptions where someone invokes a blessing upon another without using a form of ברך, so there are numerous cases where blessing and curse intentions are paired without the occurrence of ברך or ארר. Deuteronomy 30:15–20 supplies an example of such an intention (the antecedent of these verses is the extensive collection of covenant blessings and curses, chapters 27–30). In verse 15, God "sets before" the people life (חיים) and prosperity (טוב) versus death (מות) and calamity (רע). In verses 16–18, the rewards of obedience are spelled out: long life and descendants. Verse 19 ties together the parts of verse 15 and identifies "life" with "blessing," "death" with "curse."

Substantive blessings and curses are found in the following texts and are treated below: Tell Fekherye, Amman Citadel, Zakir, Sefire I and II, Hadad, and Nerab 1. The blessings and curses in all of these inscriptions are invocations of specific deities.

Tell Fekherye 7–12. The statue bearing this inscription was found just SE of Tell Fekherye which is in turn just a few kilometers NE of Tell Halaf.[53] The inscription is best dated to the mid-ninth century B.C.[54] The Neo-Assyrian version of the text consists of 38 columns while the Aramaic version is contained in 23 lines. Each version is actually comprised of two inscriptions: the first Neo-Assyrian inscription is contained in lines 1–18, the second in lines 19–38; the first Aramaic inscription ends with line 12, the second consists of lines 13–23.[55] The statue is a representation of the King Hadadyisʿi (הדיסעי) portrayed in an attitude of prayer and was intended to stand before his god Hadad of Sikan.

(7) לחיי.נבשה.ולמארכ.יומוה
(8) ולכבר.שנוה.ולשלמ.ביתה.ולשלמ.זרעה.ולשלמ
(9) אנשוה.ולמלד.מרק.מנה.ולמשמע.תצלותה.ול
(10) מלקח.אמרת.פמה.כננ.ויהב.לה.ומנ.אחר.כנ
(11) יבל.לכננה.חדס.ושמימ.לשמ.בה.וזי.ילד.שמי.מנה
(12) וישימ.שמה.הדד.גבר.להוי.קבלה

Hadadyisʿi, son of Sas-nuri, king of Gozan, dedicated the statue to his lord Hadad of Sikan (from line 1),

Chapter 5: Blessing and Curse

so that his soul may live, so that his life may be long, so that his years may be increased, so that his household may enjoy well-being, so that his descendants may enjoy well-being, so that his men may enjoy well-being, and so that illness may be removed from him, and so that his prayers may be heard, and so that his words may be accepted, this image he set up and offered (it) to him. Whoever, in the future, removes it to erect a new one, he should put my name on it. And whoever erases my name from it and puts his name instead, may Hadad the Hero be his adversary.[56]

After praising his god, Hadad, and giving his own pedigree, Hadadyis'i begins giving his reasons for constructing the statue and its inscription.[57] In lines 7b–8a, he states that he has done this "so that his soul may live and his life be long and so that his years may be increased." The phrase חי נפש occurs in 1 Sam. 20:3 where it is part of an oath paired with חי־יהוה. Here, the form is a Pa'el infinitive לחיי (cf. Ezek. 13:19 ולחיות נפשות, also Ezek. 33:12).[58] The rest of this phrase, including the parallel usage of days and years (the forms are masculine as is usual in BH, though the feminine forms are found in the inscriptions), is familiar from other inscriptions (e.g. Yeḥimilk, Eliba'al, Shipiṭba'al, Abiba'al, Tell Siran Bottle, Karatepe, Yeḥaumilk) and from biblical texts Gen. 5:5; 25:7; Deut. 32:7; etc. The formula למארכ.יומה.ולכבר.שנוה does vary in that למארכ is an infinitive[59] rather than an adjective as in the Karatepe inscription (A.iii.5–6); the verb occurs also in Nerab 2. And the word ולכבר, also an infinitive[60] or perhaps a noun "abundance, bounty,"[61] is otherwise unknown to the formula (cf. the Hadad inscription line 11 where וכברו occurs and Gibson translates it "greatness"[62]).

Lines 8b–9a contain the next part of the requested blessing. The word ולשלמ is repeated in a three-fold request for "peace" or "prosperity." The *editio princeps* treats this form as a noun "salut"[63] but Batto understands it as another Pa'el infinitive.[64] Sasson suggests that ביתה refers to the royal family, זרעה refers more specifically to his heir and a continuation of the dynasty (cf. 1 Sam. 25:6; 1 Kgs. 2:33), and אנשוה refers to those who bear arms in defense of the country.[65] Perhaps ביתה is general and the next two words are the specific objects of the blessing: the "house" is divided into the more important family or heir who will continue his dynasty and the "men" are the loyal retainers who will secure that succession. If this is the case, a better translation of שלמ is surely "prosperity" with the connotation of "success."

Line 9 contains a prayer that sickness be kept from the king. The *editio princeps* treats the verb ולמלד as a Pa'el infinitive with a prefixed מ from the root לוד[66] defined by Greenfield and Shaffer as "bend or deflect."[67] Both of these works point to Sefire for a parallel; though Sefire I.C.18 and II.C.1–2, 6 have forms of this verb, there it is used of "effacing" the treaty inscription[68] and this is also the meaning below (lines 11 and 16).[69] Some biblical texts treat removal of sickness or prevention of sickness as a sign of Yahweh's blessing or favor: the people are promised freedom from sickness as reward for obedience in Exod. 23:25 (note ברך in the immediate context) and Deut. 7:15, while 2 Kgs. 20:1–5 provides a more specific parallel in the prayer of Hezekiah for healing (cf. 2 Kgs. 1:2 the illness of Ahaziah).

The remainder of line 9 and most of line 10 contain a request that the king's words to his god (prayers) be heard and his words to his people be accepted. This request is repeated in the "B" part of the inscription (line 14). Other inscriptions were dedicated because the donor's deity heard his prayer (e.g. Barhadad, *KAI* 201) and numerous biblical texts speak of Yahweh hearing the prayer of his people (1 Kgs. 8:28; 2 Kgs. 20:5; Ps. 54:4; Neh. 1:6).[70] It seems probable that both in the inscriptions and in the biblical texts the supplicant believes that if his or her prayer is "heard" it is answered and the supplicant knows that the prayer is heard because he or she receives what is asked.[71] The king's prayers must find favor with the gods so that no natural catastrophe strikes the land and his words must find favor with his people so that his dynasty will endure (cf. I Kgs. 12:4–16). It is possible that both phrases refer to the king's prayer rather than the second referring to favor with his subjects, the phrase אמרת.פמה is also used of prayer (אמרי־פי Ps. 19:15[14]), but in light of the more explicit parallel in part "B" of the inscription (line 14), the plea for favor in both realms seems the more probable interpretation. The fifth century Phoenician Yeḥaumilk (*KAI* 10) inscription contains a request for blessing and a substantive curse very similar at points to the Tell Fekherye inscription. Particularly of interest is the request (lines 10–11) that Baʻalath give Yeḥaumilk favor with gods and men, "And may [the lady], Mistress of Byblos, give [to him] favour in the sight of the gods and favour in the sight of the people of this land!"[72] A more detailed study of this blessing element is included in the discussion of Kuntillet ʻAjrud 6.[73]

The remainder of line 10 through most of line 12 is a two-fold curse against anyone who 1) might succeed Hadadyisʻi and replace his statue with one not including Hadadyisʻi's name or 2) would efface Hadadyisʻi's name from his statue and replace it with his own. Both parts of the curse begin with an inclusive and general address ומנ.אחר (line 10) and וזי (line 11) rather than a more specific one (cf. Karatepe A.iii.12–14). The verb יבל is interpreted in different ways: Greenfield and Shaffer believe that it is a form of נבל (BH נפל) and translate "Whoever is after me, if it falls, let him rededicate it...";[74] Sasson interprets the verb as a Paʻel and translates "Whoever...removes it."[75] These two arguments illustrate the two opinions held by scholars regarding this verb: is this a warning pointed at someone who will remove the statue or someone who will repair it?[76] The verb in the second part of the curse, ילד, is a Paʻel imperfect from the root לוד or לדד which occurs with this same meaning in Sefire I.C.18 and II.C.1–2, 6, and in a different context in line 9 above. Kaufman's translation is both

Chapter 5: Blessing and Curse

consistent and adequate in each case: the king prays that sickness be "removed" from him (line 9) and threatens anyone who "removes" his name from his inscription.[77] The concern that a successor would renovate a building without giving due credit to the original builder is found frequently in the Neo-Assyrian inscriptions and was dealt with by curses against the guilty or blessings for the obedient.[78]

Hadad "the hero" (גבר)[79] is called to defend the statue by becoming the legal "adversary" of the violator. Yahweh is described as גִּבֹּר (Deut. 10:17; Ps. 24:8; Zeph. 3:17), Nergal is called "hero of the gods" in a vassal treaty of Esarhaddon,[80] and Marduk is called hero in the *Enuma Eliš*;[81] in each case military might is explicit or implied. Kaufman identifies the word קבלה as a Pe'al Participle with a pronominal suffix, the root meaning of the verb being "to bring suit, accuse."[82] Postgate notes the use elsewhere of declaring or asking a god to be legal antagonist to an oath breaker *lu bel denišú*.[83] Examples are also found in several fifth century Aramaic texts including *AP* 10:12–13, a contract for a loan; *BMAP* 1:4–6, a purchase agreement; and Hermopolis 2:10, a dispute over property.[84] While the verb קבל only takes the meaning "accuse" in Late Hebrew, the verb ריב carries the same connotations and is used in much the same way in the Bible. Several times, Yahweh stands to accuse his people (e.g., Isa. 3:13; 27:8; 57:16; Ps. 103:9; Amos 7:4 [judgment by fire]); he may also be called upon, as in the Tell Fekherye inscription, to contend in one party's behalf against another (e.g., 1 Sam. 24:15 [of David against Saul]; Micah 7:9; Jer. 50:34 [of Israel against Babylon]; and 51:36 ["Behold I will plead your cause and take vengeance for you. I will dry up her sea and make her fountain dry"]).

Tell Fekherye 12–23.

(12 צלמ.הדיסעי
(13 מלכ.גוזנ.וזי.סכנ.וזי.אזרנ.לארמ.ורדת.כרסאה
(14 ולמארכ.חיוה.ולמענ.אמרת.פמה.אל.אלהנ.ואל אנשנ
(15 תיטב.דמותא.זאת.עבד.אל.זי.קדמ.הותר.קדמ הדד
(16a יסב.סכנ.מרא.חבור.צלמה.שמ.

The statue of Hadyis'i, king of Gozan and of Sikan and of
'Azran, to establish the foundation of his throne and so that his
life may be long and so that the words of his mouth may be
pleasing to gods and to men, this image he caused to be better
than before. In the presence of Hadad who dwells in Sikan,
Lord of the Ḥabur, he has set up his statue.[85]

The "B" part of the inscription begins with Hadadyis'i's three-fold reason for improving the statue (lines 13–15). The first reason is obscured by the difficult word לארמורדת. While most scholars agree that the king is asking for a firm basis for his rule ("throne" כרסאה), they are not agreed on the division of

גמ[.]ישבת. the words, the exact reading, or the exact translation (cf. Hadad 8–9 על.משב.אבי.ונתן [ה]ד̇ד̇[.]בי[ד]י̇.חטר.חל[בבה] "Moreover, I sat on my father's throne, and Hadad gave into my hands the sceptre of authority"[86]).[87]

Second, the reason that his life will be lengthened ולמארכ.חיוה is similar to part "A" ולמארכ.יומה though in line 14 the request is more compact. Greenfield and Shaffer think that the writer lacked an Aramaic equivalent for Akkadian *pališu* "his reign, turn of office" and so used חיוה,[88] but surely חיוה would have the same meaning in a culture where the king usually ruled as long as he lived.

The third reason (lines 14–15) is also found in part "A." In lines 9–10, the king's request that he find favor with gods and men is not as clear ("so that his prayers may be heard, and so that his words may be accepted") but here it is more explicit. The verb תיטב, a Pe'al Imperfect from יטב, also occurs in Imperial Aramaic where it is a sort of technical term used to show that the party speaking was satisfied with aforementioned business terms (*AP* 2:9; 14:5; 20:9; 43:7; *BMAP* 3:6; et al.; and Aḥiqar 66[89]). Several biblical passages express the idea that an individual has the approbation of God and humans, including 1 Sam. 2:26 which also uses a form of טוב (cf. Judg. 9:9; Gen. 39:21 where favor with the keeper of the prison indicates Joseph's favor with Yahweh; Exod. 3:21; 11:3; 12:36; Prov. 3:4). For inscriptional parallels compare Yeḥaumilk lines 10–11 and Panammu line 23.[90]

The curse portion of part "B" begins at line 16. The address is again inclusive and short: מנ.ילד "whoever will remove" (cf. line 10 ומנ.אחר). The verb ילד is a Pe'al Imperfect from the root לדד or perhaps לוד (see notes on lines 9 and 11 of part "A"). The curse in this case seeks to protect cult objects מאניא זי.בת.הדד from which the monarch's name is not to be removed. The term מאניא usually refers to a vessel of metal or wood. For other uses of this word see Nerab 2:6 (vessels placed in a tomb), *AP* 20:5, and *BMAP* 7:13. For a parallel to the phrase מאניא זי.בת.הדד see Dan. 5:23 ולמאניא די־ביתה and Ezra 5:14 מאניא די־בית־אלהא.

(16b) מנ.ילד.שמי.מנ.מאניא
(17) זי.בת.הדד.מראי.מראי.הדד.לחמה.ומוה.אל.ילקח.מנ
(18) ידה.סול.מראתי.לחמה.ומוה.אל.תלקח.מנ.ידה.ול
(19) זרע.ואל.יחצד.ואלף.שערינ.לזרע ופריס.לאחז.מנה
(20) ומאה.סאונ.להינקנ.אמר.ואל.ירוה.ומאה.סור.להינקנ
(21) עגל.ואל.ירוי.ומאה.נשונ.להינקנ.אלימ.ואל.ירוי
(22) ומאה.נשונ.לפאנ.בתנור.לחמ.ואל.ימלאנה.ומנ.קלקלתא.ללקטו.אנשוה.
שערנ.לאכלו
(23) ומותנ.שבט.זי.נירגל.אל.ינתזר מנ.מתה

Whoever removes my name from the furnishings of the temple

Chapter 5: Blessing and Curse 171

of Hadad my lord, may my lord Hadad not accept his bread and his water from his hand. May my lady Sawl not accept his bread and his water from his hand. And may he sow, but not harvest. May he sow 1000 measures of barley but take only a fraction thereof. Though 100 ewes suckle one lamb, let it not be satisfied. Though 100 cows suckle one calf, let it not be satisfied. Though 100 women suckle one infant, let it not be satisfied. May 100 women bake bread in one oven but not be able to fill it. Would that his men pick up barley from the rubbish heap to eat. May death, the rod of Nergal, never cease from his land.[91]

Lines 16–18 contain the first curse: Hadadyis'i calls for Hadad and his consort Sawl not to accept sacrifices from the violator. In contrast, the willingness of the gods to accept Panammu's sacrifices was understood as a mark of their favor לִ[נ]תֹ.מת.אלהי.מנ.אשאל.ומה.ידי.מנ.יקחו.ומת.לאלהי.אהב "...]would I offer to the gods, and they used to accept (them) from my hand; and what I asked from the gods, they used always to give me" (Hadad 12–13),[92] Compare also 1 Kgs. 13:8–9, 16–17, 20–22 where the prophet from Judah is directed by Yahweh to deliver an oracle in the northern kingdom but not to accept food there (v. 17), "For it was said to me by the word of the Lord, 'You shall neither eat bread nor drink water there, nor return by the way that you came'." This may simply mean that the prophet was not to tarry in Israel (cf. 2 Kgs. 4:29, Elisha instructed Gehazi not to greet anyone along the way in his haste to heal the Shunammite's son) or, more to the point, that the prophet, as representative of Yahweh, was to refuse food in the north and thus show Yahweh's rejection of Israel. "Bread and water" are also the offerings in the Tell Fekherye text. Yahweh's displeasure with his people is marked by his refusal to "smell" their sacrifices (Lev. 26:31 contra Gen. 8:21) and on the human level acceptance of a person's gift was understood as acceptance of the person (Gen. 33:11). Kaufman points out that "bread and water of the gods" shows strong Mesopotamian connections though it is not found as part of a curse elsewhere.[93] Greenfield and Shaffer note that the Mesopotamian use of bread and water in sacrifice is limited to the meal offering and libation of the *kispu* ceremony performed at the new moon. Part of an Old Babylonian inscription from Takililišu concerning that festival reads *akalšu ellam mêšu naḫdutim uškaram u šapattam aštakkanšum* "at new moon and full moon I regularly place before him his pure bread and precious water."[94] In the Bible, bread and water represented basic sustenance needs (Exod. 34:28; Deut. 23:4–5; 1 Kgs. 22:27; Isa. 3:1; Amos 8:11; Neh. 13:2).

Lines 19–22. The curse that someone sow seed but not harvest is well-known from the Bible, Deut. 28:38 "You shall carry much seed into the field, and shall gather little in; for the locust shall consume it" (this is in a larger series of curses, vs. 30–45; also Isa. 5:10; Haggai 1:6; Micah 6:15). In contrast, Gen. 26:12 shows that abundant harvest is the mark of God's blessing, "And Isaac sowed in that land, and reaped in the same year a hundredfold. The Lord blessed him."⁹⁵ This curse is usually tied to other curses of inadequacy: 1) a lack of agricultural products, though the plants are treated and cared for as usual, the people are denied the benefits of the produce because Israel has been unfaithful to Yahweh (Haggai 1:6) and in some other references strangers will commandeer the crops (Micah 6:14–15); 2) a shortage of men, Isa. 4:1 "And seven women will take hold of one man in that day, saying, 'We will eat our own bread and wear our own clothes, only let us be called by your name; take away our reproach'";⁹⁶ 3) women becoming prostitutes "Your wife will become a prostitute in the city" (Amos 7:17; Isa. 23:15–18). Likewise, the 1000:1 ratio is familiar from the Bible: in Deut. 32:30, the Israelites are warned that their disobedience will be punished by the humiliation of cowardice: a thousand of their soldiers chased by a single enemy warrior, ten thousand by two of the enemy (cf. Isa. 30:17).⁹⁷

The curses of inadequacy are continued in lines 20–21, first with a series of curses in which a single lamb, a single calf, and a single human infant are each suckled unsuccessfully by one hundred nurses. Curses from the Sefire treaty (I.A.21–23; II.A.1–2) are very similar, though the number of nursing mothers is seven rather than one hundred:

> [and should seven nurses] anoint [*their breasts* and] nurse a young boy, may he not have his fill; and should seven mares suckle a colt, may it not be sa[ted; and should seven] cows give suck to a calf, may it not have its fill; and should seven ewes suckle a lamb, [may it not be sa]ted ⁹⁸

A very similar situation is described as a curse in the Annals of Assurbanipal,

> Every curse, written down in the oath which they took, was instantly visited (*lit.*, fated) upon them by Assur, Sin, Shamash, Adad, Bêl, Nabû, Ishtar of Nineveh, the queen of Kidmuri, Ishtar of Arbela, Urta, Nergal (and) Nusku. The young camels, asses, cattle and sheep, sucked at seven udders (*lit.*, suckling mothers) and could not satisfy their bellies with the milk. The people of Arabia asked questions, the one of the other, saying: "Why is it that such evil has befallen Arabia?" (And answered), saying: "Because we did not keep the solemn (*lit.*, great) oaths sworn to Assur; (because) we have sinned against the kindness (shown us by) Assurbanipal, the king beloved of Enlil's heart.⁹⁹

In the Bible also the failure of the available resources to satisfy the appetite is a sign of a cursed state (Lev. 26:26; Amos 4:8; Micah 6:14) and dry breasts (along with the miscarrying womb) are a mark of the Lord's judgment of his people (contrast the blessedness of Isa. 66:11 למען תינקו ושבעתם משד תנחמיה "that you may suck and be satisfied with her [Jerusalem's] consoling breasts").

Line 22a concludes this type of inadequacy curse with the curse that the combined baking demands of one hundred women would fail to fill a single oven. Leviticus 26:26 contains a very close parallel, "When I break your staff of bread,

Chapter 5: Blessing and Curse

ten women shall bake your bread in one oven, and shall deliver your bread again by weight; and you shall eat, and not be satisfied" בשברי לכם מטה־לחם ואפו עשר נשים לחמכם בתנור אחד והשיבו לחמכם במשקל ואכלתם ולא תשבעו (also compare Isa. 4:1). By means of minor reconstructions of Sefire I.A.24, Kaufman is able to provide a convincing reading that parallels this portion of the Tell Fekherye text and that of Leviticus 26:26, "And may his seven daughters bake bread in an oven(?) but not fill (it)."[100] Also compare the Esarhaddon vassal treaties, "may [there be no mill or oven] in your houses"[101] and Deut. 28:17, "Cursed shall be your basket and your kneading trough."

The last part of line 22 is likewise a futility curse but of a slightly different sort. Food will be so hard to acquire that the "men" (probably soldiers) will scavenge in the dump. The refuse dump (Neo-Assyrian *tubkinnāte*) is also referred to in Neo-Assyrian curses *ina tubkinni lū mayālšunu* "may their sleeping place be on a refuse dump" (Mati'ilu treaty IV:16).[102] The word found in the Aramaic version of the inscription, קלקלתא, also appears as a loanword in Neo-Assyrian, i.e., *issulibbi kiqiliti intathannī* "he has lifted me from the refuse dump" (*ABL* 1285).[103] This sentiment may be reflected in biblical texts, though the wording varies "He raises up the poor from the dust; he lifts the needy from the ash heap" מקים מעפר דל מאשפת ירים אביון (1 Sam. 2:8; Ps. 113:7).[104] Another likely comparison is found in Job 30:19, "He [God] has cast me into the mire, and I have become like dust and ashes" הרני לחמר ואתמשל כעפר ואפר [105] cf. Ps. 102:9(10), "For I eat ashes like bread, and mingle tears with my drink"; Isa. 44:20; Mal. 3:21). In these biblical references the context is to mourning though the association with a state of being accursed is easy to see.

The last of the curses is in line 23. Death is described as an entity in several biblical texts: e.g., Jer. 9:21; Isa. 28:15; Job 28:22; and Hosea 13:14b "O Death, where are your plagues? O Sheol, where is your destruction?" אהי דבריך מות אהי קטבך שאול.[106] Mot is also seen as a divine being bringing plague and death in such passages as 2 Sam. 24:15–25 and 1 Chr. 21:17. Sasson notes the irony that while death is the rod of punishment that Assyria's god Nergal wields, in Isa. 10:5, Assyria is termed the rod of Yahweh to punish Israel הוי אשור שבט אפי ומטה־הוא בידם זעמי.[107] Greenfield and Shaffer suggest that the phrase שבט. זי.נירגל is a word play on שבט "plague, disease," Akkadian *šibṭu*, and שבט "staff" alluding to the lion-headed staff which is the symbol of Nergal.[108] Though an exact parallel is not known, Nergal is similarly invoked against violators of contracts: 1) in one of Esarhaddon's vassal treaties (lines 455–56) "May Nergal, hero of the gods, extinguish your life with his merciless sword, and send slaughter and pes[til]ence among you" *ᵈU.GUR qarrad DINGIR ina GÍR-šú la gameli nap-šatkunu liballi šágáštú mut[ā]nu ina ŠÀ bikunu liškun;*[109] 2) in a Neo-Babylonian contract, "May Nergal, strongest of the gods...not spare him plague and defeat" *Nergal dandannu ilāni...ina šibṭu u taḫti lā igammil napšassu.*[110]

Amman Citadel.[111] This mid-ninth century inscription was discovered in 1961, during excavations of the SW corner of Jebel el-Qala'ah, the Citadel mound of Amman. The eight line text, on a roughly rectangular, white, limestone slab, was uncovered in an Iron Age stratum.[112]

1) [מ]לכמ.בנה.לך.מבאת.סבבת]
2) [.ככל.מ ס בבעלך.מתימתן]
3) [כ ח ד .א כֹהֹד-וֹ כל.מערבֹ]
4) [וֹב כל.ס.--תיל נד צדק]
5) [ל.תדל תבדלת.בטנ כרֹה]
6) [--ה.תשתע.בבנ.אלמ-]
7) [וֹשׁ.---.וֹ--------]
8) [-[--למ.לכ.וֹ------]

Mil[k]om has built for you entrances round about [...]according to all that surrounds you from *Tymtn* [to...] what had been destroyed I...throughout the west [...]and on every threshold...of the legitimate wall [...]door, at the inner door he dug [...]...fear was among the men of the portico [...]...[...]...for you...[[113]

Horn treats this as a building inscription. He understands the DN [M]ilkom to be a theophoric element in the patron's name. It is more likely this than a DN, he says, since this is the inscription of the builder claiming credit.[114] He suggests that line 8 contains the word שלמ, but is not certain.[115] Cross also treats this as a building inscription, a monumental inscription, of which this is only a part and which is addressed to the deity.[116] His reconstruction of the last line does supply a few more letters אלמ.לך.וש] but the translation remains enigmatic "...Porch to thee, and..."[117] Van Selms interprets the text as a building inscription likewise with no indication that the inscription or the building it refers to has a religious purpose. The letters -לכמ which begin line 1 are the end of a PN and the inscription glorifies him as the builder. Lines 6–8, should then be translated, "[and he mad]e the plaque outside the porch...and the p[laque]...and all...this [p]orch and..."[118]

Albright says that the text's use of the second person (לך) indicates that it contains the direct commands of the god Milkom to an unknown king of Ammon (cf. the Mesha and Zakir stelas). The god's commands regard some sort of military strategy that the king is to take and in the last lines (6–8) the king is adjured "] thou shalt trust in the Son of the Gods[...the Son of the] Gods doth [command] thee, [go and...]"[119]

In his first article on this inscription, Shea partially agrees with Albright.[120] Shea sees this inscription not as describing a building made for Milkom by the Ammonite king, but as a description of the natural defenses of Rabbath-Ammon as the gift of the god Milkom to his people.[121] Further, Milkom gives orders to the king, who is under pressure from external enemies, regarding enhancement

Chapter 5: Blessing and Curse

of the natural defenses of the Citadel and how to conduct a counter assault against the attackers. Finally, Milkom apparently promises to fight against the enemy himself (line 2).[122]

Puech and Rofé think that the inscription is virtually complete as is and understand the provenance of the stone differently. The first line "[Mi]lkom a bâti pour toi des entrés périphériques["[123] has religious implications suggesting "que l'inscription provient d'un sanctuaire plutôt que d'un palais royal."[124] The principle element of the text is a cult ordinance apparently warning the reader to show the reverence due to the sacred precincts where the inscription was posted.[125] Their reading of line 6 is similar to Albright's but their reading of the last line is fuller than that of previous scholars and their translation is the first to translate peace ("paix") in line 8 "et paix à toi et pai[x" מ]שלמ.לכ.ושלמ.[.[126] Dion agrees with this reading of line 8.[127]

Fulco agrees with Cross that each line is only part of a much longer line and he concurs that the text is a building inscription, but believes that it probably pertains to the building of a temple.[128] His translation is the first to understand the inscription in the sense of substantive blessing and curse: those who threaten the worshipper shall die but the worshipper shall have peace (שלמ).[129]

> [Mi]lkom, he has built for you the precinct entrances [(2)] that all who threaten you shall surely die [(3)] I shall surely obliterate, and all who enter [(4)] and amidst all its columns the just will lodge [(5)]L. there will hang from its doors an ornament KBH [(6)]H. will be offered within its portico [(7) ??? (8)] peace to you and pe[ace...[130]

Sasson sees the inscription as "an oracle of divine protection and assurance" which "could have emerged in time of political unrest in foreign relations or in time of imminent military threat."[131] He thinks the discovery of the tablet in the citadel is significant: it would have been affixed there as a talisman protecting the citadel and city walls from attack (cf. Ps. 48 which speaks of the inviolability of Jerusalem's walls because of God's protection).[132] He divides the text into the following overall sections: lines 1–2a are a divine command to the king of Ammon to build and strengthen the borders, lines 2b–5 are a curse on Ammon's enemies and Milkom's pledge to destroy them, line 6 is a call to stand in awe of the son of the gods (Milcom), lines 7–8 contain a promise for the security and permanency of the kingdom (cf. Ps. 110, especially vv. 5–6).[133] Sasson believes that this text contains a blessing and a curse and that it was talismanic in nature.[134] Though the works of Puech and Rofé and Fulco hinted at this interpretation, Sasson is the first to explicitly to say this and his translation is much clearer:

> The words of Mi[l]kom: Build for thyself points of entry surrounding [thy land...] on all thy surrounding fronts. They shall surely die [all thine enemies; Fear not, for] I will extirpate every one that incites ag[ainst thy land] and on all fronts justice will prevail [...fire devour]ing the innermost door [that will

not] be quenched.[...] thou shalt stand in awe of the son of the gods—[...]and d[well secure...p]eace to thee and p[eace to thy seed].[135]

According to Sasson, Milkom, by oracle, is telling the king to build up the capital's defenses in light of some impending threat. The roughly contemporaneous Zakir inscription contains a similar promise to the king from the god Ba'alshamayn transmitted by means of oracles (A:11–14).[136] Sasson suggests that the occurrence of צדק at the end of line 4 is being used here as an attribute of Milkom as it is used of Yahweh in the Bible. Line 5 shows the thoroughness of Milkom's vengeance: pursuing the enemy to the innermost hiding place (cf. Isa. 2:19, 26:20). Line 5 speaks of the complete destruction of the enemy by divine fire (cf. Isa. 30:27; Jer. 17:27; 21:12 where Yahweh's wrath blazes out and especially Amos 1:13–14 and Jer. 49:1–5 where Rabbath-Ammon itself is the target of Yahweh's anger).[137] For this protection, Milkom requires awe, reverence, and worship (line 6). In return, security is promised: Sasson restores [וש[כנת בטח "] and d[well secure]" (cf. Deut. 33:12 and Jer. 23:6). Also, peace is promised in line 8, what Sasson terms "the concluding part of a blessing" (cf. 1 Kgs. 2:33). He says that with these concluding promises, "The contractual aspect of the oracle becomes clear."[138]

As a last section to his article, Sasson suggests that the very material and size of the Amman Citadel tablet may reinforce his interpretation of it as a sort of blessing tablet.[139] As was noted above in the discussion of the Gezer Calendar, turn of the century excavations at Tell Sandaḥannah uncovered 51 limestone tablets, mostly Greek and dating from the third to second centuries B.C., that contained curses intended for a supplicated deity's perusal and action.[140] Sasson suggests that the Gezer and Amman Citadel inscriptions likewise possessed talismanic value and were placed before the gods in order to achieve the desired effect: the right progression of the seasons and defense from the enemy.

In his second article on this text, Shea has modified his interpretations to understand the defensive precautions as orders of Milkom to the king (and thus the text is an oracle) rather than mere enhancements of the natural defenses. Since this order came from the national god, it would have served as a "strong theological stimulus" to the workers building the defenses and the soldiers manning them.[141] He calls Sasson's suggestion that the text served as a talisman and was placed conspicuously for effect "interesting" and suggests himself that several copies might have been made and placed around the fortifications to ward off the enemy.[142]

The general observation may be made, that those who see the text as a building inscription void of religious context or implication say it is far from complete; while those who see some sort of religious import view it as virtually intact. The reward of the righteous is a peaceful (full? שלם) life (cf. Tell Fekherye lines 8-9 where Hadadyis'i asks שלם for himself and for everyone

Chapter 5: Blessing and Curse

attached to him). Arad 16:2 and 21:2 both contain the wish of the sender for שלמֹ on the house of the recipient as well as the recipient himself. The exact meaning of שלמֹ in this context, understood with military overtones, is probably safety from enemies or success in battle.

Zakir B 16–C2. This early to mid-eighth century stele was discovered in 1903, at Afis, 45 kilometers SW of Aleppo. The stele is in four pieces originally stacked vertically: part "A" is on the front of the lower two pieces, part "B" is along the left side of all four pieces, and part "C" is written on the right end of the top piece. The inscription thus began with the "A" section on the front of the stele and concluded with the "B" and "C" sections along the sides. A relief of the god Ilwer (mentioned in line 1), now mostly missing, once stood above part "A."[143]

16) לֹ.מֹנֹ.יהגע.אית.אֹ]שר.[
17) ידיֹ].זכר.מלכ.חמֹ]תֹ.ולֹ[
18) עֹש.מנ.נצבא.זנה.[.]ומֹנֹ.[
19) יֹ]הגע.נצבא.זנה.מנ.[קֹ
20) דֹ]מֹ.אלורֹ.ויהנסנה.[מֹנֹ.
21) אשֹ]רה.או.מנ.ישלחֹ.[בֹ]רֹ[
22) ה.] תֹהֹ[
23) יֹ]קתלי.בעֹ[לשמינ.ואלֹ
24) ור.ו[.] .ֹושמש.ושהרֹ[.]
25) ו[ ].ואלהֹי.שמיֹ[נֹ.
26) ואלה]יֹ.ארק.ובעל[.]עֹ
27) .אית.[אשא.ואית.[בֹ]
28) רה.ואית.כל.[.]שֹ[רֹ]שה.

C
1) [ ] .יהוי.עד.עלֹ[
2) מֹ[.]שמֹ].זכר.ושמֹ[.ביתה.[

Now, whoever effaces the story [of the achievements] of Zakir, king of Hamath and Luʿath, from this stele, and whoever removes this stele from Ilwer's [presence], and drags it away [from its place], or whoever sends [his son   ] let Baalshamayn and Ilwer and [ ] and Shemesh and Sahar and [   ]and the gods of heaven [and the gods] of earth and the Baal of [execute] the man and [his son and his whole] stock. [(But) for ever let] the name of Zakir and the name [of his house endure].[144]

Line 16. Gibson proposes a reading of [.וכ(16)ל.] at the end of line 15–beginning of line 16 based on a comparison with Nabatean and Palestinian Aramaic.[145] A comparison with other introductions of imprecations in contemporary inscriptions, however, shows that this reconstruction is unlikely. Both of the curse sections of the Tell Fekherye inscription, lines 10 and 16, are simply introduced by מן as here.[146] Lidzbarski says that the word יהגע is a Haphel from נוע with the sense of be squandered, wasted, removed ("verschwinden lassen, beseitigen").[147] Gibson says that the proposed derivation of יהגע from the uncontested root נוע is unconvincing as is the proposed root נגע; he suggests that it is a Pe'al rather than Haph'el form.[148]

Torrey says that the inscription describes four crimes against the monument listed in logical order: "1) obliterating the king's name; 2) destroying the monument; 3) removing it from its place; 4) injuring it in any way whatever."[149]

By effacing the record of Zakir's deeds, his enemies would deny Zakir's accomplishments. That this practice was both common and regarded as heinous in the ancient world is witnessed by the numerous inscriptions which contained provisions against it and the severity of the punishment invoked on the offender (cf. Tell Fekherye 10–11, 16; Karatepe A.iii.13–16; and C.iv.15–18 which also regards an inscription accompanying an image).[150]

Line 20. Pognon says that ויהנסנה in line 20, is from the root הנס "enlever, ôter" an imperfect third masculine singular form with a נ inserted before the third masculine singular suffix. He compares the occurrences of this verb in the Nerab inscriptions (1:6 תהנס; 2:8 לתהנס; 2:9 ותהנסני) where it is used of someone dragging away a person's grave.[151] The intention of the warning is to prevent someone from defacing the records of Zakir's achievements, "quiconque enlèvera l'ecrit (*ou* les haut faits) de Zakir (*littéralment*: la trace des mains de Zakir)."[152] And, though he does not speculate much on the meaning of the broken ending of the inscription, he does say, "Il semble qu'après avoir parlé de ceux qui détruiraient son inscription ou enlèveraient la stèle, Zakir parlait de ceux qui la feraient détruire ou enlever par d'autres."[153]

Lidzbarski says that the verb ויהנסנה is a Haphel from the root נוס "wegnehmen."[154] Koopmans gives the various options for the verb root: Haphel from נוס "fliehen, verjagen," Qal from הנס (no meaning supplied), Haphel from נשׂא=נסא "aufheben," or Haphel from אנס "es wird bedrängen" (elision of א).[155] Torrey says that the verb, whether or not a Haph'el, simply means "remove" and is not connected to later Hebrew and Aramaic אנס.[156] Gibson seems unwilling to decide on a root, but translates the word "drags."[157]

Lines 21–22. Halévy understands the last half of the line and the broken next line as a continuation of Zakir's provision against attacks on the stele. He reads the last half of the line מנ.ישלח.בה and compares it to the biblical phrase שלח ב ("exciter un homme ou une bête contre une homme ou une chose") as in Zech. 8:10 where God threatens to set every man against his neighbor or Deut. 32:24 where God threatens to send the teeth of beasts and the venom of crawling things against the people should they be disobedient. The intention is that the man with malevolent intent will fear the pain threatened to punish his sacrilege, pain

Chapter 5: Blessing and Curse   179

which will come at the hands of men or by the agency of beasts.[158] The phrase may also be understood as the attempt to enlist or command someone else to destroy the inscription and thus avoid bringing down the curse on oneself. The fifth century Phoenician Eshmunazar inscription also contains a provision against this possibility

4) קנמי את כל ממלכת וכל אדם אל יפתח אית משכב ז ו
5) אל יבקש בן מנמ כ אי שמ בן מנמ ואל ישא אית חלת משכבי ואל יעמ
6) סנ במשכב ז עלת משכב שני אפ אמ אדממ ידברנכ אל תשמע בדנמ

> Whoever you are, be you ruler or be you commoner, let none
> such open up this resting-place or seek anything in it, for they
> did not lay anything in it; and let none such lift up the box in
> which I lie or carry me away from this resting-place to another
> resting-place! Even if men speak to you, do not listen to their
> talk.[159]

He suggests that line 22 may be reconstructed as אש מדעתה או אש נכרי. The line would then be translated "un homme de sa connaissance ou un homme étranger" and this would refer to the person giving orders to disturb the grave.[160]

Regarding the interpretation possibility that someone may commission/ order someone else to destroy the inscription, compare Karatepe C.iv. 14–16, יאמ[ר] למחת שם אזתוד בסמל אל[א]ם ז or if anyone "gives orders for the name of Azitiwada to be effaced from the statue of this god…"[161] Sefire II.C.1–10 provides a fine example of someone ordering/enticing another to destroy an inscription in order to avoid a curse. It also contains a curse that carries a threat against the descendants of the violator,

> [and whoever will] give orders to efface [th]ese inscriptions
> from the bethels, where they are [wr]itten and [will] say, "I
> shall destroy the inscriptions and *with impunity* shall I destroy
> KTK and its king," should that (man) be frightened from
> effacing the inscriptions from the bethels and say to someone
> who does not understand, "I shall *reward* (you) *indeed,*" and
> (then) order (him), "Efface these inscriptions from the bethels,"
> may [he] and his son die by oppressive torment.[162]

Also compare Hadad line 34 which deals with the guilt of an individual who has caused someone to kill someone else.[163]

Gevirtz's reconstruction of line 22 varies somewhat from that of other commentators, [ ]תה[ידה.ל]. He notes Ezra 6:12 as a text with a very similar emphasis and similar language ואלהא די שכן שמה תמה ימגר כל-מלך ועמ די ישלח ידה להשניה לחבלה בית-אלהא דך די בירושלם "May the God who has caused his name to dwell there overthrow any king or people that shall put forth a hand to alter this, or to destroy this house of God which is in Jerusalem."[164] This verse is preceded by a warning of a legal penalty designed to punish anyone altering the king's decree; verse 12, then contains the curse providing further protection of the decree and the temple.

    Line 26. Torrey says that the end of the line should be read as ובעלעל (without the word divider proposed by others). This word should be translated "and with a whirlwind" and be understood as a means by which the gods will punish the violator without leaving a trace.[165]

    Line 27. Halévy says that the obscured letter just before אשא should be ל and that the resultant לאשא is a scribal error for לא.יש due to a contraction in popular pronunciation. He reconstructs the line as לחטאה.לא.יש "he will not pardon his sin" (cf. Hosea 1:6) and considers this request of the god to be a further curse on the violator.[166]

    Gevirtz proposes a reconstruction of either יהנסו or יהגעו ("cast down/remove") at the beginning of the line, thus the same should happen to the violator as he has inflicted on the stele.[167]

    Line 28. Halévy looks to the Yeḥawmilk inscription lines 15–16 in order to understand the broken ending of the Zakir inscription which he thinks carries the idea of extermination of the violator's descendants.[168] The Yeḥawmilk inscription, lines 15–16, reads

הרבת בעלת גבל אית האדם הא וזרעו את פן כל אלן ג[בל]

"May the lady, Mistress of Byblos, destroy both that man and his seed in the presence of all the gods of Byblos!"[169] The Eshmunazar inscription carries the same imprecation in lines 8–12

8) אל יכן לם משכב את רפאמ ואל יקבר בקבר ואל יכן לם בן וזרע
9) תחתנם ויסגרנם האלנמ הקדשמ את ממלכ<ת> אדר אש משל בנמ לק
10) צתנמ אית ממלכת אמ אדמ הא אש יפתח עלת משכב ז אמ אש ישא אית
11) חלת ז ואית זרע ממל<כ>ת הא אמ אדממ המת אל יכן לם שרש למט ו
12) פר למעל ותאר בחימ תחת שמש

> May they have no resting-place with the shades, and may they not be buried in a grave, and may they have no son nor seed to succeed them, but may the holy gods deliver them up to a mighty ruler who shall have dominion over them, so that they perish, both (that) ruler or that commoner who opens up (what is) over this resting-place or who lifts up this box, and the seed of that ruler or those commoners! May they have no root below nor fruit above nor renown among the living under the sun![170]

Nerab 1:11 also carries this curse ויהאבדו זרעכ "and may they cause your seed to perish!"[171] Halévy restores line 28 [זרעה.יסח.מנ.ש[ר]ה.ו] and translates it "[arrachera sa descendance de] sa racine et."[172] He says that he is able to detect a ו at the end of the line which indicates the continuation of "une phrase analogue exprimant la perdition éternelle de la race du sacrilège" (cf. Eshmunazar 22).[173]

Chapter 5: Blessing and Curse

Other examples of this curse of complete destruction include the conclusion of the Hittite vassal treaty between Muršiliš and Dupi-Tešub where specific family members as well as the vassal's house and all his property will be forfeited by his unfaithfulness.[174] In 2 Kgs. 9–10, Jehu received instructions from a prophet to wipe out Ahab's family line completely, and he did so with great enthusiasm, not only doing away with Ahab's seventy sons (2 Kgs. 10:1ff) but also his officers, friends, priests, and visitors (10:11ff).[175]

Rosenthal's reconstructive translation of line 28 is, may the aforementioned gods "[deprive him of h]ead and [and] his root and[."[176] Gevirtz opposes Rosenthal's translation based on epigraphic difficulties and because the usual semitic sequence is "root...head" not vice versa.[177] Rosenthal is apparently basing his translation on a comparison with the Eshmunazar inscription, but the order there (lines 8–12) is also "root...fruit." The root-fruit metaphor for descendants is also familiar from biblical passages like 2 Kgs. 19:30//Isa. 37:31; Ezek. 17:9; and Hosea 9:16. In his discussion of the Eshmunazar inscription, Gevirtz does point to Old Akkadian antecedents of the curse against a person and his descendants DNs *išdešu lissuhu u zeraʿšu lilqutu* "May DNs tear out his foundations and pick up his seed!"[178]

The action requested of the gods against the violator, which probably orginally began in line 23, is not preserved in the text. Gibson suggests יקתלו which he translates "execute" (cf. Nerab 1:11 יכטלוכ).[179] The proposal of Gevirtz, however, is to supply the same verb which describes the action taken against the stele[180] (cf. Karatepe A.iii.19–iv.1 which calls for the violator and that which belongs to him to be effaced [ומה] the same action the violator has taken against the inscription [ימה line 13]).

The "C" section of the inscription contains the call for blessing on Zakir and his family and the longevity of his dynasty. Though the preceding line is barely detectable and completely illegible, the reference to the family of the inscription's owner seems to indicate a call for blessing on them in distinction to both the attempt to eradicate the record of the king's achievements and the penalty of eradication called for on the violator of the inscription. Karatepe A.iii.12–iv.3 provides a strong parallel for this interpretation: after calling for the gods to punish someone for destroying the monument bearing the inscription, the section concludes with the call, אפס שם אזתוד יכן לעלמ כמ שמ שמש וירח "Only may the name of Azitiwada last forever like the name of the sun and the moon!"[181]

Sefire I. The three inscriptions designated Sefire I, Sefire II, and Sefire III were all found at the small village of Sefire about 15 miles SE of Aleppo. The first two steles each have three faces bearing writing and are roughly the form of a truncated pyramid. These faces are designated as "A," "B," and "C"; the last stele is simply a flat slab.[182] The first two steles, numbers I and

II, were purchased from villagers in 1930 and 1931[183]; the discovery of the last is more mysterious.[184] While all three of the steles record treaties between a north Syrian ruler (Matî'el) and a powerful Mesopotamian overlord (Bir Ga'yah), the relative dates of the three steles and the relationship of each to the others is far from clear.[185] The inscriptions all date from ca. 750 B.C.[186] The first two steles have resided in the Damascus Museum since 1948. The last was purchased by the Beirut Museum in 1956.

Face A. The summary of the lines preceding the first curse are as follows: lines 1–6 introduce the parties to the treaty; lines 7–13 list the gods (including Marduk, Zarpanit, Nabu, Tashmet, 'Ir, Nusk, Nergal, Laṣ, Shamash, Nur, Sin, Nikkal, NKR, KD'H, Hadad, El, and Elyon) and elements ("Heaven and Earth," "the Abyss and the Springs," and "Day and Night") who are witnesses to the treaty and who are called upon to punish the vassal, Matî'el, should he prove recalcitrant. Lines 14–16 contain the address, repeating the parties involved in the treaty. Lines 21–42 contain the curses which describe the unfaithful vassal's punishment.

(21a) [...שאת ואל תהרי

Line 21a. This line contains a curse of infertility, "[...and should seven rams cover] a ewe, may she not conceive."[187] Dupont-Sommer points out a similar curse in the Mati'ilu treaty (between this same vassal, Matî'el, and the Assyrian king Aššurnirari V) lines 11b–12a *ša ^{I}K[I.MIN-š]u kudānu aššatušu litut [u ā iši]* "*Der...des Ge[nannten] (sei) ein Maultier seine Ehefrau [soll] Nachkommenschaft nicht haben.*"[188] This is also the idea in the treaty between Šuppiluliumas and Kurtiwaza, "even so may you Kurtiwaza with a second wife that you may take, and (you) the Hurri men with your wives, your sons and your country have no seed."[189]

(21b) ושבע [מהי]נקן ימשח[ן שדיהן ו]

(22) יהינקן עלים ואל ישבע ושבע ססיה יהינקן על ואל יש[בע ושבע]

(23) שורה יהינקן עגל ואל ישבע ושבע שאן יהינקן אמר ו[אל יש]

(24a) בע ושבע בכתה יהכן בשט לחם ואל יהרגן

Line 21b–24b. These lines contain a series of curses of inadequacy: the first four concern the inability of nurses (human and animal) to feed their young,[190] the last deals with inadequate food suppply.

[and should seven nurses] anoint [*their breasts* and] nurse a young boy, may he not have his fill; and should seven mares suckle a colt, may it not be sa[ted; and should seven] cows give suckle to a calf, may it not have its fill and should seven ewes suckle a lamb, [may it not be sat]ed; and should seven *hens* go looking for food, may they not *kill* (anything)![191]

This curse is closely paralleled by lines 20–21 of the Tell Fekherye inscription, curses from the ninth campaign of Assurbanipal,[192] and the Mati'ilu treaty.[193] The various unsuccessful nurses vary: in Tell Fekherye they are sheep,

Chapter 5: Blessing and Curse 183

cows, and women; in the Assurbanipal cylinder they are camels, asses, cattle, and sheep; in the Mati'ilu treaty they are oxen, asses, sheep, and horses; and in Sefire I.A.they are women, horses, cows, and sheep (cf. Sefire II.A.1–2 where the order is horses, cows, sheep, and goats; though the text is partially restored). In the Tell Fekherye inscription one hundred nurses are insufficient to satisfy one nursling; in the Assurbanipal inscription the number is seven as it is here. The execration in *Šurpu* (III.97–98) does not mention a number but is the same curse, "The 'oath': to put the breast into the mouth of a small child. The 'oath': [to cause] the drying up of the breast."[194] Dupont-Sommer notes Hosea 9:11, 14, "Ephraim's glory shall fly away like a bird—no birth, no pregnancy, no conception!...Give them, O Lord—what wilt thou give? Give them a miscarrying womb and dry breasts," as a parallel with connections to the previous curse of barrenness as well (cf. Lam. 4:3–4 where this is an image of cruelty).[195]

Line 24b. "and should seven *hens* go looking for food, may they not *kill* (anything)!"[196] Dupont-Sommer says that this curse naturally follows the others which, after referring to humans, refers to animals domesticated by humans. Since the hen does not nurse, it must hunt food for its young.[197] The verb הרג, though unusual, is used of an animal once in the Bible (Job 20:16) though it is used only of humans in the Aramaic and Moabite texts (Hadad 26, 33, 34; Panammu 3, 5, 7; Mesha 11).[198] Likewise, the verb שוט is used in the Bible, though of humans, not animals, looking for food (Num. 11:8).[199]

But the fact that hens are not known for their predatory nature has bothered scholars from the first in commenting on this text. Bauer translated the line, "Und seine sieben Töchter mögen im Herumschweifen nach Brot gehen, und sie sollen nicht," saying that the line refers to the daughters working as prostitutes[200] and he compares the Tell Halaf inscription, "Whoever erases (my) name and puts (his) name, may he burn his seven sons before Adad, may he release his seven daughters as prostitutes for Ishtar."[201] Hillers reiterated a modified version of this theory later in some detail. He contends that the Sefire curse concerns a lack of men so, while the Tell Halaf inscription's curse is one that the daughters become prostitutes (cf. Isa. 4:1 "And seven women will take hold of one man in that day, saying 'We will eat our own bread and wear our own clothes, only let us be called by your name; take away our reproach'"), the Sefire curse goes a step further: "Mati'el's daughters will try to be prostitutes to earn a living but be unsuccessful because of the inadequate number of men" (cf. Deut. 28:68b "you will offer yourselves for sale to your enemies as male and female slaves, but no man will buy you").[202] Fitzmyer still agrees with Dupont-Sommer's reading of the word בכהה and his interpretation of the text.[203]

Kaufman suggests that since lines 20–21 of the Tell Fekherye text and 21b–24a of Sefire I parallel each other, perhaps the next portions of each text are also parallel. By means of minor reconstructions of the Sefire text (it is abraded at this point), Kaufman is able to provide a convincing reading, "And may his seven daughters bake bread in an oven(?) but not fill (it)," that parallels line 22 of the Tell Fekherye text and Lev. 26:26, "When I break your staff of bread, ten women shall bake your bread in one oven, and shall deliver your bread again by weight; and you shall eat, and not be satisfied."[204]

(24c) והנ ישקר מתע[אל ול]

(25a) ברה ולעקרה תהוי מלכתה כמלכת חל מלכת חל מזי ימלכ אשר [205]

Lines 24c–25a. The difficulty in understanding this curse comes in knowing how to divide the letters כמלכת חל מלכת חל מזי. Dupont-Sommer reads the phrase כמלכת חלמ לכת חלמ זי saying that this should be interpreted as wishing the vassal to a have an unreal kingdom, a dream kingdom which will pose no threat to Aššur (cf. Ps. 73:20; Job 20:8).[206] Fitzmyer reads the phrase כמלכת חל מלכת חל מזי and translates it, "may his kingdom become like a kingdom of sand, a kingdom of sand as long as Asshur rules!"[207] Donner and Röllig read it as כמלכת חל מלכת חלמ זי and translate, "wie ein Sandkönigreich, ein Traumkönig-reich, das..."[208] In any case, the curse seems clear: that the offending kingdom be worthless and ephemeral.

Lines 25b–35a. Under the general invocation that Hadad pour "every sort of evil" out on a rebellious Arpad, the following execrations appear: agricultural curses, curses of mourning replacing rejoicing, of devouring animals, that the city become the habitation of animals, and that the city's name not be mentioned again.

(25b) [יסכ ה]

(26a) דד כל מה לחיה בארק ובשמינ וכל מה עמל

Lines 25b–26a. "(And) [may Ha]dad [pour (over it)] every sort of evil (which exists) on earth and in heaven and every sort of trouble."[209]

(26b) ויסכ על ארפד [אבני ב]

(27a) רד

Lines 26b–27a. "And may he shower upon Arpad [ha]il-[stones]!"[210] Dupont-Sommer notes that Hadad is called upon in his character as storm-god to rain down hailstones and he notes biblical texts where Yahweh uses this weapon to fight for his people (e.g., Josh. 10:11) and to avenge his people on the Assyrians (Isa. 30:30).[211]

(27b) ושבע שנן יאכל ארבה

Line 27b. "For seven years may the locusts devour (Arpad)."[212] Locusts, ארבה, are several times used in the Bible as an image of Yahweh's judgement (e.g., Exod. 10; Deut. 28:38–39, 42; Amos 7:1; Nah. 3:15). They are also a curse in Neo-Assyrian treaties, "[may the] locust who diminishes the land [devour] your harvest."[213] Fitzmyer notes that hail and locusts also appear in this order in Exod. 10:5 and Ps. 105:32–34.[214]

(27c) ושבע שנן תאכל תולעה

Line 27c. "And for seven years may the worm eat."[215] Dupont-Sommer

Chapter 5: Blessing and Curse   185

notes that the worm, תולעה in BH, attacks the vine. Deuteronomy 28:38–39 pairs the same destroyers and in the same order as here: the locust devours the grain in the field and the worm deprives the cursed people of the fruit of the vine.[216] Compare also the worm (תולעת) which plagued Jonah's gourd (4:7).[217]

(27d) ושבע [שנן יס]
(28a) ק תוי על אפי ארקה

Line 27d–28a. "And for seven [years may] TWY come upon the face of its land!"[218] Though he admits he can decide on no certain interpretation of תוי, Dupont-Sommer says that the curses against the vegetation continue here. He proposes that this curse may be a plant sickness and תוי may be related to BH דוי "illness."[219] It might be a species of parasite that would invade the fields.[220] Or, it might refer to an inundation of the fields in a flood (cf. the similar curse in the Mati'ilu treaty, "deine Leute zur Ueberschwemmung" $niše^{meš}$-ka ana riḫṣi.[221] A similar curse appears in the Esarhaddon vassal treaties, "with a great flood (may he [Adad] submerge) your land."[222] Koopmans suggests that it may be the same thing as BH ראא in Deut. 14:5 or תוא in Isa. 51:20 "Wildschaf" or "Wildstier."[223] Fitzmyer feels that Koopmans's suggestion does not fit the context but feels the word must refer to a kind of crop blight since it comes after the locust and worm infestation and before the lack of vegetation curse.[224]

(28b) ואל יפק חצר וליתחזה ירק ולי[תחזה]
(29a) אחזה

Lines 28b–29a. "May the grass not come forth so that no green may be seen; and may its vegetation not be [seen]!"[225] Dupont-Sommer compares the first part of the line to the similar curse in the Mati'ilu treaty, "das Grün des Feldes möge nicht aufgehen, Šamaš möge (es) *vernichten*" $urqit ṣēri lu la uṣṣa^a$ $^dšamaš$ *luteši*.[226] He says the next combination of letters should be deciphered as the negation ל prefixed to the particle of existence ית (BH יש) followed by the verb חזה.[227] Fitzmyer interprets it as the negation prefixed to the Ithpe'el form of the verb and treats the whole as a result clause, "so that there may not be seen."[228] Dupont-Sommer interprets the word אחזה as a term for vegetation with a pronominal suffix appended referring to Arpad.[229]

(29b) ואל יתשמע קל כנר בארפד ובעמה המל מרק והמ[ון צע]
(30a) קה ויללה

Lines 29b–30a. "Nor may the sound of the lyre be heard in Arpad; but among its people (let there rather be) the din of *affliction* and the noi[*se of cry*]*ing* and lamentation!"[230] Dupont-Sommer notes the parallels in Ezek. 26:13 "and the sound of your lyres shall be heard no more" וקול כנוריך לא ישמע עוד and in the Mati'ilu treaty, "so soll sein Bauer auf dem Felde kein Lied anstimmen" $^{amêl}ikkaru$-$šu$ ina $ṣēri$ $ā$ $lisā$ $alala$.[231] Aššurbanipal brags of his conquest of Elam, "The noise of people, the tread of cattle and sheep, the glad shouts of rejoicing, I banished from its fields. Wild asses, gazelles and all kinds of beasts of the plain, I caused to lie down among them, as if at home."[232] In Deut. 28:47, the sound of rejoicing will be replaced by the clamorous sounds (יללה) of a conquered land. Note that the curses in lines 28b–29a and 29b–30a are also together in the Mati'ilu treaty though in reverse order.

30b) וישלחן אלהן מן כל מה אכל בארפד ובעמה [יאכל פ]

31) מ חוה ופמ עקרב ופמ דבהה ופמ נמרה וסס וקמל וא[-- יהוו]

32b) עלה קק בתנ [יש]תחט לישמנ אחוה

Lines 30b–32b.
May the gods send every sort of devourer against Arpad and against its people! [May the mo]uth of a snake [eat], the mouth of a scorpion, the mouth of the bear, the mouth of a panther! And may a moth and a louse and a [...become] to a serpent's throat! May its vegetation be destroyed unto desolation.[233]

Dupont-Sommer says the word מן introduces the enumeration of the devourers which follow: the serpent, the scorpion, the bee, the panther, the moth, and the louse. He proposes that the end of line 31 be read [וא]פ יפלנ and be translated "Et, en outre, que s'abattent..."[234] The first words of line 32 עלה קק בתנ are the preposition על with the third masculine singular pronominal suffix referring to Arpad and קקבתנ (he does not separate the letters) which refers to a bird particularly harmful to grain and fruit crops (cf. Akkadian *qaqabānu*). The word [יש]תחט he treats as an Itpaʻal imperfect (with metathesis) from the root שחת meaning "devastate, ravage."[235]

Hillers gives numerous biblical references where ravenous animals are brought against the land as punishment. Leviticus 26:22 threatens the disobedient, "And I will let loose the wild beasts among you, which shall rob you of your children, and destroy your cattle, and make you few in number, so that your ways will become desolate," while 22:6b promises the faithful, "and I will remove evil beasts from the land." Jeremiah 5:6 is a similar oracle of doom, "Therefore a lion from the forest shall slay them, a wolf from the desert shall destroy them. A leopard is watching against their cities, everyone who goes out of them shall be torn in pieces."[236] The Day of Yahweh is also described this way in Amos 5:18c–20, "It is darkness, and not light; as if a man fled from a lion, and a bear met him; or went into the house and leaned with his hand against the wall, and a serpent bit him" (cf. 2 Kgs. 2:23–24). In Jer. 8:17, the serpent is similarly referred to, "'For behold, I am sending among you serpents, adders which cannot be charmed, and they shall bite you,' says the Lord." The "mouth of the scorpion" is an unusual threat since the scorpion is known rather for the sting of its tail.[237] Likewise, the bee, with the fly, can be punishment from Yahweh (Isa. 7:18).[238] Fitzmyer thinks it more likely, however, that Dupont-Sommer's דברה "bee" is דבהה "bear" since that animal was well-known in the area and is a popular image in the Bible (1 Sam. 17:34 and Hos. 13:8).[239] The panther (נמר, RSV "leopard") is

Chapter 5: Blessing and Curse

paired as a threat with the wolf (Hab. 1:8), with the lion and wolf (Jer. 5:6), and with the lion and bear (Hos. 13:7–8).[240] The moth appearing among the predators is strange but also occurs as a description of Yahweh in Hosea 5:12, "Therefore I am like a moth to Ephraim, and like dry rot to the house of Jacob" (cf. Isa. 51:8).[241] Regarding the interpretation of the words עלה קק בתנ, Fitzmyer says they should be translated as "towards the throat of a serpent," and regard the insects mentioned last, that they may become far more vicious against Arpad than was usual.[242]

(32c) ותהוי ארפד תל ל[רבק צי ו]

(33a) צבי ושעל וארנב ושרנ וצדה ו-- ועקה

Lines 32c–33a. "And may Arpad become a mound, to [house the desert animal]: the gazelle and the fox and the hare and the wild cat and the owl and the [ ] and the magpie!"[243] For the city becoming the habitat of the animals of the "désert," Dupont-Sommer compares Isa. 13:20–22, "It will never be inhabited or dwelt in for all generations…But wild beasts will lie down there, and its houses will be full of howling creatures…(also 34:10–15; and Zeph. 2:13–15[244] and Isa. 23:13; 32:12–14; Mal. 1:3). He cites the Mati'ilu treaty, "so möge Aššur der Vater der Götter,…dein Land zur Steppe,…deine Städte zu Trümmerhügeln, dein Haus zur Ruine machen" *aššur abu ilāni^{meš}…mātaka ana tušari…ana tilê^{me} bîtika ana ḫarbati lutir.*[245] Hillers says this curse is relatively rare in other treaties but is found in other texts like the annals of Esarhaddon where Marduk is said to have cursed the land with a flood, after which birds and fish inhabited it, "Stadt liess er sein. Sumpfrohr und Weiden wuschen dort üppig und trieben Schösslinge. Vögel des Himmels und Fische der Wassertiefe gab es dort onhe Zahl."[246] The hare also appears as a symbol of ruin in the Deir 'Alla texts (Combination I, line 9b).[247] Hosea 9:6b, "Nettles shall possess their precious things of silver; thorns shall be in their tents," probably refers to abandoned habitations as well.

(33b) ואל תאמר קר[יתא הא ו]

(34) מדרא ומרבה ומזה ומבלה ושרנ ותואמ וביתאל ובינג ו[---- וא]

(35a) רנה וחזז ואדמ

Lines 33b–35a. "May [this] ci[ty] not be mentioned (any more), [nor] MDR' nor MRBH nor MZH nor MBLH nor Sharun nor Tu'im nor Bethel nor BYNN nor [...nor 'Ar]neh nor Ḥazaz nor 'Adam!"[248] Dupont-Sommer thinks that the towns listed in these lines are those dependent on Arpad as in BH a city is said to be the "mother" and the outlying towns the "daughters" (cf. Num. 21:25, 32; Josh. 15:45; Ps. 48:12[11]).[249] In contrast, in the Bible cities destroyed in the past are constantly referred to as examples to warn others against unacceptable conduct (e.g., Sodom and Gomorrah).

Lines 35b–42. These lines contain curses with accompanying rites. McCarthy traces the development of the "substitution rite" in first millenium treaties through earlier Akkadian and Hittite literature down to later Neo-Assyrian and Syrian treaties. He says the technique grew more complex but had one goal:

188   Chapter 5: Blessing and Curse

to symbolize and cause the ruin of the oath breaker.²⁵⁰ The oppressively long list of similes used to describe the treaty violator in the Esarhaddon treaties served the same purpose and was in some cases accompanied by rituals or "practical demonstrations."²⁵¹ The Hittite Soldiers' Oath also makes use of such symbols in its ritual; specific examples will be given below. Hillers treats the "simile curse" as a major form of curse and subdivides it into three groups: 1) "ritual or ceremonial curses," 2) "curses which *may* have been accompanied by a ritual," 3) curses apparently not accompanied by ritual.²⁵² The simile curses of Sefire I.A.35–42, using wax, GNB', arrows, and a calf, fall within his first group.²⁵³

(35b) איך זי תקד שעותא זא באש כן תקד ארפד ו[בנתה ר]
(36) בת ויזרע בהן הדד מלח ושחלין ואל תאמר גנבא זנה ו[נבשא זא]
(37) מתעאל ונבשה הא איכה זי תקד שעותא זא באש כן יקד מ[תעאל בא]
(38) ש ואיך זי תשבר קשתא וחציא אלן כן ישבר אנרת והדד [קשת מתעאל]
(39) וקשת רבוה ואיך זי יער גבר שעותא כן יער מתעא[ל ואיך ז]
(40) י] יגזר עגלא זנה כן יגזר מתעאל ויגזרן רבוה [ואיך זי תע]
(41) רר ז[נ]יה] כן יעררן נשי מתעאל ונשי עקרה ונשי ר[בוה ואיך ז]
(42) י תקח גברת שעותא זא] וימחא על אפיה כן יקחן [נשי מתעאל ו]

Just as this wax is burned by fire, so may Arpad be burned and [her gr]eat [daughter-cities]! May Hadad sow in them salt and *weeds*, and may it not be mentioned (again)! This GNB' and [ ] (are) Matî'el; it is his person. Just as this wax is burned by fire, so may Matî['el be burned by fi]re. Just as (this) bow and these arrows are broken, so shall 'Inurta and Hadad break [the bow of Matî'el], and the bow of his nobles! Just as a man of wax is blinded, so may Matî['el] be blinded. [Just as] this calf is cut in two, so may Matî'el be cut in two, and his nobles be cut in two! [And just as] a [ha]r[lot is stripped naked], so may the wives of Matî'el be stripped naked, and the wives of his offspring, and the wives of [his] no[bles]! And just as this wax woman is taken] and one strikes her on the face, so may the [wives of Matî'el] be taken [and]...²⁵⁴

Line 35b–36a. Dupont-Sommer reconstructs the end of line 35 as above and thus the first of the curses accompanied by "un rite magique" continues the threat that unfaithful Arpad's fate will also fall on her dependent towns.²⁵⁵ Fitzmyer believes that in lines 35b–36, the figurine was in the shape of a city, while it was in human shape, a "man of wax," in line 37.²⁵⁶ He notes Ps. 68:3(2) as "a similar expression."²⁵⁷ It could well be interpreted (with verse 2[1]) itself as referring to a rite, "Let God arise, let his enemies be scattered; let those who hate him flee before him! As smoke is driven away; as wax melts before fire, let the wicked perish before God!" (cf. Num. 5, the ritual ordeal for the woman accused of adultery). Figurines of wax or other substances were also used in magical texts like *Maqlû* (I.73–121; 135–43; II.75–102; 146–47), e.g., "Just as these figurines

Chapter 5: Blessing and Curse

melt, run, and flow away so may sorcerer and sorceress melt, run, and flow away" (II.146–47).[258]

Regarding the practice of sowing a land with salt, Dupont-Sommer cites several biblical parallels (Judg. 9:45; Deut. 29:23 [of Yahweh's judgement on the cities of the plain]; Zeph. 2:9; and Job 39:6) as well as Akkadian examples from the documents of Tiglath-pileser I, Šalmanezer I, and Adadnirāri I.[259] Fitzmyer disagrees with Dupont-Sommer's characterization of this as "un geste magique," believing rather that it is a wish that Hadad will render the land unproductive.[260] Gevirtz says that salting a devastated city was, more likely, a procedure designed to purify the site as a "preparatory or concommitant act of consecration."[261] Regarding the sowing of cress (שחלין, Mishnaic Hebrew שחלים), Dupont-Sommer finds a parallel in the Annals of Assurbanipal concerning action taken subsequent to the conquest of Elam in which salt and cress (saḫle) are sown, "je dévastai la province d'Elam, je semai dans sa campagne du sel et du cresson (?)" *nagê* ^{māt}*Elamti*^{ki} *ušaḫrib* ^{šam}*saḫlê ušappiḫa ṣîruššun*.[262] Also note near the end of the Hittite Soldiers' Oath, one of the curses is that *saḫlū* (which Goetze notes as a kind of weed commonly found on ruins) will come from his field instead of spelt and barley.[263]

Line 36b–37a. Dupont-Sommer translates the phrase, "et qu'on n'en parle plus! Ce bandit (?)-ci et [cette âme-ci], c'est Mati'el et son âme," and suggests two possible interpretations for the word גנבא: he prefers "robber" which could be used to describe an unfaithful vassal, or it might mean a "stump" to which he compares Isa. 7:4 (זנבות) describing the rulers of Aram and Israel.[264] Fitzmyer is not satisfied by Dupont-Sommer's identification. At the end of line 36 he follows Dupont-Sommer's reconstruction but leaves out the translation. Fitzmyer reserves his own reconstruction for his commentary

ואל תאמר גנבא זנה [וגנבא הא] מתאאל ונבשה הא

"And you shall not mention this GNB'; the GNB' is Matî'el and it is his person."[265]

Line 37b. This curse is the same as that in 35b except that here the name of Matî'el is substituted for that of Arpad. The vassal treaties of Esarhaddon use a similar image, "Just as they burn an image (made) of wax in the fire and dissolve one of clay in water, just so may your figure burn in the fire and sink in water" *kī šá ṣalmu šá GAB.LÀL ina IZI išárapuni [šá ṭiṭ]i ina A.MEŠ imaḫḫaḫuni [kī ḫ]ānnie lānkunu ina* ^{d}*GIŠ.BAR liqmū [ina A.]MEŠ lu ṭabú*.[266]

Lines 38a–39a. A very common curse was that the gods would break someone's weapons, usually the bow.[267] Dupont-Sommer notes a similar curse in the Mati'ilu treaty "Quant aux hommes, puisse la souveraine des femmes ôter leur arc" and in the Hittite Soldiers' Oath "Die Bogen, Schäfte (und) Waffen in ihrer Hand sollen sie zerbrechen."[268] Hillers points out another parallel in the Esarhaddon vassal treaties, "may they shatter your bow and cause you to sit beneath your enemy; may they cause the bow to come away from your hand; may they cause your chariots to be turned upside down."[269]

190   Chapter 5: Blessing and Curse

Line 39b. Dupont-Sommer speculated that the sympathetic magic at this point involved gouging out the eyes of the wax figure representing Matî'el. He compares the Mati'ilu treaty, "et puisse-t-il les aveugler"[270] and the Soldiers' Oath which contains a ceremony,

> They parade in front of them a [blind woman] and a deaf man and [you speak] as follows: "See! here is a blind woman and a deaf man. Whoever does evil to the king (and) queen, let the oaths seize him! Let them make him blind! Let them [ma]ke him [deaf]! Let them [blind] him like a blind man..."[271]

Lines 39b–40a. The calf in the text was evidently the one killed to seal the treaty; its slaughter symbolized the promised punishment for Matî'el if disloyal.[272] Dupont-Sommer again points out the Mati'ilu treaty parallel, "Dieser Kopf—nicht der kopf des Widd[ers (ist) er], der Kopf des Mati'ilu [(ist) er], der Kopf seiner Söhne, seiner Grossen, der Leute [seines] Land[es (ist) er]" *qaqqadu anniu la qaqqadu ša UDU.NI[M šūt] qaqqadu ša ¹mati'ilu [šūt] qaqqadu ša mārê^{meš}-šu rabûti^{meš}-šu nišê^{meš} māt[i-šu šūt]*.[273] The cutting of the calf as a rite in the treaty making process is reminiscent of Gen. 15:9–18 and Jer. 34:18–19.[274] Bauer thinks that this ritual involved castration of the ram, representing a like fate for Matî'el and his men;[275] an interpretation which Fitzmyer finds unlikely in light of the parallels. Note, however, that the curse of the warriors becoming women, usual for this type of document (see the discussion of the next lines), is otherwise missing in the Sefire texts; might this be that sort of curse if Bauer's suggestion is followed? The Esarhaddon vassal treaties contain a parallel use of an animal in a simile curse, but in reference to cannablism, "Just as this sheep is cut up and the flesh of her young is put in her mouth..."[276]

Lines 40b–41a. Dupont-Sommer reads the end of line 40 through the beginning of line 41 as זנה עבדא יעבד זי ואיך "Et de même que sert cet esclave-ci..." and the third full word on line 41 as יעבדן rather than יערין; he says the punishment is that these higher society women (wives of Matî'el, his sons, and his nobles) will be forced to serve (עבד) their conquerors as slaves (cf. Isa. 47:1–2, "Come down and sit in the dust, O virgin daughter of Babylon; sit on the ground without a throne...For you shall no more be called tender and delicate. Take the millstone and grind meal, put off your veil, strip off your robe, uncover your legs"). He favors this reconstruction even though he admits עבד with meaning "serve as a slave" is unattested in Aramaic.[277] McCarthy agrees with his interpretation and cites the parallels of Deut. 28:32, "Your sons and your daughters shall be given to another people, while your eyes look on and fail with longing for them all the day; and it shall not be in the power of your hand to prevent it," and the Esarhaddon treaty lines 428, 588–90.[278]

Bauer reads the third full word of line 41 as וערין and translates the curse, "es sollen entblösst werden die Frauen des Mati'el und die Frauen seiner Nachkommenschaft und die Frauen [."[279] He notes that this was the practice of the

Chapter 5: Blessing and Curse

Assyrians in conquest.²⁸⁰ The idea may be that the wives of the disobedient vassal and his people be taken and sexually assaulted by the conquerors; compare the Esarhaddon vassal-treaties (lines 428–29), "[May Venus, the brightest of the stars,] make your wives lie [in the lap of your enemy before your eyes]"²⁸¹ (cf. Amos 7:17; Isa. 13:16; Jer. 8:10; Job 31:10). Hillers agrees with this interpretation and further reads זנה as "harlot" rather than as a demonstrative pronoun. He points out that the threat of being stripped like a prostitute is not uncommon in the Bible (cf. Hos. 2:5, 12; Nah. 3:5; Jer. 13:26–27; Ezek. 16:37–38; 23:10, 29).²⁸² He notes the occurrence of a similar curse in the Matî'ilu treaty (V.9–11), "Then may the aforesaid indeed become a prostitute, and his warriors women. May they receive their hire like a prostitute in the square of the city. May land after land draw near to them."²⁸³ Fitzmyer concurs with this interpretation and with Hiller's suggestion that this curse was probably not accompanied by a ritual action.²⁸⁴ Unlike other treaties and oaths, however, the Sefire treaty does not include a curse that the warriors of the vassal be reduced to weakness "become women."²⁸⁵ This curse is clearly against the wives of Matî'el, of his sons, and of his nobles.

Lines 41b–42. Slapping the face of the wax image symbolizes the harsh treatment that Matî'el may expect for his own women (and those of his sons and nobles probably) to suffer.²⁸⁶

Face B. Lines 1–13 list the parties involved, rulers and gods, in the treaty (ועדי אלהי כתכ עמ עדי א[להי ארפד ו]עדי אלהנ המ זי שמו אלהנ, lines 5–6). Lines 21–45 (the rest of B) contain the stipulations of the treaty. The only curse in B is that should Matî'el fail in his vassalage, he "will have been false to all the gods of the treaty which is in this inscription" (lines 23 and 33).

(23) שקרתמ לכל אלהי עדיא ז]י בספרא זנה...[²⁸⁷
This form of the warning alternates with the briefer form, "you will have been false to this treaty" (line 38).

(38) שקרת בעדיא אלנ ²⁸⁸
This "no-gods" form of the phrase may also have the additional phrase appended זי בספרא זנה, (lines 27b–28a).²⁸⁹ The difference in these two warnings is probably stylistic rather than substantive; the gods are, after all, witnesses and parties to the treaty whether specifically mentioned in every phrase or not. Since the gods have been invoked as witnesses and participants in the treaty and since they have been called upon to punish the violator, it seems reasonable to think that this brief statement sums up all of the previously specified curses. Fensham notes that both brief and full maledictions may be seen in documents ranging in date from the Hittite vassal treaties to the time of Josiah. In the Bible, fuller, more specific versions are found like that of Deut. 28:15ff. which has many similarities to Sefire, but brief statements also occur, like Deut. 31:20–21

> For when I have brought them into the land flowing with milk and honey...they will turn to other gods and serve them, and

despise me and break my covenant. And when many evils and troubles have come upon them, this song shall confront them as a witness (for it will live unforgotten in the mouths of their descendants)...[290]

Fitzmyer points out the significant parallel with certain biblical passages in which the same verb (שקר) is used with reference to the covenant between Yahweh and his people, "but I will not remove from him my steadfast love, or be false (אשקר) to my faithfulness" (Pss. 89:34[33]; 44:18[17]; Isa. 63:8).[291]

Lines 23b–25. These lines contain a minimal blessing, the promise by the suzerain (Bir-Ga'yah) not to attack the vassal Matî'el should he remain compliant.

(23b) [והן

(24) תשמענ ותש[למנ עדיא אלנ ותאמר גבר עדנ הא [אנה לאכהל לא

(25) שלח יד] בכ וליכהל ברי [ל[ישלח יד בבר[כ] ועקרי בעק[רכ]

[But if you obey and ful]fill this treaty and say, "[I] am an ally," [I shall not be able to raise a hand] against you; nor will my son be able to raise a hand against [your] son, nor my offspring against [your] offspr[ing].[292]

By the words גבר עדנ הא Matî'el recognizes his obligations to Bir-Ga'yah and his status as a vassal.[293] Fensham notes that the language used to describe the vassal's faithfulness in the various suzerainty treaties is very much the same: the Hittite treaty of Muršiliš and Dupi-Tešub uses the word "honor," the treaty of Šuppiluliuma and Kurtiwaza uses "fulfill," and the Akkadian treaty of Muršiliš and Niqmepa uses "guard" or "keep" (naṣaru).[294] The Sefire steles use both of these terms: in I.B.24 the vassal is granted freedom from attack by the suzerain if he "obeys" (תשמענ) and "fulfills" (תשלמנ) the treaty; in I.C.16b–17a, the one who does not "observe" (ליצר) the words of the inscription will be punished.

Face C. This face contains the conclusion of the treaty, what remains of the blessing, and a curse to protect the inscription. At least lines 1–16 of this side appear to be the words of Matî'el himself (מה כתבת א[נה מתע[אל, lines 1–2). Lines 4–9 are a basic statement of Matî'el's reason for keeping the conditions of the treaty, "that no evil may be done against the house of Matî'el." The same word for evil (לחיה) occurs in I.A.25b–26a, and in Nerab 1:10 where it describes the kind of demise wished on the violator of a tomb (cf. Nerab 1). Surely the "evil" Matî'el hopes to avoid is that specified in the earlier passage and summarized in this one word.

(4) לתבת

(5) [א] יעבד[ו תחת] שמשא

(6) [לב[ית מ[לכי ז[י כל לח

(7) [יה לתתעבד על] בית מ

(8) תע[אל וברה ובר] ברה ע[ד

Chapter 5: Blessing and Curse

9) עלמ[-------]-ו-[---]

"May they make good relations [beneath] the sun [for (the sake of) my] ro[yal hou]se that no ev[il may be done against] the house of Matî'el and his son and] his [grand]son for[ever]."[295] Fitzmyer disagrees with Dupont-Sommer's simple translation of לטבת "pour le bien"[296] preferring to follow Moran's treatment of the word as a specialized term designating "good relationship" preserved by faithful observation of treaty stipulations (see the discussion of lines 19–20 below).[297]

Lines 15–16a. Unfortunately, lines 10–14 are missing and when the text resumes, it is at the end of what was the most complete blessing formula for the faithful vassal (cf. II.C.13–17 which is similar and in like condition).

15) יצרו אלהן מנ יו

16) מה ומנ ביתה[298]

Dupont-Sommer treats the repeated particle מנ as an interrogative-indefinite pronoun and translates, "que les dieux gardent qui ses jours et qui sa maison"; the verb used of the god's attitude toward obedient Matî'el (יצרו) is the same used for the attitude desired of people toward the inscription (יצר), i.e., that they guard it.[299] Fitzmyer, however, thinks that מנ is the preposition and that the thing ("[the evil]") from which faithful Matî'el is to be spared was in the missing line 14, so he translates, "...may (the) gods keep [all evils] away from his day and from his house" (cf. Pss. 12:8[7] "Do thou, O Lord protect us, guard us ever from the wicked generation" אתה־יהוה תשמרם תצרנו מן־הדור זו לעולם; also 32:7; 64:2; 140:2, 5).[300] The root used in the blessing of the faithful vassal (יצרו) is used for the same purpose in the treaty between Muršiliš and Niqmepa.[301]

Lines 16b–25. These lines are a curse designed to protect the inscription itself. Though the type of inscription is not the same as that of Zakir (a memorial stele), the intention of the curse and some of its substance are the same: Zakir B23–28 includes the descendants of the violator as well as the violator himself.[302]

16b) ומנ

17) ליצר מלי ספרא זי בנצבא זנה

18) ויאמר אהלד מנ מלו

19) ה או אהפכ טבתא ואשמ

20) [ל]לחית ביומ זי יעב

21) [ד] כנ יהפכו אלהן אש

22) [א ה]א וביתה וכל זי [ב]

23) ה וישמו תחתיתה [ל]

24) ע[ל]יתה ואל ירת שר

25) [ש]ה אשמ

Whoever will not observe the words of the inscription which is on this stele or will say, "I will efface some of (its) words," or "I shall upset the good relations and turn (them) [to] evil," on any day on which he will d[o] so, may the gods overturn th[at m]an and his house and all that (is) in it; and may they make its lower part its upper part! May his scio[n] *inherit* no name!³⁰³

Three situations are described that will be punished by the curse (lines 21–25): 1) someone will refuse to observe the treaty conditions (line 17), 2) someone will damage the inscription itself (line 18), or 3) someone will attempt to upset the good relationship between the vassal and his suzerain (line 19–20a). All of these violations amount to the same crime: a rupture between the vassal and suzerain and a *de facto* state of revolt.

Line 17. Fitzmyer notes the use of the cognate *naṣāru* in Esarhaddon's Vassal treaties (line 291), "Guard this treaty" *adē annute uṣra*.³⁰⁴

Line 18. Dupont-Sommer says that the verb אהלד is from the root לוד or לדד which means "effacer."³⁰⁵ Dupont-Sommer suggests a comparison to BH לוז (Prov. 3:21 paralleled to נצר; and 4:21 paralleled to שמר).³⁰⁶ Gevirtz agrees with identification of the root as לוד and favors a connection to the Mishnaic Hebrew root לוז "slander, pervert."³⁰⁷ This root is also attested in the Tell Fekherye inscription and is used in two ways, in the prayer of Hadadyisʻi (line 9) that sickness be kept (ולמלד) from him and (lines 11 and 16) describing damage to an inscription. (See the discussion of these lines of the Tell Fekherye inscription for a fuller treatment.)

Lines 19–20a. Turning (הפך) good to evil is also found in biblical texts as Yahweh's judgment (Amos 5:7; 6:12; 8:10) as well as turning evil to good is a sign of his favor (Deut. 23:5; Jer. 31:13; Ps. 30:12).³⁰⁸ The verb at the end of line 19 is ואשם, compare Mal. 1:3, "I have appointed his mountain a waste land" ואשים את־הריו שממה.³⁰⁹ Moran notes the technical use of the term *ṭābūtu* in Akkadian treaties (both vassal and parity treaties) in the sense of friendship established by treaty.³¹⁰ When the "friendship" ceases, the treaty ceases and in the case of Matîʻel, evil comes.

Lines 21–24a. The same verb is used for the action to be taken against the violator (הפך) as was used to describe his action against the treaty (line 19). For שרש as "scion" in BH, see Isa. 11:10 and 14:29.³¹¹ The overthrow (destruction) of the house of the offender is paralleled in the vassal treaties of Esarhaddon, "Do not transgress your treaty, (or) you will lose your lives, you will be turning over your dwellings to be shattered, your people to be carried off."³¹² Daniel 2:5 is also important, "The king answered the Chaldeans, 'The word from me is sure: if you do not make known to me the dream and its interpretation, you shall be torn limb from limb, and your houses shall be laid in ruins'" (cf. Amos 6:11 and Ezra 6:11). The curse idea may be a sort of play on the ideas which can be conveyed by the word שרשה, both descendant and root. If a plant is torn up from the ground, its roots exposed, it will die. If a dynasty, or any family line, is deprived of descendants, it will also cease and will eventually be forgotten (cf. Eshmunazar

Chapter 5: Blessing and Curse   195

line 11; Zakir B.28; Karatepe A.I.10; Amos 2:9).

Lines 24b–25. Dupont-Sommer says that the idea of this curse is that the house (family) of the criminal shall disappear from human memory.[313] Fensham notes that the curse of the obliteration of progeny is one of the most important and frequently occurring curses on the *kudurru* as well as in the Bible (cf. Amos 7:17; 9:1–4, 10).[314]

Sefire II.[315]

A

1) [יהינקן על ואל ישבע ושבע שורה יהינקן עגל ואן]ל ישבע ושבע
2) [שאן יהינקן אמר ואל ישבע ושבע עזן יהי]נקן גדה ואל יש
3) [בע ושבע בכתה יהכן בשט לחמ ואל יהרגן והן יש]קר לבר גאיה ול
4) [ברה ולעקרה תהוי מלכתה כמלכת חל ואשמה י]תנשי ויהוה קב
5) רה[וש]בע שנן שית שב
6)]   וש]בע שנן תהוי [ ]
7)]   -[בכל רברבי -
8)]   -[וארקה וצע   ] -[ואת
9) קה]   ויאכל] פמ אריה ופמ [ ]-[ופמ נמר]ה[--

...[and should seven mares suckle a colt, may it not be sated; and should seven cows give suck to a calf, may it n]ot have its fill; and should seven [ewes suckle a lamb, may it not be sated, and should seven goats suck]le a kid, may it not be sa[ted; and should seven hens go looking for food, may they not kill (anything). And if (Matî'el) should be un]faithful to Bir-Ga'yah and to [his son and to his offspring, may his kingdom become like a kingdom of sand; and may his name be for]gotten, and may [his grav]e be...[and for se]ven years thorns, ŠB[...and for se]ven years may there be[...] among all the nobles of...[...] and his land. And a cry [...and may] the mouth of a lion [eat] and the mouth of [a...] and the mouth of a panther [...][316]

Lines 1–2. These lines are largely restored by Dupont-Sommer (followed by Fitzmyer). As he has restored the text, the order is slightly different from that of the parallel curse in I.A.21b–24a. Here the list of unsatisfactory nurses is horses, cattle, sheep, and goats as compared to the list of women, horses, cattle, and sheep in the first stele. Each list has four members, however, since the beginning of stele II is broken, women could well have headed this list as well.[317]

Line 3a. The part of this line restored to contain the difficult passage about "hens" hunting food is actually completely obliterated. It is restored because the size of the lacuna exactly matches the space required for that formula[318] and because the previous and subsequent lines of II.A parallel I.A so closely.[319]

Lines 3b–4a. This curse is also largely restored on the basis of comparison to I.A.24b–25b, note however, that the phrase which concluded the curse in I.A ("kingdom of sand") is missing from II.A and that another phrase, "and may his name be forgotten" is substituted, though this is also largely a reconstruction.

Lines 4b–5a. Dupont-Sommer says that this curse, not included in I.A, is probably a curse aimed at the sepulcher of Matî'el.[320] Several possible parallels are found in other inscriptional curses which make clear the real peril to graves (Aḥiram, the Siloam Tomb, Nerab 1 and 2, Tabnit, Eshmunazar,[321] and Assyrian records in which a conqueror boasts of his violation of a tomb).[322]

Lines 5b–6. Dupont-Sommer thinks that these lines may be restored to contain curses like those in I.A.27–29a which begin with the formula, "For seven years..." In line 5, he understands the surviving word word שית as a type of thorn, an interpretation which strengthens the parallel to the agricultural perils of I.A. He acknowledges the very tentative nature of this reconstruction.[323] This may also be compared to the curse in I.A.36, where Hadad is called upon to sow the land of unfaithful Matî'el with salt and weeds. These weeds would choke out food crops.[324]

Line 7. All of the nobles of Arpad are, either alone or with Matî'el, the target of some lost curse.[325]

Line 8. The line contains the end of what is probably a curse against Matî'el's (literally "his") land (cf. I.A.27–29). The line ends with the word וצע[קה] which Dupont-Sommer notes may be like BH צעקה "cry of distress" (cf. I.A.29b, though he notes that this word does not appear in I.A.29b).[326]

Line 9. The threatening mouths of the lion and panther are mentioned (cf. I.A.30c–32a though the lion is not mentioned in I.A.; a further example of variants between the documents). Dupont-Sommer says that though lines 10–14 are practically illegible, surviving letters indicate a continuation of the threat of the devouring animals.[327] Hillers points out the parallel in the treaty of Esarhaddon with Ba'al of Tyre (IV.6–7), "May Bethel and Anath Bethel put you at the mercy of a devouring lion."[328] Yahweh is not infrequently pictured as a lion or compared to a lion, "So I will be to them like a lion, like a leopard I will lurk beside the way. I will fall upon them like a bear robbed of her cubs. I will tear open their breast, and there I will devour them like a lion, as a wild beast would rend them" (Hos. 13:7–8; also 5:14; Amos 3:8; Jer. 5:6; 49:19; 50:44). Lions are also seen as an image of punishment (Isa. 15:9; Jer. 2:14–15; 4:7). The promise of their removal is a sign of divine favor (Isa. 35:9).[329]

Face B.

(2 עדיא וטבתא ז[י] עבדו אלהן ב[ארפד ובעמה ולישמע מתעאל]

(2b ולישמען בנוה

(3 לישמען רבוה ולישמע עמה ולי[שמען כל מלכי ארפד [

(4 ים זי יעורן פהן תשמע נחת מ[ה]

(5 הן תאמר בנבשך ותעשת בלבב[כ גבר עדנ אנה ואשמע לבר גאיה]

(6 ובנוה ועקרה פלאכהל לאשלח י[ד בכ וברי בברכ ועקרי בעקרכ]

Chapter 5: Blessing and Curse   197

7a) ‏ולחבזתהמ ולאבדת אשמהמ‏

...the treaty and the amity whi[ch] the gods have made in [Arpad and among its people; and (if) Matî'el will not obey], and (if) his sons will not obey, (if) his nobles will not obey, and (if) his people will not obey, and (if) [all the kings of Arpad] will not o[bey...] YM who *are watchful*. But if you obey, (may) tranquillity [...And] if you say in your soul and think in your mind, ["I am an ally, and I shall obey Bir-Ga'yah] and his sons and his offspring," then I shall not be able to raise a ha[nd against you, nor my son against your son, nor my offspring against your offspring], neither to strike them nor to destroy their name.[330]

Line 4a. The first line of the text is missing but probably contained some warning formula. Dupont-Sommer says that the beginning of this line holds the end of a curse, "pour qu'ils soient aveugles" the addressees of which are unknown.[331] Fitzmyer rejects this interpretation and suggests "who are watchful."[332]

Lines 4b–7a. The phrase ‏פהן תשמע נחת מ‏] is not found in the I.B.23b–25a parallel and II.B is thus more specifically a blessing than its I.B parallel. Dupont-Sommer points out that ‏נחת‏ is precisely what will not endure in the curse of the Aḥiram sepulcher. The reward for obedience (‏תשמע‏) is tranquility, the opposite of all the violence and want threatened above.[333]

9) ‏שקרתמ לכל אלה[י עדיא זי בספרא זנה‏   [
14b) ‏שקרתמ [בעדיא אלנ‏   [
17b) [‏שק‏
18) ‏ר[ת לכל [אלהי ע]דיא זי בספר[א זנה‏   [

Lines 9b, 14b, 17b–18. Line 9b, "(then) you will have been false to all the gods [of the treaty which is in this inscription." Line 14b "you will have been false [to this treaty...]." Lines 17b–18 "[...] you [will have been false to] all the [gods of the trea]ty which is in [this] inscription."[334] The only other curses on this face of the stele are variations of the formula, "you will have been false to this treaty." The formula in line 9 is safely restored as the longer form of the formula with the plural verb. The extant part of the formula in 14b is limited to the word ‏שקרתמ‏ and thus the exact form may not be determined, though Dupont-Sommer partially restores it.[335] The formula in lines 17b–18 is complete enough that the longer variation of the formula may be restored with confidence, howbeit with the singular form of the verb.[336]

Face C.

1) --------------- [‏ומנ י‏]‏א‏
2) ‏מר להלדת ספריא [א]לנ מנ ב‏
3) ‏תי אלהיא אנ זי י[ר]שמנ ו‏

Chapter 5: Blessing and Curse

4) [ןמלו א[י]רפס דבאהא רמאי[
5) ךלמ תיאו ךתכ תיא דבהא נ
6) רפס דל נמ אה לחזיו ה
7) ל רמאיו איהלא יתב נמ א[י]
8) [י]ו רגא רגא הנא עדיל יז
9) תב נמ נלא איר[פס] דל רמא
10) [אה תמ[י בלע חצלבו איהל[א] י
11) הרבו

...[and whoever will] give orders to efface [th]ese inscriptions from the bethels, where they are [wr]itten and [will] say, "I shall destroy the inscriptions and *with impunity* shall I destroy KTK and its king," should that (man) be frightened from effacing the inscriptions from the bethels and say to someone who does not understand, "I shall *reward* (you) *indeed*," and (then) order (him), "Efface these inscriptions from the bethels," may [he] and his son die by oppressive torment.[337]

Lines 1–11. These lines are a curse designed to protect the inscription. In contrast to the protective curse at the end of stele I which deals with someone not observing the conditions of the treaty, damaging it, or otherwise disrupting relations between the vassal and suzerain, this curse is preoccupied with the idea that someone would order or pay someone else to damage the inscription.

Lines 1–3. Dupont-Sommer translates the first lines, "[Et quiconque déci]dera d'effacer [c]es inscriptions-ci des bétyles où elles sont tr[acées] et [d]ira..."[338] Dupont-Sommer notes the first non-biblical use of the term יתב איהלא which he takes to refer to the steles themselves ("la pierre sacrée") on which the inscriptions are written, indeed, the stones bearing the inscriptions "sont elle-mêmes vivantes."[339] He translates I.B.8, "Et que ne se taise aucune des paroles de cette inscription-ci!": the steles themselves will "speak" their message (the inscriptions) if not hampered by violators. Fitzmyer notes a deity in the god list of the treaty between Esarhaddon and Ba'al of Tyre named *Baiti-ilāni^{meš}.*[340] McCarthy also notes the idea of the sacred stones serving as witnesses to a treaty or covenant in the Bible, both inscribed stones (Deut. 27:2–4, 8) and uninscribed ones (Gen. 28:18–22; 31:48, 52; Josh. 24:27; Isa. 19:19–20).[341]

Line 4. Dupont-Sommer says that the use of the verb דבאהא refers not just to the destruction of another's record of achievement but here it alludes as well "au caractère magique de ces inscriptions qui, comme les bétyles eux-mêmes, sont vivantes, 'animées'."[342]

Line 5. Dupont-Sommer says that the crime mentioned in line 5 is the removal (דבהא) of the names of the overlord and his kingdom from the stele thus, magically, making them perish.[343]

Lines 7–8. He confesses that the first words are "un peu énigmatiques"

Chapter 5: Blessing and Curse 199

but says that the following words אגר אגר אנה (אגר being a first singular imperfect of גור meaning "être exilé" or "être effrayé") are the confession of the inscription's violator made to the first passerby, either, "Moi, je serai exilé, je serai exilé" or "Moi, j'ai peur, j'ai peur." The guilty one thus admits his crime and declares his readiness to be punished by exile.[344] Fitzmyer disagrees with this interpretation, saying it does not fit the context. The first אגר should rather be seen as a Pe'al Infinitive intensifying the other אגר which is a Pe'al Imperfect first singular from the root אגר related to the word for "salary."[345] Apparently, the person who wishes the inscription effaced seeks to avoid the penalty involved by bribing someone else to damage the inscription. Though the identity of the "he" mentioned (lines 10-11) is not clear, the penalty will not be avoided. Surely the instigator is the primary target of the curse. It seems unlikely that the inscriber would go to such pains to deal with this eventuality and then not provide for the party's punishment; to do so would be to suggest to the violator a way to do what he wanted with impunity.[346]

Lines 9–11. In keeping with his interpretation that this is the confession of someone who has damaged the stele, Dupont-Sommer restores the beginning of line 9 as אמר לד[ת ספ]ר[יא, "[J'ai] effacé ces [inscr]iptions-ci des bétyles." In line 10, the words ובלחץ עלב refer to the crushing fear that the guilty one feels. Fitzmyer notes that ובלחץ is otherwise unattested in Aramaic but is probably related to BH לחץ "oppression" (Exod. 3:9; Deut. 26:7). The word then describes the guilty one's death, not his guilt.[347] The last words, י[מת הא], Dupont-Sommer treats as the death penalty: only death is sufficient punishment for sacrilege.[348]

13) [ישא]ן כל אלה[י עד]יא זי בספרא
14) [זנ]ה אית מתעאל וברה ובר ברה
15) ועקרה וכל מלכי ארפד וכל רב
16) וה ועמהם מן בתיהם ומ
17) יומידם

[...] and all the gods of the [trea]ty which is in [this] inscription will [ ] Matî'el and his son and his grandson and his offspring and all the kings of Arpad and all his nobles and their people from their homes and from their days.[349]

Lines 12–17. Unfortunately, line 12 is too broken to read. Dupont-Sommer notes that the rest of line 11 is blank and that line 12 begins another curse. The missing part of 12 contained some crime against the stele or violation of the treaty. Lines 13–14a, call upon the gods to act against the violator Matî'el and his kingdom. In contrast to I.C.15–16 which called for a blessing on the faithful and their homes, this passage calls for the destruction of the guilty and their homes.[350] He quotes as evidence for his interpretation a Babylonian inscription of Sardur III:

Sardur spricht: Wer dieses Stele von (ihrem) Orte entfernt,

wer (sie) verbirgt, wer diese Inschrift beschädigt, wer an irgendeinem diese (=solche Taten) sieht (oder) sagt: "Geh! Zerstöre!", wer (als) ein anderer sagt: "Ich habe (diese Taten) gemacht", der wird von Ḫaldi, Wettergott und Sonnengott und den (anderen) Göttern ausgetilgt werden, der und (seine) Lebenskraft und sein Same werden aus dem Sonnenlicht weggebracht sein, dessen *arḫi* und *inaini* und Leben soll getötet und dem Nichts (=Vernichtung) zugeführt sein.[351]

Perhaps it makes more sense to see the empty space on line 11 as a break from the curses. Note that a blessing would be expected from a comparison to I.C where the curse against the violator is followed by a blessing on Matîʻel and his kingdom. Fitzmyer suggests that this paragraph could be understood as a blessing, though a form of the verb נצר supplied in the lacuna would not match the direct object here.[352] This pattern would also agree with that seen in the vassal treaty between the Hittite Muršiliš and the Amorite Dupi-Tešub

> The words of the treaty and the oath that are inscribed on this tablet—should Duppi-Tessub not honor these words of the treaty and the oath, may these gods of the oath destroy Duppi-Tessub together with his person, his wife, his son, his grandson, his house, his land and together with everything he owns.
>
> But if Duppi-Tessub honors these words of the treaty and the oath that are inscribed on this tablet, may these gods of the oath protect him together with his person, his wife, his son, his grandson, his house (and) his country.[353]

Based on this comparison, these lines might well be a blessing on Matîʻel and his kingdom should he prove faithful and obedient. Though the actual "blessing" part of these lines is missing, the fact that something is to be kept[354] "from (מן) their homes and from (מן) their days" seems to indicate that they will be spared (I.C.6–8, 15–16) evil (לחיה as in I.C.6), rather than that evil things will be brought upon them (על ארפד as in I.A.26). Also the flavor of this section seems similar to that of Nerab 1:5–14 which has a curse against a grave robber but a blessing should he actually change his mind and protect the tomb.

The Sefire inscriptions are treaty texts and contain the same types of curses seen in other ancient Near Eastern treaty literature, including biblical covenant solemnizations. The curse portions are highly developed and specific as are the other Aramaic curses of the target period. The blessings, however, like blessings found in the other treaty literature and the biblical covenant passages are not so extensive.[355] Indeed, the Sefire blessings are so brief as to be easily missed.

Hadad lines 15–24. This inscription (ca. 732) was written on the skirt of a four meter high statue of the god Hadad erected by Panammu I, the Panammu

Chapter 5: Blessing and Curse

of the inscription. The statue was discovered in a village just NE of Zenjirli in 1890.[356]

(15) ומן.אן.חד[.].לפנמו.בני.יאחז[.]חט[ר.וישב.על.משבי.ויסעד.אברו.ויזבח.

(16) הדד.זן.פא[.]יא[מ]ר[.].אר̇[.][.]נשי.ויזבח.]הדד.ז[.נ.פכא].י̇זבח.הדד.ויזכר.אשׁם̇.
הדד.או.

(17) כא.פא.יאמר[.]האכל.נב[ש.פנמו.עמכ.ותש̇]תי.נ[בש.].[פנמו].עמכ.עד.יזכר.נבש.
פנמו.עמ[.]

(18) [ה]ד̇[ד].בימ̇[   ]-באה[.]זא.יתנ.ל[הדד.וי]רקי.בה[.]שי[.].להדד.ולאל.
ולרכבאל.ושמ̇.נ̇

(19) ----------מ̇ג̇------בב̇----------י̇.קיצא.פבנ̇ית[ה.והושבת.בה.אלהי.ובלבבתה.
חנאת.

(20) ואלהו.[נתנו.לי.זרע.חבא]   [י--י̇ם̇].ן.חד[.].בני.יאחז.חטר.וישב.על.משבי.מלכ̇[.]

(21) על.[יאדי].ויסעד.אברו.ו̇י̇ז̇[בח.הדד.זן.ואל.הדד.זן.ואל.יזכ]ר.אשם.פנמו.יאמר[.]ת[א]כ̇ל.
נבש.פנ̇מ̇ו̇[.]

(22) עם.הדד.ותשתי.נבש.פנמו.עם.ה̇[דד.].הא[.]   [חהנ.זבחה.ואל.ירקי].בה[.]ומז.

(23) ישאל.אל.יתנ.לה.הדד.והדד.חרא.לבתכה]   א̇[ל.יתנ.לה.לאכל.ברגז.

(24) ושנה.למנע.מנה.בליל̇א.ולדלח.נתנ.ל]ה   א̇[יחי.ומ]ו̇[דדי].[מ̇ומת].[מ̇.תי.

Now, if one of Panammu's sons should grasp the sceptre and sit on my throne and maintain power and do sacrifice to this Hadad, and should say, By thee I swear, and do sacrifice to [this Hadad], whether in this way he does sacrifice to Hadad and invokes the name of Hadad or in another, let him then say, May the soul of Panammu [eat] with thee, and may the soul of Panammu drink with thee. Let him keep remembering the soul of Panammu with Hadad; in the days [ ]this[ ]let him give to [Hadad], and may he look favorably upon it as a tribute to Hadad and El and Rakkabel and Shemesh; [ ]this[ ]; so I built it, and I have made my god to dwell in it, and in his authority I have found rest. [Now the gods] have granted me a seed to cherish […] If (however), [any] of my sons should grasp the sceptre and sit on my throne as king over [Y'DY] and maintain power and do sacrifice [to this Hadad, and should not] remember the name of Panammu, saying, May the soul of Panammu eat with Hadad, and may the soul of Panammu drink with Hadad, as for him […] his sacrifice, and may he not look favourably upon it, and what he asks, may Hadad not give to him, but with wrath may Hadad confound him, […] may he not allow him to eat because of rage, and sleep may he withhold from him in the night, and may terror be given to him…[357]

Though the substantive curse of this inscription is clear, the substantive blessing is quite short and easily missed. Cooke identifies lines 15–24a as Panammu's "blessing" on his sucessor should "he be faithful to his religious and filial duty, and a curse upon him if he neglect it."[358] Cooke's reconstruction of the beginning of line 18 is יֹ[רקי]כתפ.אז הֹחבזֹ[ ]יֹ.דדֹ[ה] which he translates "[Ha]dad [ ]this his sacrifice [ ] may he look [f]avourably upon him."[359] If not for the phrase "may he look favourably" in line 18 being interpreted as referring to Hadad's approval, there would be no other explicit blessing in this inscription. The blessing, if such it is, is a call for Hadad and the other gods to regard favorably the offerings of the loving heir and bestow on him their beneficient regard described as belonging to his father Panammu. Remarkably, there is no curse included to guard the inscription itself.[360]

Indeed, an argument can be made that the substantive blessing is more subtle and does not appear at this point in the inscription at all. First, it seems a bit awkward that Hadad be interpreted as the subject of the verb when he is listed later in the same line with other gods (though, admittedly, in line 13, it is the deity's favor that matters). Secondly, seemingly, a blessing, if not as lengthy as the substantial curse section of the text, would at least take up more space that can be accounted for in the broken portions of lines 18. Thirdly, blessings, when included, usually follow the curse in Akkadian inscriptions and in Aramaic inscriptions influenced by Akkadian. Finally, I think that the real substantive blessing is found in the description of Panammu's prosperity described earlier in the inscription. The five curses in lines 22–24 are all counterparts to conditions that Panammu enjoys and the not so subtle implication is that the prosperous state that the pious Panammu revels in will pass on to his son should he show due filial honor. For example, the first curse that Hadad not accept the successor's sacrifice is in contrast to the "blessedness" that Panammu enjoys "] would I offer to the gods, and they used to accept (them) from my hand" (line 12). The second, that Hadad will not give what the disrespectful son asks but rather give him wrath should be juxtaposed to "and whatever I asked from the gods they used to give to me" (lines 4, 12). The third curse, may he not allow him to eat because of rage should be compared to "and in my days also Y'DY ate and drank" (line 9, perhaps more is in the broken lines 5–7). The fourth curse, that he not be able to sleep, may be understood as the successor's own ghost not resting quietly or it may stand in distinction to Panammu for whom, Hadad et al. "[also cut off] sword and slander from my father's house" (line 9). The fifth curse is like the fourth "and may terror be given to him" in contrast to the scepter of authority and absence of dissention that Panammu enjoyed (lines 8–9).

The curse is to come upon the successor of Panammu who fails to provide for the memory of his father and his post-mortem comfort through the *kispu* obligation.[361] The person responsible for care of the deceased's ghost (*eṭemmu* or *GIDÎM*) was usually a close relative, often the heir, and was designated as the *pāqidu* (or *LÚ.SAG.ÈN.TAR*). An uncared-for ghost would wander the earth haunting the living.[362] Essential services that the *pāqidu* was to provide include: 1) making funerary offerings (*kispa kasāpu*), 2) pouring water (*mê naqû*), and 3)

Chapter 5: Blessing and Curse

calling the name of the deceased (*šuma zakāru*).³⁶³ These concerns are all mentioned and in the same order in the Hadad inscription: 1) "May the soul of Panammu [eat] with thee" ([תאכל.נב]ש.פנמו.עמך, line 17), 2) "May the soul of Panammu drink with thee" ([ותש]תי.נ[בש].[פ]נמו.[עמך], line 17), and 3) "Let him keep remembering the soul of Panammu with Hadad" (יזכר.נבש.פנמו.עמ[ה]ד[ד]), lines 17–18). Bayliss suggests that faithfulness in the *kispu* ritual may be tied to the process of inheritance; if this is so, the threat of disinheritance may underlie all of the more obvious curses.³⁶⁴

The curse that the god(s) not accept the descendant's sacrifice is an intentional contrast to the blessed state described in lines 12–13a, ב[י.וב]ימי (12 חלבת[י.]-[.]דת.אהב.לאלהי.ומת.יקחו.מנ.ידי.ומה.אשאל.מנ.אלהי.מת.יתן[נ].[.] לי (13) "In the days of my authority [ ] would I offer to the gods, and they used to accept (them) from my hand; and what I asked from the gods, they used always to give to me."³⁶⁵ This rejection-of-sacrifice curse is also found in the Tell Fekherye inscription, lines 16–18, where it begins the penalties to be imposed on the person who removes Hadadyisʻi's name from the furnishings of Hadad's temple,

(16 ילד.שמי.מנ.מאניא
(17 ז.בת.הדד.מראי.הדד.לחמה.ומוה.ילקח.מנ
(18 ידה.סול.מראתי.לחמה.ומוה.אל.תלקח.מנ.ידה.

Whoever removes my name from the furnishings of the temple of Hadad my lord, may my lord Hadad not accept his bread and his water from his hand. May my lady Sawl not accept his bread and his water from his hand.³⁶⁶

This curse may also be found in Kilamuwa 1. Some commentators think that the first of the curses in that inscription, lines 13–15a, is actually a promise of blessing for successors who honor Kilamuwa's inscription by anointing it with oil; in contrast, the one who destroys it shall have his head destroyed.³⁶⁷ This interpretation of Kilamuwa 1, is a particularly interesting parallel since both are directed toward "sons" and both have to do with honoring their father: in the case of Kilamuwa, by honoring his inscription (cf. Zakir B16–28; also 2 Sam. 18:18; Gen. 35:20; Exod. 24:4 for the practice of erecting memorial steles), and in the case of the Hadad inscription, by honoring the deceased Panammu through the *kispu* ritual. In biblical texts, Yahweh's acceptance of sacrifice also shows his pleasure and results in his blessing (e.g., Gen. 8:20–9:7; 1 Kgs. 3: 4–14), while in his displeasure with his people he rejects their sacrifice (e.g., Amos 7:21–22).

The imprecation of lines 22b–23a should be contrasted to line 4 of the inscription. In line 4, Panammu boasts that he so enjoys the favor of the gods that, "whatever I asked from the gods they they used to give to me," ומז.אשא[ל. מנ.]אלהי.יתנו.לי (cf. 1 Kgs. 3:5; Gen. 33:10; 1 Sam. 2:26).³⁶⁸ Kuntillet ʻAjrud 6 also provides an important parallel, "Whatever he asks from a man, may it be favored…and let Yahweh give unto him as he wishes (according to his heart)"

³⁶⁹כל אשר ישאל מאש חנן...ונתן לה יהו כלבבה

The curse that the unfaithful successor not eat (line 23b) may be a curse that he not have someone to perform the *kispu* for him. The rage may well be Hadad's, in contrast to the favor bestowed on Panammu, or it may be the rage of the untended *eṭemmu* against the unfaithful heir.³⁷⁰ The curse may be intended to contrast with the state of plenty that Panammu has enjoyed (cf. lines 4–8). Panammu received all that he asked from the gods, surely including plenty of food for his people which would serve to make his reign peaceful. An inadequate supply of food would make the people restless and restless people may revolt. Civil war could certainly be seen as a mark of the deity's rage.³⁷¹

This may also be the explanation for the next curse, that he not sleep (line 24a). "Sleep" may represent peacefulness in death, a death in which the deceased is properly cared for according to ritual.³⁷² In biblical literature both the phrases "to be gathered to one's kinsmen" (יאסף אל־עמיו e.g., Gen. 25:8, 17; 35:29; Num. 20:24, 26) and "to sleep with one's fathers" (שכב עם אבתיו e.g., 1Kgs. 2:10; 15:8; 2 Kgs. 10:35) occur but are only superficially similar.³⁷³ "Sleep" as a metaphor for death is something to be avoided, to be delivered from (e.g., Ps. 13:4), except in cases where life is deemed unbearable (e.g., Job 3:11, 13, 16). Though the practice of providing sustenance for the dead may well be represented in the Bible not merely as prohibited in certain situations, but sanctioned in some cases and perhaps implicitly required in others, this interpretation does not seem to best fit the context of this inscription.³⁷⁴ "Sleep," especially since the entire phrase is "and sleep may he withhold from him in the night," may refer to the worry which would burden the ruler of a land endangered by attack from the outside or dissent from within. The curse interpreted this way would contrast to Hadad's plenty which allows him to rest, his secure position on the throne, and possession of the scepter of authority (line 9).

The last curse (line 24) is that terror be given the disrespectful heir. This curse is best seen, like the previous two, as the antithesis of Panammu's civil order: chaos as the punishment of the gods rather than a stable throne and the order portrayed by the favor of the gods (cf. lines 1–4, 8–11). Kraeling suggests that the rest of the inscription describes the chaos that will come upon the one who disregards these curses.³⁷⁵

Nerab 1. This inscription and Nerab 2 were found in 1891 at the small village of Nerab located seven kilometers SE of Aleppo. The upper part of the text surrounds the head and upraised arms of a human figure at prayer, a representation (צלמה line 3) of the priest Sin-zer-ibni named in line 1. The lower part of the inscription (lines 9–14) is written across the skirt of his robe. The theophoric element "Sin" in his name seems to indicate that he was a priest of the Babylonian moon god (though that deity is called by its Aramaic appellation Sahar here).³⁷⁶

Chapter 5: Blessing and Curse

1) שנזרבנ כמר
2) שהר בנרב מת
3) וזנה צלמה
4) וארצתה
5) מנ את
6) תהנס צלמא
7) זנה וארצתא
8) מנ אשרה
9) שהר ושמש ונכל ונשכ יסחז
10) שנכ ואשרכ מנ חינ ומות לחה
11) יכטלוכ ויהאבדו זרעכ והנ
12) תנצר צלמא וארצתא זא
13) אחרה ינצר
14) זי לכ

Sin-zer-ibni, priest of Sahar at Nerab, deceased. This is his picture and his grave. Whoever you are who drag this picture and grave away from its place, may Sahar and Shamash and Nikkal and Nusk pluck your name and your place out of life, and an evil death make you die; and may they cause your seed to perish! But if you guard this picture and grave, in the future may yours be guarded![377]

Line 4. Koopmans translates וארצתה as "und sein Bett, Ruheplatz, Begräbinsstätte, sarkophag."[378] He compares Akadian *eršu* "bed"[379] and BH ערשׂ which means "couch, bed" and may be used of a sarcophagus as a last bed in Deut. 3:11.[380]

Line 5. The address to the would-be violator is all inclusive מנ את "Whoever you (are)" like Zakir B16, 18 in contrast to Karatepe A.iii.12–13 or Aḥiram line 2 which warn only the mighty.[381]

Line 6. The verb תהנס also occurs in the Zakir stele B20–21 where it is also used with אשרה, [מנ.אש]רה ויהנסנה[382] and it occurs in Nerab 2:8–9 where it serves both to describe the prohibited action of the violator and the punishment wished upon him. Halévy says that the root is נוס and translates it here as "remuer, déplacer, retirer"[383]; Donner and Röllig's translation "fortschleppen" coincides.[384] Hoffman identifies the verb root as אנס and interprets it as "reissen" in this context.[385] Koopmans apparently agrees with this identification of the root but translates the verb as "du wirst rauben."[386]

Line 8. Gevirtz suggests a play on words between אשרה in line 8 and מנ אשרכ חינ in line 10 ("remainder of life") and another idea the word אחרת(ה) ("future") in line 13 (cf. Nerab 2:10).[387] It is simpler and more logical, however, to be content with the parallel between אשרה and אשרכ and note the reciprocal nature of the crime and its called for punishment.[388]

Line 9. Halévy posits a form of the verb יסחו in his reconstruction of the last line of the B section of the Zakir stele [מנ.ש]ר[שה]נסח.[389] He interprets it there as a curse against the family of the violator of the inscription. That interpretation fits well in the context of Nerab 1:10–11 as well, which speaks of the violator's "name" and wish that his "seed" would perish.[390] Koopmans agrees with this interpretation, translating the root נסח as "herausreissen."[391] He compares Ps. 52:7(5) where the verb occurs with שרש "But God will break you down forever; he will snatch and tear you from your tent; he will unproot you from the land of the living" גם־אל יתצך לנצח יחתך ויסחך מאהל ושרשך מארץ חיים. The emphasis, as well as the language, of the verse seems to parallel that of the Nerab inscription. He also notes Ezra 6:11–12 where the verb occurs in Darius's edict given to protect the rebuilt temple

> Also I make a decree that if anyone alters this edict, a beam shall be pulled out of his house, and he shall be impaled upon it, and his house shall be made a dunghill. May the God who has caused his name to dwell there overthrow any king or people that shall put forth a hand to alter this, or to destroy this house of God which is in Jerusalem

ומני שים טעם די כל־אנש די יהשנא פתגמא דנה יתנסח אע מן־ביתה וזקיף יתמחא עלהי וביתה נולו יתעבד על־דנה: ואלהא די שכן שמה תמה ימגר כל־מלך ועם די ישלח ידה להשניה לחבלה בית־אלהא דך די בירושלם:[392]

Line 10. Halévy supplies מחה "frappant," as the last word of line 10, saying that the first letter is uncertain. His translation of the last phrase is thus "D'une mort foudroyante."[393] Clermont-Ganneau, not satisfied with any explanations available to him for לחה, determines its general meaning to be "malemort" in comparison with יהבאשו ממתתה in Nerab 2:9–10.[394] Koopmans finds a parallel for the adjective "evil" לחה in a curse from the Sefire treaty (I.A.26); a curse that the god Hadad may bring upon the transgressor of the treaty "every sort of evil (which exists) on earth and in heaven" כל מה לחיה בארק ובשמין.[395]

Line 11a. Two verbs occur in this line of the curse יכטלוך and ויהאבדו. Donner and Röllig say that יכטלוך is the same verb as Hebrew קטל.[396] Gibson agrees and provides יקתלו in his reconstruction of the curses ending the Zakir stele (B23).[397] The verb ויהאבדו is also found in Nerab 2:10 (האבד) of action to be taken against the posterity of the violator and in Sefire II.C.4 of action against the written treaty.[398] It is used similarly in Esth. 7:4a "For we are sold, I and my people, to be destroyed, to be slain, and to be annihilated" כי נמכרנו אני ועמי להשמיד להרוג ולאבד (also compare the Moabite stele וישראל.אבד.אבד.עלמ "and Israel perished utterly for ever").[399]

Lines 11b–14. These lines contain the blessing for anyone guarding

(תנצר) the inscription, that theirs likewise be guarded. The meaning of the word אחרה is important for the understanding of this phrase; does it mean simply "in the future" or does it refer to "another" who will guard the repentant thief's grave? Clermont-Ganneau discusses both options but does not come to a conclusion.[400] In BH נצר also means "watch, guard, keep" and is used of guarding a fortification (Nah. 2:2) as well as of people observing commandments (Deut. 33:9; Pss. 25:10 and 119:2, 22, 44, 34, 56, *et passim*), and of observation of treaty conditions.[401] The Akkadian *naṣāru* means "watch over, protect." This may mean that the protector's inscription will be guarded or, it may mean that all "which (is) to you" (i.e., everything belonging to you—your family and home) will be guarded as well. If the verb ינצר is interpreted as a passive, then this is only a promise regarding the protector's memorial, as Koopmans translates, "möge beschützt werden." If the verb is treated as active, then the subject may be the unspecified protector, again according to Koopmans, "möge man schützen."[402] Donner and Röllig translate the verb as a passive ("*möge hinfort beschützt werden*"), but in italics to allow for doubt and they list both interpretation possibilities in their notes.[403]

Does the act of plucking the inscription (and tomb in some cases?) from its place somehow deny its deceased owner continued existence (is this the concern as well in Aḥiram ויגל)? If so, then the curse is for the same to come upon the violator—that the violator's name and place be taken away, that the violator suffer an evil death, and that the violator's seed perish (Nerab 1:10–11). Is אשרה (line 8) used also for "home" in the sense of a place where one lives with one's family, or only for the grave? If it is also used for home then "place and name" could refer to one's family—the corresponding curse then is, quite logically, that the violator's life, family, and therefore memory be forgotten.[404] And, logically, the one guarding the inscription and thus preserving the memory of Sin-zer-ibni should have his own guarded and thus his own memory preserved.

Summary

The substantive blessing and curse combinations in the target inscriptions date from the mid-ninth to the early seventh century and are all, with the exception of the Amman Citadel inscription, in Aramaic. These texts include blessings and curses intended to protect monumental inscriptions: a blessing on the patron of the inscription and a curse on the one damaging it (Tell Fekherye and Zakir). Other protective inscriptions are found in Sefire I and II, curses to prevent damage to the treaty document. In Nerab 1, a different combination occurs: the curse against the one violating the priest's tomb is familiar, but the text also includes a blessing for anyone protecting the inscription. Another type of blessing and curse are found in the Sefire treaties, curses for the one unfaithful to the treaty stipulations and blessings (minimal) for the faithful vassal and the gracious overlord. The Hadad inscription presents yet another type, curses for a successor who fails to show proper respect for the deceased and blessings (damaged) for a loyal heir. Finally, the Amman Citadel inscription presents the most obscure example. In it the fate of the enemies of Ammon is to be defeated by the national deity Milcom, while Milcom's people enjoy שלמ.

Conclusions

As was noted in the previous chapter, general words for "curse" are seldom found in the target inscriptions. When found, the lion's share of the occurrences are a form of the root ארר all of which appear in Hebrew inscriptions. The earlier and non-Hebrew inscriptions use specific curses instead (e.g. "May Šamaš smash his head!"). In a more limited way, the same may be said for the use of the general word for "bless." Though the root ברכ is found more often, a disproportionately large percentage of the occurrences are in Hebrew inscriptions. Phoenician inscriptions also use the root ברכ frequently, indeed, the earliest occurrences are in Phoenician inscriptions. This may testify to yet another point of contact between Phoenicia and monarchic Israel and Judah (other connections include the design of the Jerusalem Temple and intermarriage between the royal houses of these two peoples).

While general words for curse and blessing in inscriptions of the target period tend to be found in Hebrew and Hebrew and Phoenician respectively, the juxtaposition of blessing and curse are not limited to these two peoples. Both specific blessings and specific curses are frequently found in period Aramaic inscriptions. Indeed, specific (and very ingenious) curses are most abundant in the Aramaic texts. This points to culturally different ways of expressing what appear to be the same cross-cultural intentions, i.e., the intentions of the blessings and curses are the same, only the ways of expressing them differ. While this may be only implied from a comparison of the Phoenician and Hebrew inscriptions to the Aramaic inscriptions, the Bible, whose final form is from a later period, contains blessings and curses which are both general (with ברכ or ארר, etc. like the Hebrew and some Phoenician inscriptions) and specific (like the Aramaic "May one hundred nurses nurse one child and not satisfy it"). All of the inscriptions containing specific blessing and curse combinations also include deity names: none of them are vague magical imprecations, rather they are dependent for their fulfillment upon the power of the deities mentioned. Not surprisingly, the Bible, which contains materials from different periods hundreds of years apart, includes all of these types of blessings and curses.

---

[1] A near parallel does occur with another DN in a late Punic example, *CIS* i 4945 (the transliteration and translation are mine):

לאדנ לבעלחמנ ולרבת (1

Chapter 5: Blessing and Curse

2) נ לחנת פנ בעל נדר ויתנ
3) צלח בנ ברכ בנ הר השק
4) לני יברכי ואש יר
5) נו תמחנת ז וקבת
6) תנת פנ בעל

"To lord Baʻal Ḥamman and to the lady Tanit face of Baʻal ṢLḤ son of BRK son of HR the ṢQLNY vowed and he gave (this). May they bless me. But the man disturbing this gift, may Tanit face of Baʻal curse."

[2] P. Bar-Adon, "An Early Hebrew Inscription in a Judean Desert Cave," *IEJ* 25 (1975), 226–27.

[3] Ibid., p. 229. Bar-Adon's drawing (p. 227) seems to indicate another occurrence of ארר on the last line of the inscription.

[4] This is true, with the understanding that Dever's reading of the Khirbet el-Qôm inscription (Chapter 3, "ברכ Alone") is incorrect.

[5] Though the participle is also seen in ברכ בעל in Kuntillet ʻAjrud 1:2 and הברכ בעל in Karatepe A.1.1. See the discussions in Chapter 3.

[6] Garr agrees (*Dialect Geography of Syria-Palestine, 1000–586 B.C.E.* [Philadelphia: University of Pennsylvania Press, 1985], p. 131).

[7] Christopher Mitchell, *The Meaning of BRK "to Bless" in the Old Testament*, in SBL Dissertation Series, no. 95 (Atlanta: Scholars Press, 1987), p. 12.

[8] Note, though, that in other inscriptions where both יהוה and אדני occur, אדני refers to a person (e.g., Lachish 6).

[9] Bar-Adon, "Early Hebrew Inscription," p. 227.

[10] Ibid., p. 227, drawing; p. 229, transliteration. The photographs included do not show anything below line 7.

[11] Patrick Miller, Jr., "Psalms and Inscriptions," in *Supplements to Vetus Testamentum, Congress Volume Vienna: 1980*, ed. J. A. Emerton (Leiden: E. J. Brill, 1981), p. 332.

[12] François Bron, *Recherches sur les inscriptions phéniciennes de Karatepe*, II Hautes Études Orientales, no. 11 (Geneva: Libraire Droz, 1979), p. 5; and Helmut T. Bossert and U. Bahadir Alkim, *Karatepe: Kadirli ve Dolaylari*, Publications for the Institute for Research in Ancient Oriental Civilisations, no. 3, (Istanbul: Pulhan Basimevi, 1947), p. 28 and map. This inscription is from a region just outside the strict confines set out for this study; though only about thirty miles north of the modern Syrian border. However, the language, time, and contents of the inscription are such that the boundaries will be stretched in this instance.

[13] *TSSI*, vol. 3, p. 42.

[14] *KAI*, vol. 2, p. 35.

[15] It is so designated by most scholars (*KAI* 26; *TSSI*, vol. 3, p. 41; Bron, *Recherches*, p. 12) though since it was the last discovered, it is not so designated in the older treatments.

[16] *TSSI*, vol. 3, p. 41. For an overall photograph of inscription B see H. Th. Bossert et al. *Karatepe Kazilari (Birinci Ön-Rapor), die Ausgrabungen auf dem Karatepe: Erster Vorbericht*, Türk Tarih Kurumu Yayinlarindan V. Seri, no. 9 (Ankara: Türk Tarih Kurumu Basimevi, 1950), plate XIV, photograph 70.

[17] Pietro Magnanini, *Le iscrizioni fenicie dell'Oriente: testi, traduzioni, glossari* (Rome: Istituto di Studi del Vicino Oriente, 1973), p. 53. The readings are Magnanini's except for the last letter of line 9 which is read as a ר instead of ע with Gibson (*TSSI*, vol. 3, p. 50). The translation is Gibson's (Ibid., p. 51).

[18] For more detail on this substantive blessing, see the individual treatments of the inscriptions

and the Yeḥimilk inscription particularly in the "Substantive Blessing" section below.

[19] Jonas C. Greenfield, "Scripture and Inscription: The Literary and Rhetorical Element in Some Early Phoenician Inscriptions," in *Near Eastern Studies in Honor of William Foxwell Albright*, ed. Hans Goedicke (Baltimore: Johns Hopkins, 1971), p. 265.

[20] A. M. Honeyman, "Epigraphic Discoveries at Karatepe," *PEQ* 80 (1948), 27.

[21] *TSSI*, vol. 3, pp. 94–95. The text is also given below in note 72.

[22] Michael L. Barré, "An Analysis of the Royal Blessing in the Karatepe Inscription," *Maarav* 3/2 (1982), 179.

[23] Ibid., p. 178. This is actually a modified version of Barré's "Chart A." I have made the transliteration to block letters and designated the line numbers of the inscription differently as well as abbreviated DN for "divine name" and RN for "royal name."

[24] Barré, "Royal Blessing," p. 184.

[25] Ibid., pp. 184–85. Barré strengthens his point with an excellent Ugaritic parallel *UT* 1019:2–6 *ilm tǵrk tšlmk t'zzk alp ymm wrbt šnt b'd 'lm* "May the gods preserve you(r life), keep you in health, give you vigor for 1,000 days, 10,000 years, (and) forever." The only real difference being that the one word blessing section of the Ugaritic version uses verbs rather than the nouns found in the one word part of the Phoenician blessing. He also quotes a fine, contemporary (722–656 B.C.) Assyrian blessing (p. 192) providing a transliteration and transliteration from the text in *ABL* (Robert F. Harper *Assyrian and Babylonian Letters Belonging to the Kouyunjik Collections of the British Museum*, part 7 [London: The University of Chicago Press/Luzac and Co., 1902], p. 786). *ABL* 733, obv. 6–10 *Aššur LUGAL DINGIR.MEŠ ašib É.ŠÁR.RA ana LUGALEN-ia likrūb ūmē (UD.MEŠ) arkūte (GÍD.DA.MEŠ) šanate (MU.AN.NA.MEŠ) mā'du[-te] šebē littutu ana LUGAL [belī]-[iá] lid[-din]* "May Assur, the king of the gods, who dwells in Ešarra, bless the king my lord; long days, many years, (and) fullness of old age may he give to the king my lord."

[26] *TSSI*, vol. 3, pp. 94–95.

[27] Barré, "Royal Blessing," p. 187. He also makes comparisons to the Hittite portion of the text to show that all three blessings of the second panel of the Phoenician blessing have to do with longevity but confesses that the meaning of the Hittite portion cannot be certainly fixed apart from the Phoenician.

[28] Magnanini, *Le iscrizioni*, p. 53; *TSSI*, vol. 3, pp. 50–51, readings and translation.

[29] RS 24.252 verso ll. 9–12, Nougayrol, *Ugaritica V*, p. 553.

[30] *ARTU*, pp. 187, 190.

[31] Magnanini, *Le iscrizioni*, p. 53, readings; *TSSI*, vol. 3, pp. 50–53, readings and translation.

[32] *TSSI*, vol. 3, p. 14.

[33] Alan R. Millard, "Alphabetic Inscriptions on Ivories from Nimrud," *Iraq* 24 (1962), 47. The text is very fragmentary. The transliteration into block letters is mine. See the discussion in the section "ארר" in Chapter 4.

[34] Nahman Avigad, "The Epitaph of a Royal Steward from Siloam Village," *IEJ* 3 (1953), 143.

[35] Ali Abou-Assaf, Pierre Bordreuil, and Alan R. Millard, *La statue de Tell Fekherye et son inscription bilingue assyro-araméenne*, Etudes assyriologiques, vol. 7 (Paris: Editions recherche sur les civilisations, 1982), p. 23. See the discussion in this chapter in the "Substantive Blessing and Curse" section.

[36] *TSSI*, vol. 2, p. 10.

[37] Ibid., vol. 3, p. 34.

[38] Joseph A. Fitzmyer, *The Aramaic Inscriptions of Sefîre*, Biblica et Orientalia, no. 19

Chapter 5: Blessing and Curse   211

(Rome: Biblical Institute Press, 1967), p. 20.

³⁹*TSSI*, vol. 2, pp. 95, 97.

⁴⁰*DISO*, p. 147 and *TSSI*, vol. 3, p. 63 say this verb is plural in order to agree with the plural subject, but Friedrich (Johannes Friedrich, *Phönizisch-punischeGrammatik*, Analecta Orientalia, vol. 32 [Rome: Pontifical Biblical Institute, 1951], p. 174) says the form is singular (cited in *DISO*). That this practice of effacing the name of an enemy or predecessor from records of his or her accomplishments was common practice would seem to be borne out by the frequency of prohibitions against it in the inscriptions included here and in Akkadian (especially Assyrian) and Egyptian inscriptions, e.g., the name of the Eighteenth Dynasty Egyptian "pharaoh" Hatshepsut (Hashepsowe), was effaced from almost all of her monuments after her death (Alan Gardiner, *Egypt of the Pharaohs, An Introduction* [Oxford: Oxford University Press, 1961], p. 187).

⁴¹BDB, p. 562.

⁴²Magnanini, *Leiscrizioni*, p. 53; *TSSI*, vol. 3, pp. 52–53.

⁴³*TSSI*, vol. 3, p. 53.

⁴⁴Edmond Sollberger, "Samsu-Iluna's Biblingual Inscriptions C and D," *RA* 63 (1969), 37, 40.

⁴⁵Reading with the ⅏ και συμπαραμενει ויארך "and may he shine" ("live" RSV) rather than the MT's ייראוך "may they fear you."

⁴⁶*KAI* 277; *TSSI*, vol. 3, p. 151.

⁴⁷For the text of B see, H. Theodore Bossert, "Karatepe," *Turk Tarih Kurumu Belleten* 17 (1953), 143–49 (a periodical not available to me). Donner and Röllig (*KAI*, vol. 2, pp. 37–38) give B as a simple parallel to A, noting only that B breaks off at its equivalent to A.iv.

⁴⁸Magnanini, *Leiscrizioni*, pp. 53–54. The readings are Magnanini's except where Gibson's division of the words at the end of iii.20 has been followed (*TSSI*, vol. 3, p. 54).

⁴⁹*TSSI*, vol. 3, pp. 53, 55. The break in the line numbering of the C.iv portion indicates where sacrifices for the statue upon which the inscription is written were included, while the description of sacrifices precedes the relevant part of the inscription in the A section. These lines near the end of the inscription (iv.21–v.5) are either missing or only sketchily translatable.

⁵⁰Minor differences like the repetition of בעבר in C.iv.12 versus its distributive use in A.iii.11 occur but are not germane to this study.

⁵¹Fitzmyer, *Sefire*, pp. 82–83 (see the discussion of the inscription below). Actually, the one who would receive punishment—the worker or the master—is not precisely stated as the inscription is broken at this point, nor is this clear in the Sefire inscription. It seems probable that the master would still be liable and, probably, so would the servant.

⁵²Honeyman, "Epigraphic Discoveries at Karatepe," p. 35.

⁵³Abou-Assaf, et al., *La statue*, p. 1. The statue bearing this Aramaic-Assyrian bilingual inscription was discovered in February 1979. A farmer expanding his field inadvertently unearthed the statue with a piece of farm equipment.

⁵⁴The *editio princeps* dates it to 866 B.C. (Ibid., p. 103). Stephen Kaufman gives the date as 830 plus or minus thirty years ("Reflections on the Assyrian-Aramaic Bilingual from Tell Fakhariyeh," *Maarav* 3 [1982], 141).

⁵⁵For an interesting proposal as to why two inscriptions (in each language) are present on the statue, see Douglas M. Gropp and Theodore J. Lewis, "Notes on Some Problems in the Aramaic Text of the Hadd-Yithʻi Bilingual," *BASOR* 259 (1985), 55-56 and 59, n. 17.

⁵⁶Abou-Assaf, et al, *La statue*, pp. 23–24. The transliteration into block letters and the translation are mine.

⁵⁷The actual statement that he erected the "image" (line 1 דמותא, line 12 צלם) in hopes of

obtaining these attentions from his god comes in line 10.

[58] So Jonas C. Greenfield and Aaron Shaffer ("Notes on the Akkadian-Aramaic Bilingual Statue from Tell Fekherye," *Iraq* 45 [1983], 114) and Bernard C. Batto ("Book Review: La statue de Tell Fekherye et son inscription bilingue assyro-araméenne," *CBQ* 47 [1985], 502) who identify the form as a Pa'el infinitive without the expected ה.

[59] A Pe'al infinitive with מ as a stem prefix according to Abou-Assaf, et al., *La statue*, pp. 31, 55.

[60] Greenfield and Shaffer, "Notes on the Akkadian-Aramaic Bilingual Statue," p. 114; and Batto, "Book Review," p. 502.

[61] If an adjective it would be better translated "and for abundance of his years," the *editio princeps* says it is a noun (p. 31) translating it "et pour la prolongation de ses années" (p. 24).

[62] *TSSI*, vol. 2, pp. 66–67.

[63] Abou-Assaf, et al., *La statue*, p. 24; and in their "traduction synoptiques" as "prosperité" (p. 62).

[64] Batto, "Book Review," p. 502. He suggests a parallel in Sefire III.8 but that usage is surely a noun and more nearly parallels Aramaic epistolary forms שלחתי לשלמכ (cf. Hermopolis papyri 1:12; ii:17; iii:5; etc., *TSSI*, vol. 2, pp. 129–36).

[65] Victor Sasson, "The Aramaic Text of the Tell Fakhriyah Assyrian-Aramaic Bilingual Inscription," *ZAW* 97 (1985), 95.

[66] Abou-Assaf, et al., *La statue*, pp. 31, 55. Kaufman ("Reflections," p. 166) plausibly argues that the root is לדד since the imperfect form found in line 11 is written ילד rather than ילוד.

[67] Greenfield and Shaffer, "Notes on the Akkadian-Aramaic Bilingual Statue," p. 114.

[68] Abou-Assaf, et al., *La statue*, p. 31 and Greenfield and Shaffer, "Notes on the Akkadian-Aramaic Bilingual Statue," p. 114. Gevirtz ("West-Semitic Curses and the Problem of the Origins of Hebrew Law," *VT* 11 [1961], 144 and n. 2) agrees with this root identification and finds a connection with Mishnaic Hebrew לוז "slander, pervert." He translates the Sefire text "I will detract" (I.C.18).

[69] So also Gropp and Lewis ("Notes on Some Problems," pp. 49-50) who, basically agreeing with Kaufman (see the discussion of the verb in line 11), say the range of meaning for the verb is "to remove, extinguish, efface." They further note that in this text, as at Sefire, "the verb is always associated with a nominal phrase introduced by *mn*."

[70] Sasson, "The Aramaic Text," p. 96.

[71] Compare the connection between "hear" and "obey" expressed by the word שמע, e.g. Mesha stele line 28 (*TSSI*, vol. 1, p. 75; *KAI* 181).

[72] The reading and translation are from *TSSI*, vol. 3, pp. 93–94.

(8) תברכ בעלת גבל אית יחומלכ
(9) מלכ גבל ותחוו ותארכ ימו ושנתו על גבל כ מלכ צדק הא ותתנ
(10) [לו הרבת ב]עלת גבל חנ לענ אלנמ ולענ עמ ארצ ז וחנ עמ אר
(11) צ ז [קנמי את] כל ממלכת וכל אדמ אש יספ לפעל מלאכת עלת מז
(12) בח זנ [ועלת פת]ח חרצ זנ ועלת ערפת זא שמ אנכ יחומלכ
(13) מלכ גבל [תשת את]ב על מלאכת הא ואמ אבל תשת שמ אתכ ואמ תס
(14) ר מ[לאכ]בת זא [ותס]ג את ה][---] ז דל יסדה עלת מקמ ז ותגל
(15) מסתרו הסרח[ן] הרבת בעלת גבל אית האדמ הא וזרעו
(16) את פנ כל אלנ ג[בל]

May the Mistress of Byblos bless Yeḥaumilk king of Byblos and give life to him and prolong his days and his years over Byblos; for he is a lawful king! And may [the lady], Mistress of Byblos, give [to him] favour in the sight of

Chapter 5: Blessing and Curse   213

the gods and favour in the sight of the people of this land! [Whoever you are], be you ruler or be you commoner, who may do further work on this altar, [or on] this gateway of gold, or on this portico, [you shall put] my name Yeḥaumilk [beside] your own on that work; and if you do not put my name beside your own but remove this work and [shift] this [(pillar)] along with its base from this place and uncover its hiding-place, may the lady, Mistress of Byblos, destroy both that man and his seed in the presence of all the gods of Byblos!

[73] See "Substantive Blessing" in Chapter 3.

[74] Greenfield and Shaffer, "Notes on the Akkadian-Aramaic Bilingual Statue," p. 114.

[75] Sasson, "The Aramaic Text," pp. 90, 96.

[76] This point is not particularly germane to this study. Kaufman ("Reflections," pp. 166–67) believes, based on the Assyrian text and other Assyrian examples, that the reference is to a repair, *ana arkat ūmê rubû arkû enuma bītu šu ušalbaruma ēnaḫu anḫussu lūdiš* "In future times, a future prince, *when that temple has become old and worn out*, should renew its disrepair." The italics in the translation are his.

[77] Ibid., pp. 162, 166.

[78] For a fuller treatment of this element in Akkadian inscriptions see the discussion of Kilamuwa 1 below under "Substantive Curse" and these references: Ebbe E. Knudsen, "Fragments of Historical Texts from Nimrud—II," *Iraq* 29 (1967), 60–62, fragment 4378 B, col. iii, lines 20–25; Daniel D. Luckenbill, *The Annals of Sennacherib*, University of Chicago Oriental Institute Publications, vol. 2 (Chicago: The University of Chicago Press, 1924), pp. 98, 101, 116; and *AHw*, vol. 2, pp. 843–44.

[79] Compare the PN Si'-gabbari on Nerab 2 (treated below under "Substantive Curse") (*TSSI*, vol. 2, p. 97; and Stephen A. Kaufman, "Si'-Gabbari, Priest of Sahr in Nerab," *JAOS* 90 [1970], 270–71). For the word referring to a person in the inscriptions, see Hadad line 32 (*TSSI*, vol. 2, p. 68).

[80] Simo Parpola and Kazuko Watanabe, *Neo-Assyrian Treaties and Loyalty Oaths*, State Archives of Assyria, vol. 2 (Helsinki: Helsinki University Press, 1988), p. 48. Line 455 *ᵈU.GUR qarrad DINGIR*.

[81] E. A. Speiser, "The Creation Epic," in *ANET*, p. 64, tablet II, line 95.

[82] Kaufman, "Reflections," p. 167.

[83] J. Nicholas Postgate, *Fifty Neo-Assyrian Legal Documents* (Warminster, England: Aris & Phillips, Ltd., 1976), pp. 99–100, document 12, rev. line 23.

[84] For the Hermopolis papyri, *TSSI*, vol. 2, p. 133.

[85] Abou-Assaf, et al., *La statue*, p. 23. The transliteration into block letters and the translation are mine.

[86] *TSSI*, vol. 2, pp. 64–67.

[87] The *editio princeps* understands a word division between ורדת and לארם. לארם they understand as the preposition ל, prosthetic א and a noun רום translated "Pour l'exaltation (?)." They see ורדת as a transcription of the Akkadian *ridûtu* with the idea of succession, translated "et la perpétuation" but are puzzled by the absence of ל after ו (*La statue*, pp. 24, 32). Kaufman reads the words ולרדת לתרצ and translates the phrase "In order to set aright the foundation of his throne" ("Reflections," pp. 162, 167). His reading of the first word brings it into agreement with the Assyrian version. Greenfield and Shaffer ("Notes on the Akkadian-Aramaic Bilingual Statue," p. 115) agree with this reading of the first word. Sasson disagrees with the *editio princeps*, Kaufman, and Greenfield and Shaffer, he thinks that the whole is best read as one word, probably a loan word from Akkadian, having a meaning akin to the Arabic verb *wrd* "to blossom" ("The Aramaic Text," p. 97). He says the letters of the first word are clear, contra Kaufman ("Reflections," p. 138, n. 2). Gropp and

Lewis say that the "word" is a combination of "preposition *l* plus two coordinating verbal nouns in construct with *krs'h*, which together form a hendiadys roughly equivalent in meaning to the single Akkadian noun *ana tiriṣ* [found in the Neo-Assyrian version of the inscription]. The lack of word divider between *l'rm* and *wr/ddt*, rather than militating against our view actually serves to underscore the unity of the hendiadys" ("Some Problems," p. 52; the material in brackets is mine).

[88] Greenfield and Shaffer, "Notes on the Akkadian-Aramaic Bilingual Statue," p. 110.

[89] *AP*, pp. 214, 222.

[90] For a fuller treatment see part "A" lines 9–10 above and Kuntillet ʿAjrud 6 in the "Substantive Blessing" section of Chapter 3.

[91] Abou-Assaf, et al., *La statue*, p. 23. The transliteration into block letters and the translation are mine.

[92] *TSSI*, vol. 2, pp. 66–67. However, the heir who does not perform his duty to his deceased father is cursed, זבחה.ואל.ירקק.[בהן.]ומז.ישאל.אל.יתן.לה "...]his sacrifice, and may he not look favourably upon it, and what he asks, may Hadad not give to him" (lines 22–23, pp. 68–69). See the treatment of these lines under "Substantive Curse" in Chapter 4.

[93] Kaufman, "Reflections," p. 168.

[94] Jonas C. Greenfield and Aaron Shaffer, "Notes on the Curse Formulae of the Tell Fekherye Inscription," *RB* 92 (1985), 52. They do note (p. 53) that a hieroglyphic Luwian text from Carchemish has a very similar curse "with him may the gods DN^1, DN^2, DN^3 be angry and from him [the malefactor] may they [the gods] not accept bread and (liquid) offering!"

[95] Greenfield and Shaffer, "Notes on the Curse Formulae," p. 54.

[96] Compare troublesome Sefire I.A.24 and Hillers's treatment of it (Delbert R. Hillers, *Treaty Curses and the Old Testament Prophets*, in Biblica et Orientalia, vol. 16 [Rome: Pontifical Biblical Institute, 1964] pp. 71–74).

[97] The corresponding blessing for obedience is found in Lev. 26:8.

[98] Fitzmyer, *Sefire*, pp. 14–15, 80–81. Compare Tell Fekherye, line 20:

ומאה.סאון.להינקן.אמר.ואל.ירוה

and Sefire I.A.23–24: ‏.ושבע שאן יהינקן אמר ו֯אל יש]בע‏.

[99] Daniel D. Luckenbill, *Ancient Records of Assyria and Babylonia*, vol. 2 (Chicago: The University of Chicago Press, 1927), p. 318, §828.

[100] Kaufman, "Reflections," pp. 170–72. His reasoning and reconstruction (in conjunction with Professor Zuckerman, p. 170, n. 77) are very detailed and logical. For a further look at his treatment, see the discussion of Sefire I.A.24 below.

[101] Donald J. Wiseman, *The Vassal-Treaties of Esarhaddon* (London: The British School of Archaeology in Iraq, 1958), pp. 61–62, lines 443–44.

[102] Greenfield and Shaffer, "Notes on the Akkadian-Aramaic Bilingual Statue," p. 116.

[103] Ibid. In their later article on this inscription ("Notes on the Curse Formulae," p. 56), they describe this word's use in later Aramaic and give other examples of Neo-Assyrian use.

[104] Note that this was where Job went to sit in his uncleanness, Job 2:8. Compare also Lam. 5:4.

[105] The RSV emends the first word of the verse to אל הן.

[106] The text is emended by the editors of the RSV, אהי to איה each time.

[107] Sasson, "The Aramaic Text," p. 102.

[108] Greenfield and Shaffer, "Notes on the Akkadian-Aramaic Bilingual Statue," p. 116. They also note that Akkadian *ḫaṭṭu* has a similar semantic range, from "staff" to "illness."

Chapter 5: Blessing and Curse

[109] Noted by Greenfield and Shaffer, "Notes on the Curse Formulae," p. 59; this fuller quotation is from Parpola and Watanabe, *Neo-Assyrian Treaties and Loyalty Oaths*, p. 48.

[110] Greenfield and Shaffer, "Notes on the Curse Formulae," p. 59.

[111] This text is being treated in this study for two reasons: 1) several of the commentators (as will be seen below) draw attention to the contrast between the fate of the enemies of Milkom versus the good things that will be the lot of Milkom's people; 2) several of the students of this inscription reconstruct the word שלם twice in the last, extant line.

[112] Siegfried H. Horn, "The Ammān Citadel Inscription," *BASOR* 193 (1969), 2. Horn dates the inscription to the first half of the eighth century (p. 8 and n. 14). Cross ("Epigraphic Notes on the Amman Citadel Inscription," *BASOR* 193 [1969], 17) dates it ca. 850. The lower right hand corner of the tablet is missing, leaving it not quite rectangular.

[113] Horn, "The Ammān Citadel Inscription," p. 8.

[114] Ibid., p. 9.

[115] Ibid., p. 13.

[116] Cross, "Amman Citadel Inscription," pp. 18–19. He emphasizes that none of the lines is complete; he does not even offer a translation of the text as a whole.

[117] Ibid., p. 19.

[118] A. van Selms, "Some Remarks on the 'Amman Citadel Inscriptions," *BO* 32 (1975), 8.

[119] William F. Albright, "Some Comments on the 'Amman Citadel Inscription," *BASOR* 198 (1970), 38–39.

[120] William H. Shea, "Milkom as the Architect of Rabbath-Ammon's Natural Defences in the Amman Citadel Inscription," *PEQ* 111 (1979), 18.

[121] Ibid., pp. 20–22. Shea says that Milkom is recognized here not as a creator god but as a "landscape-architect god" who is interested in shaping only this little corner of the world for his people (p. 23).

[122] Ibid., pp. 23–24.

[123] Émile Puech and Alexander Rofé, "L'inscription de la Citadelle d'Amman," *RB* 80 (1973), 534.

[124] Ibid., p. 531.

[125] Ibid., pp. 532, 534.

[126] Ibid.

[127] Paul. E. Dion, "Notes d'épigraphie ammonite," *RB* 82 (1975), 32–33. His translation is the same though his reading is a bit more conservative: [ש]לם.לכ.וש[למ].

[128] William J. Fulco, "The 'Amman Citadel Inscription: A New Collation (with Text)," *BASOR* 230 (1978), 41.

[129] Fulco does not actually say that the tablet includes a substantive blessing and curse; this is my interpretation based on his translation.

[130] Ibid.

[131] Victor Sasson, "The 'Amman Citadel Inscription as an Oracle promising Divine Protection: Philological and Literary Comments," *PEQ* 111 (1979), 117.

[132] Ibid., p. 118.

[133] Ibid. Sasson's reading of the text differs significantly from that of Horn, but only the differences in those parts relevant to this discussion will be noted.

[134] Ibid., p. 125.

[135] Ibid., p. 118.

[136] Ibid., p. 120. See *TSSI*, vol. 2, pp. 8–9.

[137] Sasson, "The 'Amman Citadel Inscription," p. 123.

[138] Ibid., p. 124.

[139] For the idea that the material, limestone, was significant, see Wolf Wirgin, "The Calendar Tablet from Gezer," in *Eretz-Irael: Archaeological, Historical and Geographical Studies* vol. 6, ed. M. Avi-Yonah, et al. (Jerusalem: Israel Exploration Society, 1960), p. 11. Of course, Sasson's idea that the size is significant depends on his belief that the Amman Citadel inscription is almost complete as is. The tablets treated by Wünsch actually vary in size fairly significantly. In fact the Gezer tablet is a mere four-and-a-half inches long by two-and-three-quarters inches wide, while the Amman Citadel inscription is approximately ten inches tall and eight inches wide at its widest point, making it a good deal larger.

[140] See the discussion above (under "Substantive Blessing") for more detail and the work of R. Wünsch "Limestone Inscriptions," pp. 156–87, especially pp. 186–87.

[141] William H. Shea, "The Amman Citadel Inscription Again," *PEQ* 113 (1981), 109.

[142] Ibid., p. 110.

[143] This information is available from several sources. Pognon (H. Pognon, *Inscriptions sémitiques de la Syrie, de la Mesopotamie et de la Mossoul* [Paris: Imprimerie Nationale, 1907], pp. 156–58) provides a detailed account of the expedition which led to the discovery of the inscription, omitting only the exact find site which he revealed later. Information regarding the relations of the fragments of the inscriptions to one another and their positions on the stone was gleaned mainly from a study of photographs.

Though several other vocalizations of the RN זכר have been suggested, including Zakar and Zakkur (the last being widely accepted), I shall leave that debate to others and simply use the older standard Zakir.

[144] *TSSI*, vol. 2, pp. 10, 12. Gibson's readings of the lines discussed here agree with those of Pognon (*Inscriptions sémitiques*, p. 175) except in the following cases: Pognon does not provide the ל at the beginning of line 16, likewise, he does not read the ר at the end of line 21, nor the ה at the beginning of line 22, nor the word יקתלו in line 23, nor reconstructions by Gibson in line 28. Pognon provides no reconstruction of C1 (though he does recognize the existence of the line). Halévy's reading (Jacob Halévy, "Nouvelles remarques sur l'inscription de Zakir," *RS* 16 [1908], 370) differs significantly from line 21 on and so is given here:

(21) אש[ר]ה.או.מנ.ישלח.בה
(22) אש.מדע[ת]ה.או.אש.נכ
(23) רי.בעלשמין.וא[ל
(24) ור.ואל.ורשפ.ושמש.ושהר
(25) והדד.ואלהי.שמי[נ.
(26) ואלה[י].ארק.ובעלעי[נ
(27) לחטאה.ל[אשא.ואית.
(28) זרעה.יסח.מנ.ש[ר]ש[ה.ו
(29) [יאבדה.לעלמ]

(Notes to Halévy's reading of the inscription: in line 23, the material before the ל in the word [בע]לשמין is a reconstruction and should be marked as such; in line 24, the stone is broken before the word ושמש, thus the words preceding it should be marked as a reconstruction, the same may be said for the material preceding ש[ר]ש in line 28.)

[145] *TSSI*, vol. 2, pp. 10, 16.

[146] Abou-Assaf, et al., *Lastatue*, p. 23. Cf. Sefire I.C.16 where this is also the case (Fitzmyer, *Sefire*, p. 20).

Chapter 5: Blessing and Curse 217

[147] Mark Lidzbarski, *Ephemeris für Semitische Epigraphik*, vol. 3 (Giessen: Alfred Töpelmann, 1915), p. 10.

[148] *TSSI*, vol. 2, p. 16. Gibson does not propose a meaning for the root הנע, though he does suggest a parallel in a curse on a later Aramaic boundary marker from Bahadirli near Karatepe (Ibid., p. 156).

[149] Charles C. Torrey, "The Zakar and Kalamu Inscriptions," *JAOS* 35 (1915), 363.

[150] This practice was apparently common in the ancient world, e.g. Hatshepsut (Ḥashepsowe), daughter of Thutmose I and wife of Thutmose II, who succeeded her husband, serving as Pharaoh during Egypt's 18th dynasty. Her successor, Thutmose III, son of her husband's concubine, effaced her name from her monuments and replaced it with his own or that of his father. Actually, the person responsible for this change is not known with certainty, but this conclusion seems likely. (See Gardiner, *Egypt of the Pharaohs*, pp. 181–88.)

[151] Pognon, *Inscription sémitiques*, p. 172.

[152] Ibid., p. 176.

[153] Ibid., n. 7.

[154] Lidzbarski, *Ephemeris*, vol. 3, p. 10. G. Hoffmann ("Aramaische Inschriften aus Nêrab bei Aleppo, neue und alte Götter," *ZA* 11 [1896], 212) in comparison with Targumic Aramic, locates this same verb in the Nerab inscriptions (1:6; 2:8, 9) as from the root אוס "rauben."

[155] J. J. Koopmans, *Aramäische Chrestomathie, ausgewählte Texte (Inschriften, Ostraka und Papyri) bis zum 3. Jahrhundert n. Chr.* (Leiden: Nederlands Instituut voor Het Nabije Oosten, 1962), p. 29.

[156] Torrey, "The Zakar and Kalamu Inscriptions," p. 363.

[157] *TSSI*, vol. 2, pp. 11, 16–17.

[158] Halévy, "Nouvelles remarques," p. 368.

[159] *TSSI*, vol. 3, pp. 106–07. Halévy points to Hadad lines 27–31 which he says describe punishments to be inflicted for "la profonation accomplie par la main d'autres personnes" ("Nouvelles remarques," p. 368).

[160] Halévy, "Nouvelles remarques," p. 368

[161] *TSSI*, vol. 3, p. 54–55.

[162] Fitzmyer, *Sefire*, p. 83. This inscription will be discussed and the text given below.

[163] *TSSI*, vol. 2, pp. 68–69.

[164] Gevirtz, "West-Semitic Curses," p. 144 and n. 6.

[165] Torrey, "The Zakar and Kalamu Inscriptions," p. 364. He cites late parallels from the Koran and Midrash.

[166] Ibid., p. 369.

[167] Gevirtz, "West-Semitic Curses," pp. 144–45. Note, however, that he has translated [י]הגע at the beginning of line 19 simply as "destroy."

[168] Ibid.

[169] *TSSI*, vol. 3, pp. 94–95.

[170] The reading and translation are from *TSSI*, vol. 3, pp. 106–07.

[171] Ibid., vol. 2, pp. 95–96.

[172] Halévy, "Nouvelles remarques," pp. 370–71. Torrey agrees with this interpretative reconstruction ("The Zakar and Kalamu Inscriptions," p. 364.) A similar curse occurs in a late Neo-Assyrian inscription, "angrily may they [the above mentioned gods] curse him and eradicate from all lands

his name, his seed, his offspring, (and) his progeny" (A. K. Grayson, "Cylinder C of Sin-šarra-iškun, a New Text from Baghdad," in *Studies on the Ancient Palestinian World*, ed. J. W. Wevers and D. B. Redford [Toronto: University of Toronto Press, 1971], p. 166).

[173] Ibid., p. 370. Interestingly, Halevy does not deal with the "C" part of the inscription even though Pognon had reported it and translated it (*Inscriptions sémitiques*, pp. 175-76).

[174] Albrecht Goetze, "Treaty Between Mursilis and Duppi-Tessub of Amurru," in *ANET*, p. 205.

[175] This may be simply a religious pogrom or it may be an attempt to wipe Ahab from memory. Notice too, that Ahab had killed not only Naboth, in order to appropriate his vineyard, but his sons as well (2 Kgs. 9:26).

[176] Franz Rosenthal, "Zakir of Hamath and Lu'ath," in *ANET*, p. 656.

[177] Gevirtz, "West-Semitic Curses," p. 145, n. 1.

[178] Ibid., p. 150, n. 2. The gods are called upon to act as harvesters or gleaners. For BH לקט as "glean, harvest" see Ruth 2:2ff.

[179] *TSSI*, vol. 2, pp. 12–13.

[180] יהגעו/יהנסו cf. line 16, "West-Semitic Curses," pp. 144-45.

[181] Ibid., vol. 3, pp. 52–53.

[182] Fitzmyer, *Sefîre*, pp. 2, 9 and n. 12. This scheme replaces the more awkward one used in earlier studies.

[183] Ibid., p. 1

[184] A. Dupont-Sommer and J. Starcky, "Une inscription araméenne inédite de Sfiré," *BMB* 13 (1956), 24.

[185] Fitzmyer, *Sefîre*, p. 2. The name of the vassal in the first two steles is Matî'el; in the extant portion of the third he is unnamed. Fitzmyer notes that all scholars do not accept this document as a vassal treaty, some viewing it rather as a parity treaty: a treaty between equals each possessing a version that stipulates his own obligations (Martin Noth, "Der historische Hintergrund der Inschriften von Sefire," *ZDPV* 77 [1961], 118–72; and *KAI*, vol. 2, pp. 271–72). Fitzmyer shows, however, the unsatisfactory points of this proposal (pp. 124–25). Fitzmyer also provides (p. 121) an outline of Korošeç's analysis of elements in Hittite vassal treaties (V. Korošeç, *Hethitische Staatsverträge: Ein Beitrage zu ihrer juristischen Wertung*, in Leipziger rechtswissenschaftliche Studien, vol. 60 [Leipzig: T. Weicher, 1931]). For a full treatment of these treaties—their forms, contents, and relationship to biblical materials—see Dennis J. McCarthy, *Treaty and Covenant: A Study in Form in the Ancient Oriental Documents and in the Old Testament* in Analecta Biblica, vol. 21 (Rome: Pontifical Biblical Institute, 1963). In a recent treatment, H. F. van Rooy ("The Structure of the Aramaic Treaties of Sefire," *JS* 1 [1989], 133–39) summarizes the major views and offers a hypothesis of the relationships of the segments; also, as he points out, the faces of each stele may be better read in a different order, i.e., ADBC instead of ABCD for stele I (p. 135).

[186] Fitzmyer, *Sefîre*, pp. 2–3. He fixes the *terminus ad quem* as 740 B.C. and notes that Matî'el was already king in Arpad as earlier as 754 B.C. when Aššurnirāri V established his own vassal treaty with Matî'el (Mati'ilu in the Neo-Assyrian text). The treaty between Aššurnirāri V and Mati'ilu bears many strong parallels to the Sefire treaties and will be compared throughout this study. Though translations may be found in McCarthy (*Treaty and Covenant*, pp. 195–97) and Erica Reiner ("Treaty Between Ashurnarari V of Assyria and Mati'ilu of Arpad" in *ANET*, pp. 532–33), the treatment by Ernest Weidner ("Der Staatsvertrag Aššurnirâris VI. [=V] von Assyrien mit Mati'ilu von Bit-Agusi," *AfO* 8 [1932], 17–34), providing transliterations and translations on facing pages, remains the standard.

[187] Fitzmyer, *Sefîre*, pp. 12–17. The readings of the inscription and the translations of the Sefire materials given below are Fitzmyer's unless otherwise specified. With regard to both the Aramaic readings and the English translations, Fitzmyer has the following note (p. 13, n. 13):

Chapter 5: Blessing and Curse 219

italics indicate uncertain readings; words in parentheses () are inserted for the sake of English idiom; square brackets [] indicate editorial restorations of lacunae; angular brackets < > indicate editorial additions to the text.

[188] André Dupont-Sommer, "Les inscriptions araméennes de Sfiré (stèles I et II)" *MPAIBL* 15 (1958), 38. Text quoted from Weidner ("Der Staatsvertrag," pp. 22–23). The italics are his. Luckenbill's translation is, "As for [Mati'ilu's] wife,—may she, [like] the mule, [be useless] for bearing children" (*Ancient Records*, vol. 1, p. 268). Dupont-Sommer refers to a Hittite text which is also to the point, "Und seine Gattinnen mögen Söhne (und) Töchter nicht gebären...seine Rinder (und) seine Schafe mögen Jungstier (und) Bock nicht gebären" (p. 38; quoted from Johannes Friedrich, "Der hethitische Soldateneid," *ZA* 35 [1924], 165). Fitzmyer also compares the treaty of Šamši-adad V with Marduk-zākir-šumi I (*AfO* 8 [1932-33], 27-29; Fitzmyer, *Sefîre*, p. 41).

[189] Albrecht Goetze, "God List, Blessings and Curses of the Treaty Between Suppiluliumas and Kurtiwaza," in *ANET*, p. 206.

[190] Hillers (*Treaty-Curses*, p. 28) terms these first four curses "futility curses."

[191] Fitzmyer, *Sefîre*, pp. 14–15.

[192] Luckenbill, *Ancient Records*, vol. 2, p. 318, §828.

[193] Reiner, "Treaty Between Ashurnirari V of Assyria and Mati'ilu of Arpad," p. 533. This version of the curse is more concise that the others, "Let there be no milk to suck for the oxen, asses, sheep, and horses in his land."

[194] Šurpu text as quoted by Hillers (*Treaty-Curses*, p. 22) from Erica Reiner, *Šurpu: A Collection of Sumerian and Akkadian Incantations*, AfO supplement 11 (Berlin: Ernst Weidner, 1958), the standard text. Hillers also cites the Era Epic IV.121 where the god Era says, "I will make the breast dry up, so that the baby shall not live" (*Treaty-Curses*, p. 62).

[195] Dupont-Sommer, "Les inscriptions araméennes de Sfiré," pp. 38–39. On the mention of the nurse anointing her breast, he says that this was done because the women would then give more milk. Fitzmyer disagrees, noting that he gives no evidence of this belief in antiquity and prefers Hiller's explanation (*Treaty-Curses*, p. 61, n. 52) that the breasts were anointed to prevent soreness and cracking (*Sefîre*, p. 42).

[196] Fitzmyer, *Sefîre*, p. 15.

[197] Dupont-Sommer, "Les inscriptions araméennes de Sfiré," pp. 40–41.

[198] Fitzmyer, *Sefîre*, p. 44.

[199] Hillers, *Treaty-Curses*, p. 73

[200] Hans Bauer, "Ein aramäischer Staatsvertrag aus dem 8. Jahrhundert v. Chr., Die Inschrift der Stele von Sudschin," *AfO* 8 (1932), 7.

[201] Ibid. Fitzmyer also notes this and translates it thus (*Sefîre*, p. 43).

[202] Hillers, *Treaty-Curses*, pp. 73–74.

[203] Fitzmyer, *Sefîre*, p. 44. He gives a full summary of the various interpretations up to the time (1967) of his work (pp. 43–45). Hillers suggests that the stone cutter made a mistake in writing this word and even if the reading is actually בכתה, it is an error for בנתה (*Treaty-Curses*, p. 72).

[204] Kaufman, "Reflections," pp. 170–72. His reasoning and reconstruction (in conjunction with Professor Zuckerman, p. 170, n. 77) are very detailed and logical. Fitzmyer's reading of I.A.24a is ושבע בכתה יהכן בשט לחם ואל יהרגן (*Sefîre*, p. 14). Tell Fekherye line 22a reads ומאה נשון לאפן בתנור לחם ואל ימלאנה (my transliteration). Kaufman proposes a reading of בנתה for the second word which provides a logical alternative to נשון. He says that Fitzmyer's יהכן is better read as יאפן and his בשט is actually ב(?)ט(?) (to which he compares Late Akkadian *nappaṭu* "brazier." The last word (יהרגן) can be read as ימלאן with the resulting reading ושבע בנתה יאפן ב| ט לחם ואל ימלאן being an almost exact parallel to the line in Tell Fekherye. Though Dupont-Sommer says the

ב in בכתה is certain, it is quite hard to read (Dupont-Sommer, "Les inscriptions araméennes de Sfiré," p. 40). However, based on a comparison of the photographs in Dupont-Sommer (plate VI) and Fitzmyer (Sefîre, plate XIV), the ה in יהכנ does not look like an א, and the reading בשׁ appears correct; the letters of the last word are very difficult to read because of the pitted surface of the stone.

[205] Fitzmyer, Sefîre, p. 14.

[206] Dupont-Sommer, "Les inscriptions araméennes de Sfiré," pp. 41–42.

[207] Fitzmyer, Sefîre, pp. 14–15. The readings and translation are his. There is also the issue of who or what is referred to by אשׁר. Does this refer to the god? If so, why is he not mentioned in the list of deities? Does this refer to the country? If so, what relationship is implied between Bir-Ga'yah and Assyria? (p. 45) Either interpretation is acceptable for the understanding of the curse given above.

[208] KAI, vol. 2, p. 248.

[209] Fitzmyer, Sefîre, pp. 14–15. The readings and translation are his.

[210] Ibid. The readings and translation are his.

[211] Dupont-Sommer, "Les inscriptions araméennes de Sfiré," p. 43.

[212] Fitzmyer, Sefîre, pp. 14–15. The readings and translation are his.

[213] Wiseman, *The Vassal-Treaties of Esarhaddon*, pp. 61–62, lines 442–43.

[214] Fitzmyer, Sefîre, p. 46.

[215] Ibid., pp. 14–15. The readings and translation are his.

[216] Dupont-Sommer, "Les inscriptions araméennes de Sfiré," p. 43.

[217] Fitzmyer, Sefîre, p. 46.

[218] Ibid., pp. 14–15. The readings and translation are his.

[219] Dupont-Sommer, "Les inscriptions araméennes de Sfiré," p. 44. Though this word is rare and used only of humans (BDB, p. 118).

[220] Dupont-Sommer, "Les inscriptions araméennes de Sfiré," p. 44.

[221] Ibid. The text is quoted from Weidner ("Der Staatsvertrag," pp. 22–23, V.6).

[222] Wiseman, *The Vassal-Treaties of Esarhaddon*, pp. 61–62, line 442; also lines 488–89, "[May an irresistible flood come up from the earth and devastate you]."

[223] Koopmans, *Aramäische Chrestomathie*, vol. 2, p. 51.

[224] Fitzmyer, Sefîre, pp. 46–47.

[225] Ibid., pp. 14–15. The readings and translation are his.

[226] Dupont-Sommer, "Les inscriptions araméennes de Sfiré," p. 45. The text is quoted from Weidner ("Der Staatsvertrag," pp. 20–21, IV.20). Italics in the translation are Weidner's.

[227] Dupont-Sommer, "Les inscriptions araméennes de Sfiré," p. 45.

[228] Fitzmyer, Sefîre, p. 47.

[229] Dupont-Sommer, "Les inscriptions araméennes de Sfiré," p. 45.

[230] Fitzmyer, Sefîre, pp. 14–15. The readings and translation are his.

[231] Dupont-Sommer, "Les inscriptions araméennes de Sfiré," p. 45. The text is quoted from Weidner ("Der Staatsvertrag," pp. 20–21, IV.19).

[232] Luckenbill, *Ancient Records*, vol. 2, p. 311, §811; noted by Hillers, *Treaty-Curses*, p. 57.

[233] Fitzmyer, Sefîre, pp. 14–15. The readings and translation are his.

Chapter 5: Blessing and Curse 221

[234] Dupont-Sommer, "Les inscriptions araméennes de Sfiré," p. 46.

[235] Ibid., p. 47. He notes שחם as an alternate root for שחת comparing Syriac. He is unable to provide a meaning for the rest of the word (בחן).

[236] Hillers, *Treaty-Curses*, pp. 54–55.

[237] Fitzmyer, *Sefîre*, p. 48.

[238] Hillers, *Treaty-Curses*, p. 56.

[239] Fitzmyer, *Sefîre*, pp. 48–49. Hillers agrees (*Treaty-Curses*, p. 55, n. 34).

[240] Fitzmyer, *Sefîre*, p. 49.

[241] Hillers, *Treaty-Curses*, p. 56. Fitzmyer notes a reference in the Gilgamesh Epic, tablet 12, lines 93–94 (*Sefîre*, p. 49).

[242] Fitzmyer, *Sefîre*, p. 49.

[243] Ibid., pp. 14–15. The readings and translation are his.

[244] Dupont-Sommer, "Les inscriptions araméennes de Sfiré," pp. 47–48. He deals with the animals mentioned in some detail (pp. 48–49).

[245] Ibid., p. 47. The text and translation are from Weidner ("Der Staatsvertrag," pp. 22–23, V.5–6).

[246] Hillers, *Treaty-Curses*, p. 44; quoted from Rykle Borger, *Die Inscriften Asarhaddons Königs von Assyrien*, AfO supplement 9 (Graz: Ernst Weidner, 1956), p. 14, §11:A–G, Episode 7, Fassung c:G.

[247] See the comments on that text below ("Substantive Curse"). The line number given is that of the modified text changed after Hoftijzer's study; he calls this line 11.

[248] Fitzmyer, *Sefîre*, pp. 14–15. The readings and translation are his.

[249] Dupont-Sommer, "Les inscriptions araméennes de Sfiré," pp. 48–49. He treats the possible identities and locations of these towns in the following pages (pp. 49–50). Fitzmyer agrees with this interpretation of the relationship between the cities (*Sefîre*, p. 51).

[250] McCarthy, *Treaty and Covenant*, pp. 103–04.

[251] Wiseman (*The Vassal-Treaties of Esarhaddon*, p. 26) lists over forty examples in lines 526–658; as noted in McCarthy (*Treaty and Covenant*, p. 104).

[252] Hillers, *Treaty-Curses*, p. 19. The italics are his.

[253] Ibid.

[254] Fitzmyer, *Sefîre*, pp. 14–17. The readings and translation are his.

[255] Dupont-Sommer, "Les inscriptions araméennes de Sfiré," pp. 50–51. He compares use of magical rites in the Hittite "Soldier's Oath,"

> Then he places wax and mutton fat in their hands. He throws them on a flame and says: "Just as this wax melts, and just as the mutton fat dissolves,—whoever breaks these oaths, [shows disrespect to the king] of the Hatti [land], let [him] melt lik[e wax], let him dissolve like [mutton fat]!" [The me]n declare: "So be it!"

(translation from Albrecht Goetze, "The Soldiers' Oath" in *ANET*, p. 353; text and another translation in Friedrich, "Der hethitische Soldateneid," pp. 162–63).

[256] Fitzmyer, *Sefîre*, p. 52.

[257] Ibid., p. 53.

[258] Hiller, *Treaty-Curses*, pp. 20–21. The standard for *Maqlû* is Gerhard Meier, *Die assyrische Beschwörungssammlung Maqlû*, AfO supplement 2 (Berlin: Ernest Weidner, 1937).

²⁵⁹ Dupont-Sommer, "Les inscriptions araméennes de Sfiré," pp. 52. He provides full references for these texts.

²⁶⁰ Fitzmyer, *Sefire*, p. 53.

²⁶¹ Stanley Gevirtz, "Jericho and Shechem: A Religio-Literary Aspect of City Destruction," *VT* 13 (1963), 62. He compares curses against conquered sites from Hittite Ḫunusa to Jericho and Shechem to Carthage (pp. 60–61). In every case except Shechem, a curse against the one who rebuilt or reinhabited the city is accompanied by sowing the site with spice (salt, cress, or other). Though he is unable to "categorically rule out" the possibility that salt or cress may have served as "symbols of perpetual ruin and infertility," he maintains that is less likely (p. 62, n. 2).

²⁶² Dupont-Sommer, "Les inscriptions araméennes de Sfiré," pp. 52–53. He does not actually explain the detrimental effect that this was thought to have though he goes into some detail as to the actual species of plant. Luckenbill simply transliterates it and says in parenthesis "some prickly plant" (*Ancient Records*, vol. 2, p. 310, §811). Also compare the *Šurpu* curses where salt and cress (*saḫlu*) are mentioned several times in ceremonial curses (e.g., III.95), though the ritual may differ from that in the Sefire treaty (Hillers, *Treaty-Curses*, p. 23).

²⁶³ Goetze, "The Soldiers' Oath," p. 354.

²⁶⁴ Dupont-Sommer, "Les inscriptions araméennes de Sfiré," p. 54.

²⁶⁵ Fitzmyer, *Sefire*, p. 54.

²⁶⁶ Wiseman, *The Vassal-Treaties of Esarhaddon*, pp. 75–76, lines 608–11.

²⁶⁷ Hillers, *Treaty-Curses*, p. 60.

²⁶⁸ Dupont-Sommer, "Les inscriptions araméennes de Sfiré," pp. 56–57, quoting the Mati'ilu treaty, V.12–13; and the Soldiers' Oath, quoted from Friedrich, "Der hethitische Soldanteneid," p. 167, II.51.

²⁶⁹ Hillers, *Treaty-Curses*, p. 60; quoted from Wiseman, *The Vassal-Treaties of Esarhaddon*, pp. 71–72, lines 573–75. Also compare line 453 of the Esarhaddon treaties (pp. 63–64).

²⁷⁰ Dupont-Sommer, "Les inscriptions araméennes de Sfiré," p. 57, quoting the Mati'ilu treaty, VI.2.

²⁷¹ Ibid. He does not give the text; it is quoted here from Goetze, "The Soldiers' Oath," p. 354, III.2–9

²⁷² In the Mati'ilu treaty (I.10–14) it is specified that the lamb is not brought forth for sacrifice, slaughter, or divination, but "um die Vertragsbestimmungen Aššurnirāris, des Königs des Landes [Assyrien], *mit* Mati'ilu abzuschliessen, ist er her[aufgebracht]" (Weidner, "Der Staatsvertrag," p. 19).

²⁷³ Dupont-Sommer, "Les inscriptions araméennes de Sfiré," p. 57. The transliteration and translation are from Weidner ("Der Staatsvertrag," pp. 18–19, I.21–23). Further lines, through line 35, specify how the shoulder of the ram, torn off, also represents that of Matî'el his sons, his nobles, and his people.

²⁷⁴ Fitzmyer, *Sefire*, p. 56.

²⁷⁵ Bauer, "Ein Aramäischer Staatsvertrag," p. 10.

²⁷⁶ So Hillers (*Treaty-Curses*, pp. 19–20) and Erica Reiner ("The Vassal-Treaties of Esarhaddon," in *ANET*, p. 539); note, however, that Wiseman does not understand the ewe to be slain (*The Vassal-Treaties of Esarhaddon*, p. 70).

²⁷⁷ Dupont-Sommer, "Les inscriptions araméennes de Sfiré," pp. 58–59. He notes that there is inadequate room for עבדא but says, "le mot עבדא aurait été omis par haplography après יעבד." Though Dupont-Sommer's biblical parallel is apt, he stops too soon; the further context reveals that the women will be stripped naked and this agrees with the other reading of the text (see Bauer, just below).

Chapter 5: Blessing and Curse 223

[278] McCarthy, *Treaty and Covenant*, p. 122. Line 428 is quoted in the next paragraph; lines 588–90 appear ambiguous.

[279] Bauer, "Ein aramäischer Staatsvertrag," p. 10.

[280] Ibid. He cites the depictions on the Bronze Gates of Balawat.

[281] Wiseman, *The Vassal-Treaties of Esarhaddon*, pp. 61–62, lines 428–29.

[282] Hillers, *Treaty-Curses*, pp. 58–60.

[283] Ibid., p. 58.

[284] Fitzmyer, *Sefire*, p. 57.

[285] Compare not only the Mati'ilu treaty (as quoted above), but also the Hittite Soldiers' Oath (*ANET*, p. 354) and Esarhaddon (Borger, *Die Inschriften Asarhaddons*, p. 99, lines 55–56) among others (see Hillers, *Treaty-Curses*, pp. 66–68) as well as the Bible (Isa. 19:16; Jer. 50: 35–38; 51:30; Nah. 3:13; et al.).

[286] Dupont-Sommer, "Les inscriptions araméennes de Sfiré," pp. 59–60. He plausibly reconstructs the end of the line to match the end of the previous curse.

[287] Fitzmyer, *Sefire*, pp. 16–19. The transliteration and translation are Fitzmyer's. In line 33, the longer form of the formula also occurs, howbeit, the first part שקרתם לכל אלהי is missing and is restored thus [שקרת לא]להי by Fitzmyer (p. 18). Note that the singular rather than the plural form of the verb is used, the usual לכל is missing, and instead the ל is attached to the noun אלהי. These formulas are quite irregular in I.B though apparently the differences are stylistic rather than substantive (see the treatment of Sefire III in Chapter 4 "Substantive Curse").

[288] Ibid., p. 18.

[289] An even briefer version appears in lines 36b–37a. As Fitzmyer states, there is really not enough room for any of the extant versions of the formula to fit in the lacuna (*Sefire*, p. 71). Therefore, either an otherwise unknown version was used, or the engraver skipped part of one of the usual versions. Fitzmyer gives the reading שק[רת ז]נה in his overall reading of the text (p. 19) and suggests the reconstruction שק[רת בעדיא]<זי בספרא>[ז]נה in his notes (p. 71).

[290] F. Charles Fensham, "Malediction and Benediction in Ancient Near Eastern Vassal-Treaties and the Old Testament," *ZAW* 74 (1962), 6.

[291] Fitzmyer, *Sefire*, p. 107.

[292] Ibid., pp. 16–17. The transliteration and translation are his.

[293] Dupont-Sommer, "Les inscriptions araméennes de Sfiré," p. 77.

[294] Fensham, "Malediction and Benediction," pp. 6–7. A translation of the Muršiliš—Dupi-Tešub treaty is found in *ANET*, pp. 203–05; and of the Šuppiluliuma—Kurtiwaza treaty in *ANET*, pp. 205–06. The spellings "Dupi-Tešub" and "Šuppiluliuma" are used except in quotations.

[295] Fitzmyer, *Sefire*, pp. 18–19.

[296] Dupont-Sommer, "Les inscriptions araméennes de Sfiré," p. 90.

[297] Fitzmyer, *Sefire*, p. 74. Moran's article is cited below.

[298] Fitzmyer, *Sefire*, pp. 20–21.

[299] Dupont-Sommer, "Les inscriptions araméennes de Sfiré," pp. 88, 91. This is stated positively; the inscription states it negatively, "Whoever will not guard/observe the words of this inscription " ומנ ליצר מלי ספרא זי.

[300] Fitzmyer, *Sefire*, p. 75.

[301] Fensham, "Malediction and Benediction," pp. 6–7.

[302] See the discussion of these lines in the treatment of the Zakir inscription above.

303 Fitzmyer, *Sefîre*, pp. 18–19. The readings and translation are his.

304 Ibid., p. 75; Wiseman, *The Vassal-Treaties of Esarhaddon*, pp. 51–52. Also compare the ending of the treaty between Muršiliš and Dupi-Tešub (*ANET*, p. 205) and quoted below in the discussion of II.C which has a very similar conclusion.

305 Dupont-Sommer, "Les inscriptions araméennes de Sfiré," pp. 92–93.

306 Ibid., p. 93. Fitzmyer favors the BH cognate idea over the Mishnaic Hebrew connection (*Sefîre*, p. 76). The meaning, "Let them [the words of the father] not escape from your sight" (Prov. 4:21a), seems a bit remote, though.

307 Gevirtz, "West-Semitic Curses, p. 144, n. 2.

308 Dupont-Sommer, "Les inscriptions araméennes de Sfiré," p. 93.

309 The translation is mine.

310 W. L. Moran, "A Note on the Treaty Terminology of the Sefîre Stelas," *JNES* 22 (1963), 174–75. He cites several examples from Mari and El-Amarna.

311 Fitzmyer, *Sefîre*, p. 76. He notes that Akkadian *šuršu* may also mean "sprout, scion."

312 Wiseman, *The Vassal-Treaties of Esarhaddon*, pp. 51–52, lines 292–95.

313 Dupont-Sommer, "Les inscriptions araméennes de Sfiré," p. 95.

314 F. Charles Fensham, "Common Trends in Curses of the Near Eastern Treaties and *Kudurru*-Inscriptions Compared with Maledictions of Amos and Isaiah," *ZAW* 75 (1963), 159–60.

315 For information on the discovery and arrangement of the Sefire steles see the beginning of the treatment of Sefire I.

316 Fitzmyer, *Sefîre*, pp. 80–81.

317 See the discussion of I.A.21b–23 in this section.

318 Dupont-Sommer, "Les inscriptions araméennes de Sfiré," p. 101.

319 See the discussion of line I.A.24 in this section.

320 Dupont-Sommer, "Les inscriptions araméennes de Sfiré," p. 101.

321 The first four inscriptions mentioned are treated in this study. The texts of the Tabnit and Eshmunazar inscriptions may be found in *TSSI*, vol. 3, pp. 103, 106, and 108 respectively; also Magnanini, *Le iscrizioni*, pp. 6 and 3–4 respectively.

322 See Assurbanipal's record of his conquest of Elam (Luckenbill, *Ancient Records*, vol. 2, p. 310, §810, quoted below in the discussion of the Aḥiram inscription).

323 Dupont-Sommer, "Les inscriptions araméennes de Sfiré," p. 102.

324 See the discussion of I.A.36 above and the Akkadian parallels mentioned.

325 Dupont-Sommer, "Les inscriptions araméennes de Sfiré," p. 102. Cf. I.A.38–42.

326 Ibid., pp. 102–03.

327 Ibid., p. 103.

328 Hillers, *Treaty-Curses*, p. 55. The standard text for this inscription is Borger, *Die Inschriften Asarhaddons*, p. 109, §99, ^{d}Baltili^{mes} ^{d}Ana(?)ti Bā[til]i^{mes}

329 Hillers, *Treaty-Curses*, p. 56.

330 Fitzmyer, *Sefîre*, pp. 80–81. The readings and translation are his.

331 Dupont-Sommer, "Les inscriptions araméennes de Sfiré," p. 108.

332 Fitzmyer, *Sefîre*, pp. 86–87.

Chapter 5: Blessing and Curse 225

[333] Dupont-Sommer, "Les inscriptions araméennes de Sfiré," pp. 108–09. Fitzmyer agrees (*Sefire*, p. 87).

[334] Fitzmyer, *Sefire*, pp. 80–83.

[335] Dupont-Sommer, "Les inscriptions araméennes de Sfiré," pp. 105, 113.

[336] Ibid., p. 114. See the fuller treatment of this formula in the discussion of Sefire III in the section "Substantive Curse" of Chapter 4.

[337] Fitzmyer, *Sefire*, pp. 82–83.

[338] Dupont-Sommer, "Les inscriptions araméennes de Sfiré," p. 116.

[339] Ibid., p. 119.

[340] Fitzmyer, *Sefire*, p. 90.

[341] McCarthy, *Treaty and Covenant*, p. 126. Cf. Hab. 2:11 and Luke 19:40.

[342] Dupont-Sommer, "Les inscriptions araméennes de Sfiré," p. 120.

[343] Ibid., pp. 120–21. He says that the unusual spelling of אהבד for אהאבד is "purement accidentelle" but indicates that א is quiescent. Fitzmyer agrees (*Sefire*, p. 91).

[344] Dupont-Sommer, "Les inscriptions araméennes de Sfiré," pp. 121–22.

[345] Fitzmyer, *Sefire*, p. 91. He also says that attempts to relate it to גור "be afraid" are "unconvincing."

[346] Compare Karatepe C.iv.13–21 in this chapter in the section "ברכ and שלמ with Substantive Curse."

[347] Fitzmyer, *Sefire*, p. 92.

[348] Dupont-Sommer, "Les inscriptions araméennes de Sfiré," p. 122.

[349] Fitzmyer, *Sefire*, pp. 82–83.

[350] Dupont-Sommer, "Les inscriptions araméennes de Sfiré," p. 123.

[351] Ibid., p. 124. Quoted from Friedrich W. König, *Handbuch der chaldischen Inschriften*, AfO supplement 8, part 2 (Graz, Austria: Ernest Weidner, 1957), p. 115, §102, section IV.

[352] Fitzmyer, *Sefire*, p. 92.

[353] Goetze, "Treaty Between Mursilis and Duppi-Tessub of Amurru" in *ANET*, p. 205. Compare also the treaty between Šuppiluliumas and Kurtiwaza (Goetze, "God List, Blessings, and Curses of the Treaty Between Suppiluliumas and Kurtiwaza," *ANET*, p. 206) which concludes with a very similar protective oath, curse, and blessing.

[354] As Fitzmyer says above, no verb is provided. Perhaps good things might be withdrawn from him and his people all their days.

[355] Though in the biblical passages the blessings are much more extensive than in other ancient Near Eastern treaty texts.

[356] *TSSI*, vol. 2, p. 60.

[357] *TSSI*, vol. 2, pp. 66–69. Koopmans's readings (*Aramaïsche Chrestomathie*, vol. 1, p. 6) are the same as those of Gibson in all of the firm readings and most of the reconstructions. Koopmans differs in his reconstruction of the missing text of line 21 where he reads ויזבח|.הדד.זנ.יזכר not supplying the negation which Gibson provides.

[358] *NSI*, p. 167.

[359] Ibid., pp. 160, 162.

[360] Gibson notes that some commentators think the inscription does contain a warning against tampering with it (*TSSI*, vol. 2, p. 61). He says that this position is based on a misreading of one

word and a misinterpretation of another word both in line 28. Lidzbarski (*Handbuch der nordsemitischen Epigraphik*, vol. 1 [1962; rpt. Hildesheim: Georg Olms, 1898], p. 442); Cooke (*NSI*, pp. 162, 170); and Donner and Röllig (*KAI*, vol 2, p. 222) read ינבּ "stehlen" where Gibson reads ינמר (marking the ר as questionable). These scholars also interpret זכרי as "my memorial" (*NSI*, p. 162) whereas Gibson translates it as "males" (*TSSI*, vol. 2, p. 75). Cooke acknowledges that the details and context are uncertain and translates it with a question mark (*NSI*, p. 170). Donner and Röllig translate ינבּ with italics (*KAI*, vol. 2, p. 216). The reading of the word in question is difficult but Gibson's interpretation makes more sense in the context.

[361] Cooke says that the food and drink are actually offered to the god on behalf of the dead rather than to the dead (*NSI*, p. 168). Kraeling, however, agrees with the above interpretation,

> Every sacrifice or libation at the tombs of the fathers brings food to their languishing spirits and cheers for a little their awful gloom. And the dread that impious descendants might forget this duty causes him to shudder. He dwells at length, therefore, on the theme of the obligations of his descendants toward himself and his god Hadad.

(Emil Kraeling, *Aram and Israel or the Aramaeans in Syria and Mesopotamia*, Columbia University Oriental Studies, vol. 13, New York: Columbia University, 1918, p. 124).

[362] Miranda Bayliss, "The Cult of the Dead Kin in Assyria and Babylonia," *Iraq* 35 (1973), 116. Though Bayliss nowhere refers to the Aramaic texts, her discussion obviously applies to the Hadad inscription as well. For more references see "*kispu*," *CAD* K, vol. 8, pp. 425–27.

[363] Ibid. Bayliss gives as an example a Babylonian incantation text against harmful ghosts who have not had these services provided for them (*CT*, vol. 16, 10, verso 5–14). Potential offspring were sometimes referred to as *nāq mê* and *zākir šumi*, ones who would continue the name of the deceased through these ceremonies (p. 117). Notice Assurbanipal's action against Elamite rulers who had not feared Assur and Ishtar, "I laid restlessness upon their shades. I deprived them of food-offerings and libations of water" (Luckenbill, *Ancient Records*, vol. 2, p. 310, §810) and the curse from the Esarhaddon vassal treaties, "May your [ghost] have none appointed as funeral-libation pourer" (Wiseman, *The Vassal-Treaties of Esarhaddon*, pp. 63–64, line 452). Ancient Hebrew and Ugaritic cults of the dead are the study of a recent monograph by Theodore J. Lewis (*Cults of the Dead in Ancient Israel and Ugarit*, Harvard Semitic Monographs, no. 39 [Atlanta: Scholars Press, 1989]); see especially Lewis's summary of the Ugaritic cult and the tomb structure designed for maintenance of the cult (pp. 95–98). Expressing a contrasting view is Wayne Pitard ("The Tombs of Ugarit and the Tombs of the Dead," a paper [group S114] read at the SBL 1990 Annual Meeting in New Orleans, November 18, 1990) who convincingly argues that Claude Schaeffer's "channels and openings for the insertion of water for the dead" were not used for that purpose. Pitard says that the channels were for drainage, the openings caused by robbers, and the "windows" are simply niches some of which have no closing blocks at their backs. As regards Judah, however, Elizabeth Block-Smith has recently strengthened the case for continuing post-mortem care and the importance of the dead in ancient Judah ("The Cult of the Dead in Judah," a paper [group S114] read at the SBL 1990 Annual Meeting in New ORleans, November 18, 1990).

[364] Ibid., pp. 119–20. Evidence of the heir serving as *pāqidu* is seen in a curse on a late second millennium *kudurru*, *aplām nāq mē likimšuma* "May (Ninurta) deprive him of an heir, a pourer of water" and in a Babylonian omen text *NÍG.TUK*^{meš} *IBILA zakir MU TUK-ši* "It means wealth—he will have an heir, one to call his name." Certain Middle Babylonian and Nuzi adoption texts make inheritance contigent on observance of this ritual. The duty to perform the ritual is heavier on adjacent generations (rarely extends past duty to grandparents). The moral obligation seems heaviest on the son who will inherit all his father has worked for (p. 121).

[365] *TSSI*, vol. 2, pp. 66–67.

[366] The text is my transliteration of the readings of Abou-Assaf, et al. *La statue*, p. 23; the translation is mine.

[367] See the discussion of Kilamuwa 1 in this section above.

[368] *TSSI*, vol. 2, pp. 64–65.

Chapter 5: Blessing and Curse

[369] Judith M. Hadley, "Some Drawings and Inscriptions on Two Pithoi from Kuntillet 'Ajrud," *VT* 37 (1987), 187; and Moshe Weinfeld, "Kuntillet 'Ajrud Inscriptions and Their Significance," *SEL* 1 (1984), 125. It is not clear from Hadley's article how much space occurs between the two segments of the inscription.

[370] See observations from Bayliss's article above (n. 359) on the restlessness and troublesomeness of the untended spirit of the deceased (*eṭemmu*). Also see the quote from Assurbanipal just below (n. 369).

[371] *Vox populi vox dei*? Compare the blessing of Kuntillet 'Ajrud 6 and Tell Fekherye lines 9–10, 14. The explanation could be even more basic, a severe famine would affect even the king's food supply.

[372] Note the activity of Assurbanipal against the graves of the Elamite kings, "Their bones (members) I carried off to Assyria. I laid restlessness upon their shades. I deprived them of food-offerings and libations of water" (Luckenbill, *Ancient Records*, vol. 2, p. 310, §810).

[373] Nicholas Tromp (*Primitve Conceptions of Death and the Nether World in the Old Testament*, Biblica et Orientalia, no. 21 [Rome: Pontifical Biblical Institute, 1969], pp. 168–71) says that the first phrase means that the deceased is united with forefathers in Sheol, while the second phrase refers to the peaceful nature of a king's death (contra "and he died" which signifies a violent death as apropriate for a wicked king).

[374] The subject is far too complex to deal with in this study but these brief notes will attempt to verify the above statement and lead the reader to fuller treatments. Sprock (*Beatific Afterlife in Ancient Israel and in the Ancient Near East*, Alter Orient und Altes Testament, vol. 219 [Neukirchener-Vluyn: Butzon & Bercker Kevelaer, 1986], p. 248) says that references to providing for the dead do not point to more than "normal care for the dead, because they were believed to live on in more or less the same way as before death." Brichto ("Kin, Cult, Land and Afterlife—A Biblical Complex," *HUCA* 44 [1973], 29) says the declaration of the worshipper in Deut. 26:14—that none of the tithe was eaten while in mourning, nor while unclean, nor was any offered to the dead—not only attests "to the practice, as late as the time of Deuteronomy, of offerings made to the dead; it attests that normative biblical religion accorded them the sanction of toleration." And though he says, "Sanction does not...equal prescription," he makes a strong case for an understanding of the fifth commandment that expects proper funeral rituals to be performed (pp. 29–31, and n. 49). Elizabeth Block-Smith agrees ("The Cult of the Dead in Judah"). Sprock also discusses the Ugaritic cult of the dead, the *mrz'* or *mrzḥ*, and notes possible parallels in biblical literature, e.g. Jer. 16:7 (pp. 196–202, 248–49).

[375] Kraeling, *Aram and Israel*, p. 124. The intention of these lines is difficult to decipher but may describe means for dealing with dynastic grievances (*TSSI*, vol. 2, pp. 60–61).

[376] *TSSI*, vol. 2, pp. 93–94. Dupont-Sommer includes good photographs of both the entire Nerab stelas (*Les Araméens*, L'Orient ancien illustré, vol. 2 [Paris: L'Orient ancien illustré, 1949], p. 87).

[377] *TSSI*, vol. 2, pp. 95–96, readings and translation.

[378] Koopmans, *Aramäische Chrestomathie*, vol. 1, p. 92.

[379] *CAD* E, vol. 4, pp. 315–18. Apparently *eršu* is not used for sarcophagus. Akkadian *erṣetu* "earth, land, nether" can also refer to the netherworld but apparently not to the grave (Ibid., pp. 310–11).

[380] So BDB, p. 793.

[381] See the comparison of opening formulas in the discussion of the Karatepe inscription above ("ברכ and שלמ with Substantive Curse").

[382] *TSSI*, vol. 2, p. 11.

[383] Jacob Halévy, "Les deux stèles de Nerab," *RS* 4 (1896), 281–82.

[384] *KAI*, vol. 2, p. 275.

[385] Hoffman, "Aramäische Inschriften," p. 209.

[386] Koopmans, *AramäischeChrestomathie*, vol. 1, p. 93. See also his study of the possible roots in his discussion of the Zakir stele B20 (p. 29).

[387] Gevirtz, "West-Semitic Curses," p. 148. He notes parallels from Ugaritic literature (2 Aqht VI: 35–37) where the word *'uḥryt* means "future" and is in parallel to *aṯryt* meaning something like "what is left over, remains": "As for man, what does he get as his destiny? What does man get as his fate?" *mt uḥryt mh yqḥ mh yqḥ mt aṯryt* (translation *UL*, p. 90; transliteration *UT*, pp. 248–49); the answer is death. He approves of Ginsberg's interpretation of that Ugaritic passage (line 35) "How can a mortal acquire a latter estate?" Man's "latter estate" is his "future" (H. L. Ginsberg, "The North-Canaanite Myth of Anath and Aqhat," *BASOR* 98 [1945], 21).

[388] Halévy agrees with this interpretation, translating אשרכ as "ta place," which means the person's very existence "c'est-à-dire ton existence" ("Les deux stèles," p. 282).

[389] Halévy, "Nouvelles Remarques...Zakir," p. 370.

[390] See the discussion of the Zakir inscription above. Worthy of note at this point is the fifth-fourth century Aramaic inscription from Guzneh (Gözen) in Persian Asia Minor. This short boundary stone inscription contains the same curse against both the perpetrator and his family.

(1 עד תנה תחום ד--[-]
(2 ומן זי א תתב ויב[ע]
(3 ון לה בעלשמין
(4 רבא שהר ושמש
(5 ולזרעא זילה

"To here (extends) the boundary of D[ ] If any of you turn (it) back, then may great Baalshamayn, Sahar, and Shamash seek him out, and his seed!" (*TSSI*, vol. 2, pp. 153–54). Richard S. Hanson's translation is similar, "...may the great Baal Shamen, Sahar and Shamash hold both him and his seed responsible" ("Aramaic Funerary and Boundary Inscriptions from Asia Minor," *BASOR* 192 [1968], 11).

[391] Donner and Röllig agree and suggest "ausrotten" as a translation (*KAI*, vol. 2, p. 275).

[392] Koopmans, *AramäischeChrestomathie*, vol. 1, p. 93. He also compares a fifth century inscription from Tema (Tema i) in which the person harming the monument will be "removed" along with his posterity:

(12b וגבר
(13 זי יחבל סותא זא אלהי תימא
(14 ינסחוהי וזרעה ושמה מן אנפי
(15 תימא...

"If any man harms this monument, let the gods of Tema remove him and his seed and his posterity from Tema" (*TSSI*, vol. 2, pp. 148–50).

[393] Jacob Halevy, "Deux notes épigraphiques: Un dernier mot sur les inscriptions de Nêrab," *RS* 5 (1897), 189.

[394] Charles Clermont-Ganneau, "Les stèles araméennes de Neîrab," *Études d'archéologieorientale* 2 (1897), 199.

[395] Koopmans, *Aramäische Chrestomathie*, vol. 1, p. 93. The quotations are from Fitzmyer, *Sefire*, pp. 14–15. The adjective also occurs in Sefire I.C.6–7, 20 and III.2 (Fitzmyer, *Sefire*, pp. 18, 20, 97) where the contexts are not curses. Fitzmyer says that the term usually means "evil" in a moral sense and gives other references, all later (e.g., *AP* 30:7; 31:6; 32:6; et al.) (p. 105). Gevirtz disagrees with a similar statement on the subject by Fitzmyer in an earlier work ("The Aramaic Suzerainty Treaty from Sefire in the Museum of Beirut," *CBQ* 20 [1958], 453); he says that Fitzmyer's examples are inadequate for determining the meaning of לחיה since "they appear after the names of persons and may conceivably be titles totally unrelated to our form" ("West-Semitic

Chapter 5: Blessing and Curse

Curses," p. 148, n. 4). Gevirtz's criticism is true in several of the cases (*AP* 30:7; 31:6; 32:6; *AD* 5:7) but not in all; certainly not here in Nerab 1:10 nor in any of the Sefire examples.

[396] *KAI*, vol. 2, p. 275.

[397] *TSSI*, vol. 2, p. 12. The ק to כ and ט to ת shifts are not misprints.

[398] Fitzmyer, *Sefire*, p. 82.

[399] *TSSI*, vol. 1, pp. 74, 76.

[400] Clermont-Ganneau, "Les stèles," p. 200.

[401] See the discussion of Sefire I.C.17 above in this section.

[402] Koopmans, *Aramäische Chrestomathie*, vol. 1, p. 93. Is the verb passive or active? Is the נ always assimilated in the imperfect of the active stems, or only sometimes? In line 12, תנצר occurs and must be interpreted as active but in line 9, יסחו occurs which is interpreted as active and from the root נסח with the נ assimilated. But compare the Tema i inscription cited above (footnote 390) where, in line 14, the active form ינסחוהי occurs with נ unassimilated. Gibson interprets it as passive (*TSSI*, vol. 2, p. 97). Randall Garr notes these possible exceptions to the rule of assimilation ("If *peal* [rather than pael], תנצר/ף would be the only instance where *nun* failed to assimilate to the following consonant"), but says they may be explained by the presence of a vowel between the נ and the next letter preventing assimilation (*Dialect Geography*, p. 42; the italics are Garr's, the material in brackets is my insertion).

[403] *KAI*, vol. 2, pp. 275–76.

[404] See the interpretations of the Zakir inscription above in this section.

Chapter 6

Conclusions

The semantic survey of terms for blessing and curse, the history of interpretation of biblical blessing and curse, and the presentation of the inscriptions have revealed some significant information about the use of blessing and curse in the inscriptions. Important points of contact between the cultures of the ancient Near East have been shown by this study. Similarities and dissimilarities in cultural expressions of blessing and curse have emerged.

The semantic survey in Chapter 2 showed not only that similar roots for general words meaning "blessing" and "curse" are present in most of the languages surveyed, but also that the intentions of those words in the different languages are the same. This includes the biblical literature. Based on available inscriptions (evidence is by no means equal from all of the languages surveyed), general words for blessing and curse are used in the same kinds of literature (letters, votive inscriptions, tomb inscriptions, religious texts—where available) with what appear to be the same intent: the wish that the deity invoked bestow the desirable things of life (progeny, food, security, etc.) on the favored one or deprive the despised one of the same things.

The views of the scholars summarized in the second half of Chapter 2 show agreement concerning the meaning of blessing and curse in the Bible. As in the other literatures, blessing consists of a wish for someone to receive the good things: land, numerous progeny, sufficient food, clothing, safety, etc. Curse is the wish that someone be deprived of these same things. Scholars have disagreed, however, over how blessing and curse operate. Pedersen thought that an individual's "soul power" enabled him or her to bless or curse another at will. Other scholars evolved the view that the Old Testament manifests a theologized view of older magical ideas of blessing and curse. Other scholars say what appears to be a "magical word" is really illocutionary utterance, blessings and curses are dependent on the action of the correct person, at the correct time, by means of the correct formula (e.g., Thistleton). Still others say that while society accepted these forms as valid, ultimately the blessing or curse is only Yahweh's to give (e.g., Mitchell).

The inscriptions yield no evidence for the evolutionary development of the concept of the power to bless and curse posited by the earlier scholars in Chapter 2.[1] Though the extant material is not vast, the range of the material, from the beginning of the tenth century to the beginning of the sixth century, is significant. In the complete Hebrew inscriptions of the target period, a deity is always invoked for blessing. In the vast majority of cases that deity is Yahweh alone but notable exceptions like Kuntillet ʻAjrud 3 and Khirbet el-Qôm occur where Asherah/an asherah is also invoked and Kuntillet ʻAjrud 1 where Baʻal is petitioned. These "extra-Yahweh" invocations are vigorously debated but quite clear enough to be striking exceptions.[2] Mitchell's argument that Yahweh is consistently the source of blessing in the Bible can also be said to apply to the extant Hebrew inscriptions.[3] And the same may be said to the inscriptions in the other languages. The power of the various deities, explicitly invoked or strongly implied, is depended on by those other peoples for the blessings or curses to have effect.

Non-Hebrew inscriptional blessings from the target period also include DNs in their blessings. There is only one non-Hebrew inscription which does not invoke a DN explicitly for blessing, the Tell Siran Bottle. The Ḥorvat ʻUza, Karatepe, Ivory Box from Ur, Yeḥimilk, Abibaʻal, Elibaʻal, Shipiṭbaʻal, and Kilamuwa 2 texts are Phoenician and Edomite inscriptions which contain a blessing without curse and all invoke a deity to bring about that blessing. The Hebrew kingdoms share with the lands with which they had contact the characteristic of including a DN in blessing.

The investigation of blessing in Chapter 3 revealed that in target inscriptions the root ברכ was used only in Hebrew, Phoenician, and Edomite.[4] The earliest of these texts is Kuntillet ʻAjrud 4 (ca. 800); the latest texts in this category are Arad 16 and 21 (ca. 595). This use of the root ברכ in Phoenician and Hebrew inscriptions, and especially its appearance several times in the Phoenician/Hebrew of mysterious Kuntillet ʻAjrud, show another sign of the close relationship between Israel/Judah and Phoenicia.[5] It would also seem that ברכ in these inscriptions is used as a compact and general way of expressing all of the more specific blessing (i.e., substantive blessing) ideas seen in the longer and/or earlier inscriptions.

Hebrew inscriptions express the wish for blessing by using ברכ, שלמ, or a combination of the two. Substantive or specific blessings do not occur in the Hebrew inscriptions.[6] שלמ without ברכ does not occur with a DN other than Yahweh or in letters which are not Judean Hebrew. Based on a comparison with biblical greetings, this sort of greeting may be a development of the fuller greeting form.

The numerous contacts between Hebrew and Phoenician culture lend weight to the theory that the use of ברכ for blessing at this early period is evidence of Hebrew-Phoenician sharing of religious concepts. The use of the general root ברכ for blessing breaks down along geographical lines. No Aramaic inscriptions of the period use the root ברכ, or any specific word for bless. Aramaic uses specific (substantive) blessings. Hebrew inscriptions prefer the general word ברכ in blessings. Phoenician inscriptions, from the geographical area between the Hebrews and Arameans, use ברכ (the Ivory Box) and substantive blessings (Yeḥimilk, Abibaʻal, Elibaʻal, Shipiṭbaʻal, and Kilamuwa 2) and even combine them in the same inscriptions (Karatepe). Hebrew inscriptions, which date to the last two centuries of the target period (ca. 800 to 595 B.C.), prefer the

Chapter 6: Conclusions

general word ברכ.

To a lesser extent, a difference in form of blessing based on chronology may be observed. Tenth to mid-eighth century blessings are more likely to use specific blessings—invocations for life, health, fertility, food, etc. while the mid-eighth century or later inscriptions (but not Aramaic) tend to use the general term ברכ. In inscriptions of the eighth century either may be found. The oldest of the Israelite inscriptions, Gezer (if interpreted as a blessing tablet), shows the older tendency to use substantive blessing rather than the typical Hebrew blessing with ברכ.[7]

The study of curses in the Hebrew inscriptions reveals that, apart from Torczyner's interpretation of Lachish 5, there are no explicit epigraphic examples in which Yahweh is invoked for a curse. The presence of several forms of ארר in the Khirbet Beit Lei burial cave along with prayers to Yahweh seem to imply that Yahweh was the one invoked to bring about the curse. The presence of a name with the theophoric element -יהו in the Siloam Tomb inscription curse points in the same direction. The curse found in the En Gedi cave, though broken, makes the same implication.[8] Biblical examples of ארר with Yahweh are infrequent but not rare.[9] All of the inscriptions with a form of ארר date from the time between the Siloam Tomb (ca. 700) and the Lachish letters (ca. 595) and are Judean Hebrew. Interestingly, the only other language using this root is Akkadian, which uses it both alone and with other substantive curses. Apart from the Akkadian usages, the evidence seems to indicate that ארר came to be used late in the target period and was endemic to Judah.

Non-Hebrew inscriptions also tend to invoke a deity in their curses.[10] The Tell Fekherye, Amman Citadel, Kilamuwa 1, Zakir, Sefire I, II, and III, Panammu, Hadad, and Nerab 1 and 2 inscriptions all invoke deities. Non-Hebrew inscriptions rarely use a general word for curse.[11] Instead, they use a specific (substantive) curse.

Further, this information lends weight to the theory that the inscriptions show a developmental trend from earlier periods where specific curses were used to later periods where a more general word was used, perhaps with a certain well-known (but not stated) stock of curses implied by that one word.

Chapter five reveals that blessing and curse combinations are rare in Hebrew (En Gedi) and Phoenician inscriptions, but more abundant in Aramaic texts. Both ברכ and substantive blessing (one of which is שלמ) and curse are present in the mid-eighth century Karatepe inscriptions. These three exemplars of the same text (A, B, C) combine elements found both in early Byblian Phoenician inscriptions requesting long life (Yeḥimilk, etc.) and the later Hebrew and Edomite more general requests for blessing which use ברכ (Kuntillet 'Ajrud, Arad, Lachish, Ḥorvat 'Uza, etc.).

The substantive blessings and curses date from the mid-ninth to the early seventh century and are predominantly Aramaic (Tell Fekherye, Amman

Citadel, Zakir, Sefire I and II, Hadad, and Nerab 1).

It remains to be seen whether future inscriptional finds will corroborate or invalidate the conclusions of this study.

Suggestions for Further Study

The inscriptions provide abundant material for further study. The examination of blessing and curse in inscriptions of the Persian period (possibly even the Hellenistic period and the Punic inscriptions of the first century A.D.) would be a useful study. The number of letters and other types of inscriptions becomes more abundant the later the study is continued with the fifth and fourth centuries being particularly well-represented by letters. The other religious elements which were to be part of this work (oaths, sacred precincts, vessels, and personnel, Yahweh and his asherah, and the localization of Yahweh and other religious elements scattered throughout the inscriptions) remain a mine of material unexamined as a group. In addition, Akkadian general words for blessing and curse as well as substantive blessings and curses should be examined for meaning and chronological frequency. Determining how these ideas appeared in different stages of Akkadian (especially Neo-Assyrian and Neo-Babylonian which were contemporary with the time period of this inscription) would be an important contribution.

---

[1] McCarter concurs, "To speak of theologizing an older, magical concept by bringing all blessing under Yahweh's control...is probably a mistake. We're dealing with *very* old concepts, not likely to have changed significantly in a few centuries. What is likely is that the attitude towards blessing and curse and the understanding of the nature of their efficacy were complex. Magical and theologized concepts probably existed side by side, quite possibly in the thinking of a single individual." (Personal communication.)

[2] Incomplete Hebrew inscriptions which might have originally borne contrary evidence are Samaria C1101 and 1220, the Wadi Muraba'ât papyrus, and the Gezer Calendar. The first three are incomplete and the function of the last is debated.

[3] Christopher W. Mitchell, *The Meaning of BRK "To Bless" in the Old Testament*, (SBL Dissertation Series, vol. 95. Atlanta: Scholars Press, 1987), p. 171. I say "blessing" only here because he dealt only with blessing. However, the same may be said for curse, see below.

[4] Post-period inscriptions in Aramaic and Phoenician contain the root as well.

[5] See the comments of Peckham just below. Meshel terms the script "Phoenician," but also says there are Hebrew inscriptions and does not say which inscriptions are in which script ("Did Yahweh Have a Consort? The New Religious Inscriptions from the Sinai," *BAR* 5/2 [1979], 27, 34). The script in the photograph on page 35 (pithos 1) looks as much like the Hebrew of the Siloam Tunnel inscription as it does like the Phoenician Kilamuwa inscription.

[6] Wadi Muraba'ât, Kuntillet 'Ajrud 6, and the Gezer Calendar are the only inscriptional Hebrew blessings without ברכ. The first is a poorly preserved letter, the second is also broken, and the intention of the third is uncertain.

[7] This could, however, be due to the type of literature of the Gezer Calendar, rather than a chronological tendency.

[8] See "ברכ and "ארר" in Chapter 5.

[9] For example, Josh. 6:26 "Joshua laid an oath upon them at that time saying, 'Cursed

Chapter 6: Conclusions

before the Lord be the man (ארור האיש לפני יהוה) that rises up and rebuilds this city, Jericho. At the cost of his first-born shall he lay its foundation, at the cost of his youngest son shall he set up its gates'." Note that the oath is laid on them by Joshua, but it is attributed to Yahweh and thus brought to pass by him 1 Kgs. 16:34 (cf. Gen. 3:14, 17; 4:11; 1 Sam. 26:19; Jer. 11:3; Mal. 1:14).

[10] The Aḥiram, Tell Siran Bottle, and Deir 'Alla inscriptions do not, in their current forms, contain DNs. Of these, only the Aḥiram and Tell Siran Bottle inscriptions are complete. The Aḥiram and Tell Siran inscriptions may well have implied references to deities. The extant forms of the Deir 'Alla inscriptions are thought by some commentators to have a DN. They do have a reference to a council of the gods in another part of the inscriptions.

[11] אלה is used in the Panammu and Arslan Tash 1 inscriptions with treaty overtones. The beginning of the Panammu inscription may describe a situation where the oath has already been broken (the specific oath situation is not known), the curse already let loose upon the violator. In the Arslan Tash inscription אלה refers to an oath whose conditions are not stated and therefore whose force can only be guessed at. The occurrences of the root קבב in Deir 'Alla Combinations IX and X is difficult to interpret because of the broken contexts; the occurrence in Combination II.17 is debated.

APPENDIX

1	2	3	4	5	6
Abibaʿal	ca. 925	1905	Ph	Dussaud, *Syria* 5 (1924)	same, plate XLII
Aḥiram	ca.1000	1923	Ph	Dussaud, *Syria* 5 (1924)	Vincent, *RB* 34 (1925), pls VI, VII Driver, *Semitic Writing*, plates 51 (2), 52 *ANEP*, no. 456
Amman Citadel	ca. 800	1961	Am	Horn, *BASOR* 193 (1969)	same, fig. 1
Arad Ostraca 16, 18, 21	605-595 Str. VI	1962-67	JH	Aharoni, *Arad Inscriptions*	same
Arad Ostraca 40	ca. 701 Str. VIII	1962-67	JH	Aharoni, *Arad Inscriptions*	same
Arslan Tash	7th c.	1933	Ph(?)	du Mesnil, *Mélanges Syrien*	same, p. 422 *ANEP*, no. 662
Deir ʿAlla	7th c.	1967	Am/Ar	*ATDA*	same, plates 1-28
Elibaʿal	ca. 914	1925	Ph	Dussaud, *Syria* 6 (1925)	same, plate XXV Montet, *Byblos et l'Égypt*
En Gedi	ca. 700	1974	JH	Bar-Adon, *IEJ* 25 (1975)	same, plate 25:B,C
Tell Fekherye	ca. 850	1979	Ar	Abou-Assaf, et al. *La Statue*	same, pls. XII-XIV

Gezer Calendar	10th c.	JH	Lidzbarski, *PEFQS* 41 (1909)	same Birnbaum, *Scripts*, no. 2
Hadad	ca. 750	Ar	von Luschan & Sachau, *Ausgrabungen in Sendschirli I*	Birnbaum, *Scripts*, no. 016
Ḥorvat ʿUza	ca. 587	Ed	Beit-Arieh & Cresson, *TA* 12 (1985)	same, plate 12.2
Ketef Hinnom	ca. 600	H	Barkay, *Ketef Hinnom* Yardeni, *VT* 41 (1991)	same, pp. 29-30
Khirbet Beit Lei	ca. 700	JH	Naveh, *IEJ* 13 (1963)	same, plates 11-13
Khirbet el-Qôm	ca. 725	JH	Dever, *HUCA* 40/41 (1969-70)	same, plates VI:B, VII
Kilamuwa 1	ca. 825	Ph	von Luschan, *Ausgrabungen in Sendschirli IV*	*KAI*, vol. 3, plate XXVII
Kilamuwa 2	ca. 825	Ph	W. Andrae, *Ausgrabungen in Sendschirli V*	same, plate 47
Kuntillet ʿAjrud	ca. 800	H/Ph	Meshel, *Kuntillet ʿAjrud*	same
Lachish Ostraca 2, 3, 5, 6, 7, 9	ca. 595	JH	Torczyner, *Lachish Letters*	same
Lachish Ostraca	ca. 595	JH	Ussishkin, *TA* 10 (1983)	same, plate 41:1

31					
Nerab 1	ca. 690	1890	Ar	Clermont-Ganneau, *Études d'archéologie orientale* 2 (1897)	*ANEP*, no. 280 Dupont-Sommer, *Les araméens*, p. 87, fig. 12
Nerab 2	ca. 690	1890	Ar	same	*ANEP*, no. 635 Dupont-Sommer, *Les araméens*, p. 87, fig. 13
Nimrud Hebrew	ca. 740	1961	HIs	Millard, *Iraq* 24 (1962)	same, plate XXIVa *TSSI*, vol. 2, fig. 22(?)
Panammu	ca. 731	1888	Ar	von Luschan and Sachau, *Ausgrabungen in Sendschirli I*	Birnbaum, *Scripts*, no. 017 *TSSI*, vol. 2, plate IV
Samaria C1101	ca. 725	1932	HIs	Sukenik, *PEQ* 65 (1933), 152ff	same, plate III, fig. 2 Birnbaum, *Scripts*, no. 13 *Objects from Samaria*, plate I.1
Samaria C1220	ca. 735	1932	HIs	Birnbaum, *Objects from Samaria*	same, plate II.7
Sefire I, II	ca. 740	1931	Ar	Dupont-Sommer, "Les inscriptions araméennes"	same, pls. I-XXVIII Fitzmyer, *Séfire*, pls XIV-XVII
Sefire III	ca. 740	1931	Ar	Dupont-Sommer, "Une inscriptions araméenne"	same, plates I-V
Shipiṭbaʻal	ca. 900	1935	Ph	Dunand, *Byblia Grammata*	same, plates XVb, XVI Birnbaum, *Scripts*, no. 07

Siloam Tomb	ca. 700	1870	JH	Avigad, *IEJ* 3 (1953)	Driver, *Semitic Writing*, pl. 50.2
Tell Siran Bottle	ca. 600	1972	Am	Zayadine & Thompson, *Berytus* 22 (1973)	same, plate 8 *ANEP*, no. 811
Ur, Ivory Box	ca. 690	1927	Ph	Burrows, *JRAS* (1927)	same
W. Murabbaʿât	ca. 650	1952	JH	Benoit, et al. DJD II	same, plate VIII
Yeḥimilk	ca. 950	1929	Ph	Dunand, *RB* 39 (1930)	same, pt. 2, plate XXVIII
Zakir	ca. 775	1903	Ar	Pognon, *Inscriptions sémitiques*	Birnbaum, *Scripts*, no. 04 Driver, *Semitic Writing*, pl. 53.1 same, pls. IX, X, XXXV, XXXVI Birnbaum, *Scripts*, no. 015 (part A only)

Explanation of columns:
1 - location/name of inscription
2 - date of inscription
3 - date of discovery
4 - language of inscription: Am = Ammonite; Ar = Aramaic; Ed = Edomite; HIs = Hebrew, Israelite; JH = Hebrew, Judean; Ph = Phoenician
5 - best/most complete publication
6 - publication of photographs and/or drawings

BIBLIOGRAPHY

Books

Abou-Assaf, Ali, Pierre Bordrueil, and Alan. R. Millard. *La statue de Tell Fekerye et son inscription bilingue assyro-arameene*. Paris: Editions Recherche sur les civilisations, 1982.

Aharoni, Yohanan. *Arad Inscriptions*. Ed. Anson F. Rainey. Trans. Judith Ben-Or. Jerusalem: The Israel Exploration Society, 1981.

_____. *The Archaeology of the Land of Israel*. Trans. Anson F. Rainey. Philadelphia: Westminster, 1982.

_____. *Beer-Sheba I: Excavations at Tel Beer-Sheba 1969–1971*. Tel Aviv: Institute of Archaeology, University of Tel Aviv, 1973.

_____. *The Land of the Bible, A Historical Geography*. 2nd ed. Philadelphia: Westminster Press, 1979.

Aistleitner, J. *Wörterbuch der Ugaritischen Sprache*. 2nd ed. Ed. O. Eissfeldt. Berlin: Akademie-Verlag, 1965.

Albright, William Foxwell and George E. Mendenhall, "Taanach 1." In *Ancient Near Eastern Texts Relating to the Old Testament*. Ed. James B. Pritchard. Princeton: Princeton University Press, 1969.

"*arallu*." *Chicago Assyrian Dictionary*. A, vol. 1, part 2. Ed. A. Leo Oppenheim. Chicago: The University of Chicago Press, 1968.

"*arāru*." *Chicago Assyrian Dictionary*. A, vol. 1, part 2. Ed. A. Leo Oppenheim. Chicago: The University of Chicago Press, 1968.

"*arru*." *Chicago Assyrian Dictionary*. A, vol. 1, part 2. Ed. A. Leo Oppenheim. Chicago: The University of Chicago Press, 1968.

"*ba'āšu*." *Chicago Assyrian Dictionary*. B, vol. 2. Ed. A. Leo Oppenheim. Chicago: The University of Chicago Press, 1965.

Barkay, Gabriel. *Ketef Hinnom: A Treasure Facing Jerusalem's Walls*. Israel Museum Catalogue no. 274. Jerusalem: Israel Museum, 1986.

Barnett, R. D. *A Catalogue of the Nimrud Ivories, with Other Examples of Ancient Near Eastern Ivories in the British Museum*. London: The Trustees of the British Museum, 1957.

_____. "Layard's Nimrud Bronzes and Their Inscriptions." In *Eretz Israel: Archaeological, Historical, and Geographical Studies*, vol. 8. Ed. N. Avigad, M. Avi-Yonah, H. Z. Hirschberg, and B. Mazar. Jerusalem: Israel Exploration Society, 1967.

Benoit, Pierre, J. T. Milik, and Roland de Vaux. *Les Grottes de Murabba'at*. Discoveries in the Judaean Desert. vol. II, 2 parts. Oxford: The Clarendon Press, 1961.

"*birku*." *Chicago Assyrian Dictionary*. B, vol. 2. Ed. A. Leo Oppenheim. Chicago: The University of Chicago Press, 1965.

Birnbaum, S. A. *The Hebrew Scripts.* 2 parts. Leiden: E.J. Brill, 1971.

———. "The Sherds." In *The Objects from Samaria.* Ed. J. W. Crowfoot, G. M. Crowfoot, and Kathleen M. Kenyon. London: Palestine Exploration Fund, 1957.

"bîšu." *Chicago Assyrian Dictionary.* B, vol. 2. Ed. A. Leo Oppenheim. Chicago: The University of Chicago Press, 1965.

Bliss, Frederick Jones and R. A. Stewart MacAlister. *Excavations in Palestine During the Years 1898–1900.* London: Palestine Exploration Fund, 1902.

Borger, Rykle. *Die Inscriften Asarhaddons Königs von Assyrien.* Archiv für Orientforschung, Supplement 9. Graz: Ernst Weidner, 1956.

Bossert, H. Theodore and U. Bahadir Alkim. *Karatepe: Kadirli ve Dolyari, (Second Preliminary Report).* Publications for the Institute for Research in Ancients Oriental Civilisations, no. 3. Istanbul: Pulhan Basimevi, 1947.

——— et al. *Karatepe Kazilari (Birinci Ön-Rapor), die Ausgrabungen auf dem Karatepe: Erster Vorbericht.* Türk Tarih Kurumu Yayinlarindan V. Seri, no. 9. Ankara: Türk Tarih Kurumu Basimevi, 1950.

Brichto, Herbert Chanan. *The Problem of "Curse" in the Hebrew Bible.* Journal of Biblical Literature Monograph Series, vol. 13. 1963; rpt. Philadelphia: Society of Biblical Literature, 1968.

Bron, François. *Recherches sur les inscriptions phéniciennes de Karatepe.* II Hautes Études Orientales, no. 11. Geneva: Libraire Droz, 1979.

Carroll, Robert P. *The Book of Jeremiah: A Commentary.* Old Testament Library. Philadelphia: Westminster, 1986.

Cooke, George A. *A Textbook of North-Semitic Inscriptions.* Oxford: Clarendon Press, 1903.

Cowley, A. *Aramaic Papyri of the Fifth Century B. C.* 1923; rpt. Osnabrück: Otto Zeller, 1967.

Cross, Frank M. Jr. "The Cave Inscriptions from Khirbet Beit Lei." In *Near Eastern Archaeology in the Twentieth Century.* Ed. James A. Sanders. Garden City, New York: Doubleday & Co., 1970.

Cuneiform Texts from Babylonian Tablets in the British Museum. Part 16. 1963; rpt. London: The Trustees of the British Museum, 1901.

Cverný, Jaroslav. *Late Ramesside Letters.* Bibliotheca Aegyptiaca, vol 9. Bruxelles: Édition de la Fondation Égyptologique Reine Élisabeth, 1939.

Dalley, Stephanie, C. B. F. Walker, and J. D. Hawkins. *The Old Babylonian Tablets from Tell al Rimah.* Baghdad: The British School of Archaeology in Iraq, 1976.

Dalman, Gustav H. *Aramäische-Neuhebräisches Handwörterbuch zu Targum, Talmud und Midrasch.* 1938; rpt. Hildesheim, W. Germany: Georg Olms Verlag, 1987.

De Moor, Johannes C. *An Anthology of Religious Texts from Ugarit.* NISABA, vol. 16. Leiden: E. J. Brill, 1987.

———. "A Note on CTA 19 (1Aqht): 1.39–42." In *Ugarit-Forschungen*, vol. 6. Ed. Kurt Bergerhof, Manfried Dietrich, Oswald Loretz. Neukirchener-Vluyn, W. Germany: Verlag Butzon & Berker Kevelaer, 1974.

_____. "Studies in the New Alphabetic Texts from Ras Shamra I." In *Ugarit-Forschungen*, vol. 1. Ed. Kurt Bergerhof, Manfried Dietrich, Oswald Loretz, Johannes C. de Moor. Neukirchener-Vluyn, W. Germany: Verlag Butzon & Berker Kevelaer, 1969.

Dietrich, Manfred and Oswald Loretz. "UG. *BŠ, *T*BŠ, Hebr. *ŠBS (Am 5,11) sowie UG IŠY und ŠBŠ." In *Ugarit-Forschungen*, vol. 10. Ed. Kurt Bergerhof, Manfried Dietrich, Oswald Loretz. Neukirchener-Vluyn, W. Germany: Verlag Butzon & Berker Kevelaer, 1978.

Dietrich, Manfred, Oswald Loretz, and J. Sanmartín. "Die Ugaritischen Totengeister RPU(M) und die biblischen Rephaim." In *Ugarit-Forschungen*, vol. 8. Ed. Kurt Bergerhof, Manfried Dietrich, Oswald Loretz. Neukirchener-Vluyn, W. Germany: Verlag Butzon & Berker Kevelaer, 1976.

Diringer, David. "Early Hebrew Inscriptions." *Lachish III: The Iron Age*, pt. 1. London: Oxford University Press, 1953.

_____. *Le iscrizioni antico-ebraiche palestinesi*. Firenze: Felice Le Monnier, 1934.

Donner, Herbert and Wolfgang Röllig. *Kanaanäische und aramäische Inscriften*. 3 vols. Wiesbaden, W. Germany: Otto Harrassowitz, 1962–64.

Driver, Godfrey R. *Aramaic Documents of the Fifth Century B.C.* Oxford: Clarendon, 1954.

Dunand, Maurice. *Byblia Grammata: Documents et recherches sur le développement de l'écriture en Phénicie*. In Études et Documents d'Archaeologie, vol. 2. Beirut: République Libanaise, 1945.

Dupont-Sommer, André. *Les Araméens*. L'Orient ancien illustré, vol. 2. Paris: L'Orient ancien illustré, 1949.

Dussaud, René. *Les monuments palestiniens et judaïques*. Paris: Musée du Louvre, 1912.

Emerton, J.A. "The Meaning of the Ammonite Inscription from Tell Siran." In *Von Kanaan bis Kerala*. Ed. W.C. Delsman, et al. Neukirchener-Vluyn, W. Germany: Verlag Butzon & Berker Kevelaer, 1982.

"eršu." *Chicago Assyrian Dictionary*. E, vol. 4. Chicago: The University of Chicago Press, 1958.

Fisher, Loren F., ed. *Ras Shamra Parallels: The Texts from Ugarit and the Hebrew Bible*. 2 vols. Analecta Orientalia, vol. 49. Rome: Pontificium Institutum Biblicum, 1972.

Fitzmyer, Joseph A. *The Aramaic Inscriptions of Sefire*. Biblica et Orientalia, no. 19. Rome: Biblical Institute Press, 1967.

_____. *The Genesis Apocryphon of Qumran Cave 1*. Biblica et Orientalia, vol. 18a. 2nd ed. Rome: Biblical Institute Press, 1971.

_____. *A Wandering Aramean: Collected Aramaic Essays*. Missoula, MT: Scholars Press, 1979.

Fowler, Jeaneane D. *Theophoric Personal Names in Ancient Hebrew, A Comparative Study*. Journal for the Study of the Old Testament Supplement Series, no. 49. Sheffield: Journal for the Study of the Old Testament, 1988.

Friedrich, Johannes. *Phönizisch-punische Grammatik*. Analecta Orientalia, vol.

32. Rome: Pontifical Biblical Institute, 1951.
Gardiner, Alan. *Egypt of the Pharaohs, An Introduction.* Oxford: Oxford University Press, 1961.
Garr, W. Randall. *Dialect Geography of Syria-Palestine, 1000–586 B.C.E.* Philadelphia: University of Pennsylvania Press, 1985.
Geraty, L. T. "The Historical, Linguistic, and Biblical Significance of the Khirbet el-Qôm Ostraca." In *The Word of the Lord Shall Go Forth.* Ed. Carol Meyers and M. O'Connor. Winona Lake, Indiana: Eisenbrauns, 1983.
Gibson, John C. L. *Canaanite Myths and Legends.* 2nd ed. Edinburgh: T. & T. Clark, 1978.
_____. *Textbook of Syrian Semitic Inscriptions: Volume 1 Hebrew and Moabite Inscriptions.* Oxford: Clarendon Press, 1971.
_____. *Textbook of Syrian Semitic Inscriptions: Volume 2 Aramaic Inscriptions.* Oxford: Clarendon Press, 1975.
_____. *Textbook of Syrian Semitic Inscriptions: Volume 3 Phoenician Inscriptions.* Oxford: Clarendon Press, 1982.
Goetze, Albrecht. "God List, Blessings and Curses of the Treaty Between Suppiluliumas and Kurtiwaza." In *Ancient Near Eastern Texts Relating to the Old Testament.* 3rd ed. Ed. James B. Pritchard. Princeton: Princeton University Press, 1969.
_____. "The Soldiers' Oath." In *Ancient Near Eastern Texts Relating to the Old Testament.* 3rd ed. Ed. James B. Pritchard. Princeton: Princeton University Press, 1969.
_____. "Treaty Between Mursilis and Duppi-Tessub of Amurru." In *Ancient Near Eastern Texts Relating to the Old Testament.* 3rd ed. Ed. James B. Pritchard. Princeton: Princeton University Press, 1969.
_____. "Treaty Between Suppiluliuma and Kurtiwaza." In *Ancient Near Eastern Texts Relating to the Old Testament.* 3rd ed. Ed. James B. Pritchard. Princeton: Princeton University Press, 1969.
Gordon, Cyrus. *Ugaritic Literature: A Comprehensive Translation of the Poetic and Prose Texts.* Roma: Pontificium Institutum Biblicum, 1949.
_____. *Ugaritic Texbook.* Analect Orientalia, vol. 38. Rome: Ponificium Institutum Biblicum, 1967.
Grayson, A. K. "Cylinder C of Sin-šarra-iškun, a New Text from Baghdad." In *Studies on the Ancient Palestinian World.* Ed. J. W. Wevers and D. B. Redford. Toronto: University of Toronto Press, 1971.
Greenfield, Jonas C. "Aspects of Aramean Religion." In *Ancient Israelite Religion.* Eds. Patrick D. Miller, Jr., Paul D. Hanson, and S. Dean McBride. Philadelphia: Fortress Press, 1987.
_____. "Scripture and Inscription: The Literary and Rhetorical Element in Some Early Phoenician Inscriptions." In *Near Eastern Studies in Honor of William Foxwell Albright.* Ed. Hans Goedicke. Baltimore: Johns Hopkins, 1971.
Grönbech, Vilhelm. *The Culture of the Teutons.* 3 vols. Trans. W. Worster. Copenhagen: Jespersen Og Pios Forlag, 1931.

"*gullulu.*" *Chicago Assyrian Dictionary.* G, vol. 5. Ed. A. Leo Oppenheim. Chicago: The University of Chicago Press, 1956.

Hackett, Jo Ann. *The Balaam Text from Deir 'Allā.* Harvard Semitic Monographs, vol. 31. Chico, CA.: Scholars Press, 1984.

_____. "Religious Traditions in Israelite Transjordan." In *Ancient Israelite Religion.* Eds. Patrick D. Miller, Jr., Paul D. Hanson, and S. Dean McBride. Philadelphia: Fortress Press, 1987.

Harper, Robert F. *Assyrian and Babylonian Letters Belonging to the Kouyunjik Collections of the British Museum.* Part 7. London: The University of Chicago Press/Luzac and Co., 1902.

Harris, Zellig S. *A Grammar of the Phoenician Language.* American Oriental Series, vol. 8. New Haven, CT: American Oriental Society, 1936.

Heidel, Alexander. *The Gilgamesh Epic and Old Testament Parallels.* 2nd ed. Chicago: University of Chicago Press, 1949.

Herr, Larry G. *The Scripts of Ancient Northwest Semitic Seals.* Harvard Semitic Monographs, vol. 18. Missoula, MT: Scholars Press, 1978.

Hillers, Delbert R. *Treaty Curses and the Old Testament Prophets.* Biblica et Orientalia, vol. 16. Rome: Pontifical Biblical Institute, 1964.

Jackson, Kent. *The Ammonite Language of the Iron Age.* Harvard Semitic Monographs, no. 27. Chico, CA.: Scholars Press, 1983.

Jean, Charles F. and Jacob Hoftijzer. *Dictionaire des inscriptions sémitiques de l'ouest.* Leiden: Brill, 1965.

Joines, K. Randolph. *The Incomparable Divine Kinsman of Second Isaiah.* Haddonfield, NJ: Haddonfield House, 1976.

Hoftijzer, Jacob and G. van der Kooij, eds. *Aramaic Texts from Deir 'Alla.* Documenta et Monumenta Orientis Antiqui, no. 19. Leiden: E. J. Brill, 1976.

_____ eds. *The Balaam Text from Deir 'Alla Reevaluated: Proceedings of the International Symposium Held at Leiden 21-24 August 1989* (Leiden: E.J. Brill, 1991)

Huehnergard, John. *Ugaritic Vocabulary in Syllabic Transcription.* Harvard Semitic Studies, vol. 32. Atlanta: Scholars Press, 1987.

"*karābu.*" *Chicago Assyrian Dictionary.* K, vol. 8. Ed. A. Leo Oppenheim. Chicago: The University of Chicago Press, 1968.

"*kispu.*" *Chicago Assyrian Dictionary.* K, vol. 8. Ed. A. Leo Oppenheim. Chicago: The University of Chicago Press, 1968.

Knudtzon, J. A., ed. *Die El-Amarna-Tafeln mit Einleitung und Erläuterungen.* 2 vols. 1964; rpt. Aalen: Otto Zeller Verlagsbuchhandlung, 1915.

Koehler, Ludwig and Walter Baumgartner. *Lexicon in Veteris Testamenti Libros.* Leiden: E. J. Brill, 1958.

König, Friedrich W. *Handbuch der chaldischen Inschriften.* Archiv für Orientforschung supplement 8. Graz, Austria: Ernest Weidner, 1957.

Koopmans, J. J. *Aramäische Chrestomathie, ausgewählte Texte (Inschriften, Ostraka und Papyri) bis zum 3. Jahrhundert n. Chr.* Leiden: Nederlands Instituut voor Het Nabije Oosten, 1962.

Korošeç, V. *Hethitische Staatsverträge: Ein Beitrage zu ihrer juristischen Wertung.*

Leipziger rechtswissenschaftliche Studien, vol. 60. Leipzig: T. Weicher, 1931.
Kraeling, Emil. *Aram and Israel or the Aramaeans in Syria and Mesopotamia.* Columbia University Oriental Studies, vol. 13. New York: Columbia University, 1918.
_____. *The Brooklyn Museum Aramaic Papyri.* 1969; rpt. New Haven, Conn.: Yale University Press, 1953.
Kutscher, Eduard Y. "Aramaic." In *Linguistics in South West Asia and North Africa.* Current Trends in Linquistics, vol. 6. Ed. Thomas A Sebeok. The Hague: Mouton, 1970.
Kutler, Laurence. "A 'Strong' Case for Hebrew *mar.*" In *Ugarit-Forschungen,* vol. 16. Ed. Kurt Bergerhof, Manfried Dietrich, Oswald Loretz. Neukirchener-Vluyn, W. Germany: Verlag Butzon & Berker Kevelaer, 1984.
Landsberger, B. *Sam'al: Studien zur Entdeckung der Ruinenstäette Karatepe.* Veröffentlichungen der Türkischen Historischen Gesellschaft VII, series no. 16. Ankara: Türkischen Historischen Gesellschaft, 1948.
Lang, Bernhard. *Monotheism and the Prophetic Minority, An Essay in Biblical History and Sociology.* The Social World in Biblical Antiquity, no. 1. Sheffield: Almond Press, 1983.
Lemaire, André. *Les écoles et la formation de la Bible dans l'ancien Israël.* Orbis Biblicus et Orientalis, vol. 39. Göttingen: Éditions Universitaires Fribourg Suisse Vandenhoeck & Ruprecht, 1981.
_____. *Inscriptions Hébraïques, Tome I: Les ostraca.* Paris: Les Éditions du Cerf, 1977.
Lewis, Theodore J. *Cults of the Dead in Ancient Israel and Ugarit.* Harvard Semitic Monographs, no. 39. Atlanta: Scholars Press, 1989.
Lichtheim, Miriam. *Ancient Egyptian Literature.* 3 vols. Berkeley, CA: University of California Press, 1973.
Lidzbarski, Mark. *Altaramäische Urkunden aus Assur.* Wissenschaftliche Veröffentlichung der Deutschen Orient-Gesselschaft, no. 38. Leipzig: J. C. Hinrichs'sche Buchhandlung, 1921.
_____. *Ephemeris für semitische Epigraphik.* 3 vols. Giessen: Verlag von Alfred Töpelmann, 1915.
_____. *Handbuch der nordsemitischen Epigraphik.* 2 vols. 1962; rpt. Hildesheim: Georg Olms, 1898.
Loretz, Oswald. "Die ammonitische Inscrift von Tell Siran." in *Ugarit-Forchung,* vol. 9. Ed. Kurt Bergerhof, Manfried Dietrich, Oswald Loretz. Neukirchener-Vluyn, W. Germany: Verlag Butzon & Berker Kevelaer, 1977.
Luckenbill, Daniel D. *Ancient Records of Assyria and Babylonia.* 2 vols. Chicago: The University of Chicago Press, 1927.
_____. *The Annals of Sennacherib.* University of Chicago Oriental Institute Publications, vol. 2. Chicago: The University of Chicago Press, 1924.
Macalister, R. A. Stewart. *A Century of Excavation in Palestine.* London: The Religious Tract Society, 1925.
_____. *The Excavation of Gezer 1902–1905 and 1907–1909.* 2 vols. London: John Murray, 1912.

Magnanini, Pietro. *Le iscrizioni fenicie dell'Oriente: testi, traduzioni, glossari.* Rome: Istituto di Studi del Vicino Oriente, 1973.

Maier, Walter A., III. *'Ašerah: Extra Biblical Evidence.* Harvard Semitic Monographs, no. 37 Atlanta: Scholars Press, 1986.

Margalit, Baruch. "Lexicographical Notes on the AQHT Epic (Part I:KTU 1.17–18)." In *Ugarit-Forschungen*, vol. 15. Ed. Kurt Bergerhof, Manfried Dietrich, Oswald Loretz. Neukirchener-Vluyn, W. Germany: Verlag Butzon & Berker Kevelaer, 1983.

Masson, Olivier and Maurice Sznycer. *Recherches sur les Phéniciens à Chypre.* II Hautes Études Orientales, no. 3. Paris: Libraire Droz, 1972.

Matthews, Victor H. and Don C. Benjamin. *Old Testament Parallels, Laws and Stories from the Ancient Near East.* Mahwah, NJ: Paulist Press, 1991.

McCarter, P. Kyle. "Aspects of the Religion of the Israelite Monarchy: Biblical and Epigraphic Data." In *Ancient Israelite Religion.* Eds. Patrick D. Miller, Jr., Paul D. Hanson, and S. Dean McBride. Philadelphia: Fortress Press, 1987.

McCarthy, Dennis J. *Treaty and Covenant: A Study in Form in the Ancient Oriental Documents and in the Old Testament.* Analecta Biblica, vol. 21. Rome: Pontifical Biblical Institute, 1963.

Meier, Gerhard. *Die assyrische Beschwörungssammlung Maqlû.* Archiv für Orientforschung supplement 2. Berlin: Ernest Weidner, 1937.

Meshel, Ze'ev. *Kuntillet 'Ajrud: A Religious Centre from the Time of the Judean Monarchy on the Border of Sinai.* Israel Museum Catalogue, No. 175. Jerusalem: Israel Museum, 1978.

du Mesnil du Buisson. "Une tablette magique de la région du moyen Euphrate." In *Mélanges Syriens offerts a Monsieur René Dussaud.* Paris: Librairie Orientaliste Paul Geuthner, 1939.

Michaud, Henri. *Sur la pierre et l'argile.* Paris: Delachaux et Niestlé, 1958.

Miller, Patrick Jr. "Psalms and Inscriptions." In *Supplements to Vetus Testamentum, Congress Volume Vienna: 1980.* Ed. J. A. Emerton. Leiden: E. J. Brill, 1981.

Mitchell, Christopher W. *The Meaning of BRK "To Bless" in the Old Testament.* Society of Biblical Literature Dissertation Series, vol. 95. Atlanta: Scholars Press, 1987.

Montet, Pierre. *Byblos et l'Egypte, quatre campagnes de fouilles a Gebeil 1921–1922–1923–1924.* Bibliothèque archaéologique et historique, vol. 11. Paris: Librairie Orientaliste Paul Geuthner, 1928.

Moscati, Sabatino. *L'epigrafia ebraica antica, 1935–1950.* Biblica et Orientalia, vol. 15. Roma: Pontificio Istituto Biblico, 1951.

Mowinckel, Sigmund. *The Psalms in Israel's Worship.* 2 vols. Trans. D. R. Ap-Thomas. Oxford: Basil Blackwell, 1962.

Naveh, Joseph. *The Development of the Aramaic Scripts.* The Israel Academy of Sciences and Humanities Proceedings, vol. 5/1. Jerusalem: The Israel Academy of Sciences and Humanities, 1970.

Nougayrol, Jean. *Le Palais Royal d'Ugarit.* vol. 3, part 1. Mission de Ras Shamra, vol. 6. Ed. Claude F. A. Schaeffer. Paris: Imprimerie Nationale, 1955.

_____, et al. *Ugaritica V.* Mission de Ras Shamra, vol. 16. Paris: Imprimerie Nationale, 1968.

Olyan, Saul. *Asherah and the Cult of Yahweh in Israel.* Society of Biblical Literature Monograph Series, no. 34. Atlanta: Scholars Press, 1988.

Pardee, Dennis. "Letters from Tel Arad." In *Ugarit Forschungen,* vol. 10. Ed. Kurt Bergerhof, Manfried Dietrich, and Oswald Loretz. Neukirchener-Vluyn, W. Germany: Verlag Butzon & Bercker Kevelear, 1978.

_____. "The Semitic Root *mrr* and the Etymology of Ugaritic *mrr∥brk.*" In *Ugarit-Forschungen*, vol. 10. Ed. Kurt Bergerhof, Manfried Dietrich, Oswald Loretz. Neukirchener-Vluyn, W. Germany: Verlag Butzon & Berker Kevelaer, 1978.

_____ et al. *Handbook of Ancient Hebrew Letters.* Society of Biblical Literature Sources for Biblical Study, vol. 15. Chico, CA.: Scholars Press, 1982.

Parpola, Simo and Kazuko Watanabe. *Neo-Assyrian Treaties and Loyalty Oaths.* State Archives of Assyria, vol. 2. Helsinki: Helsinki University Press, 1988.

Parrot, André. *Maledictions et Violations de Tombes.* Paris: Librairie Orientaliste Paul Geuthner, 1939.

Peckham, Brian. "Phoenicia and the Religion of Israel." In *Ancient Israelite Religion.* Ed. Patrick D. Miller, Jr., Paul D. Hanson, and S. Dean McBride. Philadelphia: Fortress Press, 1987.

Pedersen, Johannes. *Israel: Its Life and Culture.* 4 vols in 2. Copenhagen: Branner Og Korch, 1926.

Pognon, H. *Inscriptions sémitiques de la Syrie, de la Mesopotamie et de la Mossoul.* Paris: Imprimerie Nationale, 1907.

Postgate, J. Nicholas. *Fifty Neo-Assyrian Legal Documents.* Warminster, England: Aris & Phillips, Ltd., 1976.

Procksch, O. "λεγω." *Theological Dictionary of the New Testament.* vol. 4. Ed. Gerhard Kittel. Trans. Geoffrey W. Bromiley. Grand Rapids: Wm. B. Eerdmans, 1967.

"*qallalu.*" *Chicago Assyrian Dictionary.* Q, vol. 13. Ed. A. Leo Oppenheim. Chicago: The University of Chicago Press, 1968.

Reiner, Erica. *Šurpu: A Collection of Sumerian and Akkadian Incantations.* Archiv für Orientforschung, supplement 11. Berlin: Ernst Weidner, 1958.

_____. "Treaty Between Ashurnairari V of Assyria and Mati'ilu of Arpad." In *Ancient Near Eastern Texts Relating to The Old Testament.* 3rd ed. Ed. James B. Pritchard. Princeton: Princeton University Press, 1969.

_____. "The Vassal-Treaties of Esarhaddon." In *Ancient Near Eastern Texts Relating to The Old Testament.* 3rd ed. Ed. James B. Pritchard. Princeton: Princeton University Press, 1969.

Rosenthal, Franz. "Incantations: The Amulette from Arslan Tash." In *Ancient Near Eastern Texts Relating to The Old Testament.* 3rd ed. Ed. James B. Pritchard. Princeton: Princeton University Press, 1969.

_____. "Azitawadda of Adana." In *Ancient Near Eastern Texts Relating to The Old Testament.* 3rd ed. Ed. James B. Pritchard. Princeton: Princeton University Press, 1969.

Bibliography

    _____. "Kilamuwa of Y'DY-SAM'AL." In *Ancient Near Eastern Texts Relating to The Old Testament*. 3rd ed. Ed. James B. Pritchard. Princeton: Princeton University Press, 1969.

    _____. "Zakir of Hamath and Lu'ath." In *Ancient Near Eastern Texts Relating to The Old Testament*. 3rd ed. Ed. James B. Pritchard. Princeton: Princeton University Press, 1969.

Scharbert, Josef. "אלה." *Theological Dictionary of the Old Testament*, vol. 1. rev. Ed. G. Johannes Botterweck and H. Ringgren. Trans. John T. Willis. Grand Rapids: Wm. B. Eerdmans, 1975.

    _____. "ארר." *Theological Dictionary of the Old Testament*, vol. 1. rev. Ed. G. Johannes Botterweck and H. Ringgren. Trans. John T. Willis. Grand Rapids: Wm. B. Eerdmans, 1975.

    _____. "ברך." *Theological Dictionary of the Old Testament*, vol. 2. Ed. G. Johannes Botterweck and Helmer Ringgren. Trans. John T. Willis. Grans Rapids: Wm. B. Eerdmans, 1975.

Schrader, Eberhard, ed. *Keilinschriftliche Bibliothek, Sammlung von assyrichen und babylonischen Texten*. 3 vols. in 1. Berlin: H. Reuther's Verlagsbuchhandlung, 1889.

Schroer, S. "Zur Deutung der Hand unter der Grabinschrift von Chirbet el Qôm." *Ugarit-Forschung*, vol. 15. Ed. Kurt Bergerhof, Manfred Dietrich, and Oswald Loretz. Neukirchen-Vluyn, W. Germany: Verlag Butzon & Bercker Kevelaer, 1983.

Segert, Stanislav. *A Grammar of Phoenician and Punic*. München: Verlag C. H. Beck, 1976.

Spronk, Klaas. *Beatific Afterlife in Ancient Israel and in the Ancient Near East*. Alter Orient und Altes Testament, vol. 219. Neukirchener-Vluyn: Butzon & Bercker Kevelaer, 1986.

Tigay, Jeffrey H. *You Shall Have No Other Gods: Israelite Religion in the Light of Hebrew Inscriptions*. Harvard Semitic Studies, no. 31. Atlanta: Scholars Press, 1986.

Torczyner, Harry, et al. *Lachish I (Tell ed Duweir): The Lachish Letters*. London: Oxford University Press, 1938.

Tromp, Nicholas. *Primitve Conceptions of Death and the Nether World in the Old Testament*. Biblica et Orientalia, no. 21. Rome: Pontifical Biblical Institute, 1969.

Ungnad, Arthur. *Aramäische Papyrus aus Elephantine*. Leipzig: J. C. Hinrichs'sche Buchhandlung, 1911.

von Rad, Gerhard. *Old Testament Theology*. 2 vols. Trans. D. M. G. Stalker. New York: Harper & Row, 1965.

von Soden, Wolfram, ed. *Akkadische Handwörterbuch*. 3 vols. Wiesbaden: Otto Harrassowitz, 1965–1981.

Wehmeier, Gerhard. *Der Segen im Alten Testament*. Theologisches Disertationen, no. 6. Basel: Friedrich Reinhardt Kommissionsverlag, 1970.

Weinfeld, Moshe. "'You Will Find Favor...in the Sight of God and Man' (Proverbs 3:4)—The History of An Idea." In *Eretz Israel: Archaeological, Historical and Geographical Studies*, vol. 16. Ed. B. A. Levine and A. Malamat. Jerusalem: Israel Exploration Society, 1982.

Wente, Edward F. *Late Ramesside Letters.* Studies in Ancient Oriental Civilization, vol. 33. Chicago: University of Chicago Press, 1967.

Westermann, Claus. *Blessing in the Bible and the Life of the Church.* Trans. Keith Crim. Philadelphia: Fortress Press, 1978.

Wilson, John A. "Egyptian and Hittite Treaties." In *Ancient Near Eastern Texts Relating to The Old Testament.* 3rd ed. Ed. James B. Pritchard. Princeton: Princeton University Press, 1969.

_____. "Magical Protection for a Child." In *Ancient Near Eastern Texts Relating to The Old Testament.* 3rd ed. Ed. James B. Pritchard. Princeton: Princeton University Press, 1969.

Wirgin, Wolf. "The Calendar Tablet from Gezer." In *Eretz-Irael: Archaeological, Historical and Geographical Studies,* vol. 6. Ed. M. Avi-Yonah, et al. Jerusalem: Israel Exploration Society, 1960.

Wiseman, Donald J. *The Vassal-Treaties of Esarhaddon.* London: The British School of Archaeology in Iraq, 1958.

Wünsch, R. "The Limestone Inscriptions of Tell Sandaḥannah." In *Excavations in Palestine During the Years 1898–1900.* London: Palestine Exploration Fund, 1902.

Yadin, Yigael. *Hazor: The Rediscovery of a Great Citadel of the Bible.* London: Weidenfeld and Nicholson, 1975.

_____. "Symbols of Deities at Zinjirli, Carthage, and Hazor." In *Near Eastern Archaeology in the Twentieth Century.* Ed. James A Sanders. New York: Doubleday & Company, 1970.

_____, et al. *Hazor III–IV: An Account of the Third and Fourth Seasons of Excavations, 1957–1958, Plates.* Jerusalem: Magnes Press, 1961.

Journals

Ahlström, G. W. "The Tell Siran Bottle Inscription." *Palestine Exploration Quarterly* 116 (1984), 12–15.

Aimé-Giron, Noël. "Adversaria Semitica (III): VII—Baʻal Ṣaphon et les dieux de Taḥpanhès dans un nounveau papyrus phénicien." *Annales du service des antiquitiés de l'Egpte* 40 (1940), 433–60.

Albright, William F. "An Aramaean Magical Text in Hebrew from the Seventh Century B.C." *Bulletin of the American Schools of Oriental Research* 76 (1939), 5–11.

_____. "The Gezer Calendar." *Bulletin of the American Schools of Oriental Research* 92 (1943), 16–26.

_____. "The Lachish Letters After Five Years." *Bulletin of the American Schools of Oriental Research* 82 (1941), 18–24.

_____. "Notes on Early Hebrew and Aramaic Epigraphy." *Journal of the Palestine Oriental Society* 6 (1926), 75–102.

_____. "The Oldest Hebrew Letters: The Lachish Ostraca." *Bulletin of the American Schools of Oriental Research* 70 (1938), 11–17.

_____. "Ostracon C 1101 of Samaria." *Palestine Exploration Quarterly* 68 (1936), 211–15.

_____. "The Phoenician Inscriptions of the Tenth Century B.C. from Byblus." *Journal of the American Oriental Society* 67 (1947), 153–60.

_____. "A Prince of Taanach in the Fifteenth Century B.C." *Bulletin of the American Schools of Oriental Research* 94 (1944), 12–27.

_____. "Some Comments on the 'Amman Citadel Inscription." *Bulletin of the American Schools of Oriental Research* 198 (1970), 38–40.

Amiet, Pierre. "Observations sur les 'Tablettes magiques' d'Arslan Tash." *Aula Orientalis* 1 (1983), 109.

Avigad, Nahman. "The Epitaph of a Royal Steward from Siloam Village." *Israel Exploration Journal* 3 (1953), 137–52.

_____. "The Second Tomb-Inscription of the Royal Steward." *Israel Exploration Journal* 5 (1955), 163–66.

Baldacci, M. "The Ammonite Text from Tell Siran and North-West Semitic Philology." *Vetus Testamentum* 31 (1981), 363–68.

Bar-Adon, P. "An Early Hebrew Inscription in a Judean Desert Cave." *Israel Exploration Journal* 25 (1975), 226–32.

Barré, Michael L. "An Analysis of the Royal Blessing in the Karatepe Inscription." *Maarav* 3/2 (1982), 177–94.

Batto, Bernard C. "Book Review: La statue de Tell Fekherye et son inscription bilingue assyro-araméenne." *Catholic Biblical Quarterly* 47 (1985), 501–03.

Bauer, Hans. "Ein aramäischer Staatsvertrag aus dem 8. Jahrhundert v. Chr., Die Inschrift der Stele von Sudschin." *Archiv für Orientforschung* 8 (1932), 1–16.

_____. "Eine phönikische Inschrift aus dem 13. Jahrh.." *Orientalische Literaturezeitung* 28 (1925), 130–39.

_____. "Die כלמו-Inschrift aus Sendschirli." *Zeitschrift der Deutschen Morganländische Gesellschaft* 67 (1913), 684–91.

Bayliss, Miranda. "The Cult of the Dead Kin in Assyria and Babylonia." *Iraq* 35 (1973), 115–25.

Beck, Pirhiya. "The Drawings from Horvat Teiman (Kuntillet 'Ajrud)." *Tel Aviv* 9 (1982), 3–68.

Becking, B.E.J.H. "Zur Interpretation der ammonitische Inschrift vom Tell Siran," *Bibliotheca Orientalis* 38 (1981), 273-76.

Beit-Arieh, Itzaq. "New Light on the Edomites." *Biblical Archaeology Review* 14/2 (1988), 29–41.

_____ and Bruce Cresson. "An Edomite Ostracon from Horvat 'Uza." *Tel Aviv* 12 (1985), 96–101.

_____. "Horvat 'Uza," *Excavations and Surveys in Israel* 7-8 (1988/89), 181.

_____. "Horvat 'Uza: A Fortified Outpost on the Eastern Negev Border." *Biblical Archeologist* 54/3 (1991), 126–35.

_____. "Notes and News: Horvat 'Uza, 1982." *Israel Exploration Journal* 32

(1982), 262–63.
Bresciani, Edda. "Papiri aramaici egiziani di epoca persiana presso il Museo Civico di Padova." *Rivista degli studi orientali* 35 (1960), 11–24.
Brichto, Herbert C. "Kin, Cult, Land and Afterlife—A Biblical Complex." *Hebrew Union College Annual* 44 (1973), 1–54.
Burrows, Eric. "Phoenician Inscription from Ur." *Journal of the Royal Asiatic Society* [no vol.] (1927), 791–94.
Caquot, André. "Observations sur la première tablette magique d'Arslan Tash." *Journal of the Ancient Near Eastern Society of Columbia University* 5 (1973), 45–51.
_____ and André Lemaire. "Les Textes Araméens de Deir 'Alla." *Syria* 54 (1977), 189–208.
Chase, Debra. "A Note on an Inscription from Kuntillet 'Ajrud." *Bulletin of the American Schools of Oriental Research* 246 (1982), 63–67.
Clermont-Ganneau, Charles. "Les stèles araméennes de Neîrab." *Études d'archéologie orientale* 2 (1897), 182–223.
Cooke, Stanley A. "The Old Hebrew Alphabet and the Gezer Tablet." *Palestine Exploration Fund Quarterly Statement* 41 (1909), 284–309.
Coote, Robert B. "The Tell Siran Bottle Inscription." *Bulletin of the American Schools of Oriental Research* 240 (1980), 93.
Cowley, A. "Two Aramaic Ostraka." *Journal of the Royal Asiatic Society* [no vol.] (1929), 107–12.
Cross, Frank M., Jr. "Epigraphic Notes on the Amman Citadel Inscription." *Bulletin of the American Schools of Oriental Research* 193 (1969), 13–19.
_____. "Epigraphic Notes on Hebrew Documents of the Eighth-Sixth Centuries B.C.: II. The Muraba'ât Papyrus and the Letter Found Near Yabneh-Yam." *Bulletin of the American Schools of Oriental Research* 165 (1962), 34–46.
_____. "Notes on the Ammonite Inscription from Tell Siran." *Bulletin of the American Schools of Oriental Research* 212 (1973), 12–15.
_____ and Richard J. Saley, "Phoenician Incantations on a Plaque of the Seventh Century B.C. from Arslan Tash in Upper Syria." *Bulletin of the American Schools of Oriental Research* 197 (1970), 42–49.
Daiches, Samuel. "Notes on the Gezer Calendar and Some Babylonian Parallels." *Palestine Exploration Fund Quarterly Statement* 41 (1909), 113–18.
Dever, William G. "Asherah, Consort of Yahweh: New Evidence from Kuntillet 'Ajrud." *Bulletin of the American Schools of Oriental Research* 255 (1984), 21–37.
_____. "Iron Age Epigraphic Material from the Area of Khirbet el-Kôm." *Hebrew Union College Annual* 40–41 (1969–70), 139–204.
_____. "Recent Archaeological Confirmation of the Cult of Asherah in Ancient Israel." *Hebrew Studies* 23 (1982), 37–43.
de Vaux, Roland. "Les ostraka de Lachis." *Revue biblique* 48 (1939), 181–206.
Dhorme, Paul. "La fille de Nabonide." *Revue d'assyriologie et d'archaéologie orientale* 11 (1914), 105–17.
Dion, Paul. E. "Notes d'épigraphie ammonite." *Revue biblique* 82 (1975), 24–33.

Dunand, Maurice. "Nouvelle inscription phénicienne archaïque." *Revue biblique* 39 (1930), 321–31.
Dupont-Sommer, André. "L'inscription de l'amulette d'Arslan-Tash." *Revue de l'histoire des religions* 120 (1939), 133–59.
_____. "Les inscriptions araméennes de Sfiré (stèles I et II)." *Mémoires présentés à l'académie des inscriptions et belles-lettres* 15 (1958), 197–351. Seperately bound edition, pp. 1–151.
_____. "Une inscription araméenne inédite de Sfiré." *Bulletin du Musée de Beyrouth* 13 (1956), 23–41.
_____. "Une inscription nouvelle du roi Kilamou et le dieu Rekoub-el." *Revue de l'histoire des religions* 133 (1947–48), 19–33.
_____. "Note on a Phoenician Papyrus from Saqqara." *Palestine Exploration Quarterly* 81 (1949), 52–57.
_____. "Le syncrétisme religieux des Juifs d'Éléphantine d'après un ostracon araméen inédit." *Revue de l'histoire des religions* 130 (1945), 17–28.
Dussaud, René. "Dédicace d'une statue d'Orsokon I par Eliba'al, Roi de Byblos." *Syria* 6 (1925), 101–17.
_____. "Une inscription phénicienne découverte à Our en Chaldée." *Syria* 9 (1928), 267–68.
_____. "Les inscriptions phéniciennes du tombeau d'Aḥiram, roi de Byblos." *Syria* 5 (1924), 135–57.
_____. "Le prophète Jérémie et les lettres de Lakish." *Syria* 19 (1938), 256–71.
Emerton, John A. "New Light on Israelite Religion: the Implications of the Inscriptions from Kuntillet 'Ajrud." *Zeitschrift für die Alttestamentliche Wissenschaft* 94 (1982), 2–20.
Fensham, F. Charles. "Common Trends in Curses of the Near Eastern Treaties and *Kudurru-* Inscriptions Compared with Maledictions of Amos and Isaiah." *Zeitschrift für die Alttestamentliche Wissenschaft* 75 (1963), 155–75.
_____. "Malediction and Benediction in Ancient Near Eastern Vassal-Treaties and the Old Testament." *Zeitschrift für die Alttestamentliche Wissenschaft* 74 (1962), 1–9.
Fitzmyer, J. A. "The Aramaic Letter of King Adon to the Egyptian Pharaoh." *Biblica* 46 (1965), 41–59.
_____. "The Aramaic Suzerainty Treaty from Sefire in the Museum of Beirut." *Catholic Biblical Quarterly* 20 (1958), 444–76.
_____. "The Padua Aramaic Papyrus Letters." *Journal of Near Eastern Studies* 21 (1962), 15–24.
Foresti, Fabrizio. "Characteristic Literary Expressions in the Arad Inscriptions Compared with the Language of the Hebrew Bible." *Ephemerides Carmeliticae* 32 (1981), 327–341.
Friedrich, Johannes. "Der hethitische Soldateneid." *Zeitschrift für Assyriologie und vorderasiatische Archäologie* 35 (1923), 161–91.
Fulco, W. "The Amman Citadel Inscription: A New Collation (with Text)." *Bulletin of the American Schools of Oriental Research* 230 (1978), 39–43.

Galling, Kurt. "Beschriftete Bildsiegel des ersten Jahrtausends v. Chr. vornehmlich aus Syrien und Palastina." *Zeitschrift des Deutschen Palästinavereins* 64 (1941), 121–202.

_____. "Ein Ostrakon aus Samaria als Rechtsurkunde." *Zeitschrift des Deutschen Palästinavereins* 77 (1961), 173–85.

_____. "The Scepter of Wisdom, A Note on the Gold Sheath of Zendjirli and Ecclesiates 12:11." *Bulletin of the American Schools of Oriental Research* 119 (1950), 15–18.

Ganor, N. R. "The Lachish Letters." *Palestine Exploration Quarterly* 99 (1967), 74–77.

Gaster, Theodor. "A Canaanite Magical Text." *Orientalia* 11 (1942), 41–79.

Gevirtz, Stanley. "Jericho and Shechem: A Religio-Literary Aspect of City Destruction." *Vetus Testamentum* 13 (1963), 52–62.

_____. "A Spindle Whorl with Phoenician Inscription." *Journal of Near Eastern Studies* 26 (1967), 13–16.

_____. "West-Semitic Curses and the Problem of the Origins of Hebrew Law." *Vetus Testamentum* 11 (1961), 137–58.

Gilula, M. "To Yahweh Shomron and His Asherah." (Hebrew, Eng. summ.) *Shnaton* 3 (1978–79), 129–137.

Ginsberg, H. L. "Lachish Notes." *Bulletin of the American Schools of Oriental Research* 71 (1938), 24–27.

_____. "Lachish Ostraca New and Old." *Bulletin of the American Schools of Oriental Research* 80 (1940), 10–13.

_____. "The North-Canaanite Myth of Anath and Aqhat." *Bulletin of the American Schools of Oriental Research* 98 (1945), 15–23.

_____. "Ugaritico-Phoenicia." *Journal of the Ancient Near Eastern Society of Columbia University* 5 (1973), 131–47.

Glueck, Nelson. "Ostraca from Elath." *Bulletin of the American Schools of Oriental Research* 80 (1940), 3–10.

_____. "Ostraca from Elath." *Bulletin of the American Schools of Oriental Research* 82 (1941), 3–11.

Gordon, Cyrus H. "Notes on the Lachish Letters." *Bulletin of the American Schools of Oriental Research* 70 (1938), 17–18.

Greenfield, Jonas C. "Some Phoenician Words." *Semitica* 38 (1990), 155-58.

_____ and Aaron Shaffer. "Notes on the Akkadian-Aramaic Bilingual Statue from Tell Fekherye." *Iraq* 45 (1983), 109–16.

_____. "Notes on the Curse Formulae of the Tell Fekherye Inscription." *Revue biblique* 92 (1985), 47–59.

Gropp, Douglas M. and Theodore J. Lewis. "Notes on Some Problems in the Aramaic Text of the Hadd-Yithʻi Bilingual." *Bulletin of the American Schools of Oriental Research* 259 (1985), 45-61.

Hadley, Judith M. "The Khirbeth el-Qôm Inscription." *Vetus Testamentum* 37 (1987), 50–62.

_____. "Some Drawings and Inscriptions on Two Pithoi from Kuntillet ʻAjrud." *Vetus Testamentum* 37 (1987), 180–213.

Halévy, Jacob. "Deux notes épigraphiques: Un dernier mot sur les inscriptions de Nêrab." *Revue semitique* 5 (1897), 189–92.
_____. "Les deux stéles de Nerab." *Revue semitique* 4 (1896), 279–85, 369–73.
_____. "Les inscriptions du Roi Kalumu." *Revue semitique* 20 (1912), 19–30.
_____. "Nouvelles remarques sur l'inscription de Zakir." *Revue semitique* 16 (1908), 243–46, 357–76.
Hanson, Richard S. "Aramaic Funerary and Boundary Inscriptions from Asia Minor." *Bulletin of the American Schools of Oriental Research* 192 (1968), 3–11.
Heltzer, M. "Eighth Century B.C. Inscriptions from Kalakh (Nimrud)." *Palestine Exploration Quarterly* 110 (1978), 3–9.
Hempel, Johannes. "Die israelitische Anschauungen von Segen und Fluch im Lichte altorientalisher Parallelen." *Beiheft zur Zeitschrit für die Alttestamentliche Wissenschaft* 81 (1961), 30–113.
_____. "Die Ostraka von Lakiš." *Zeitschrift für die Alttestamentliche Wissenschaft* 15 (1938), 126–39.
Herr, Larry G. "The Formal Scripts of Iron Age Transjordan." *Bulletin of the American Schools of Oriental Research* 238 (1980), 21–34.
Herzog, Z.; Aharoni, M.; Rainey, A.; and Moshkovitz, S. "The Israelite Fortress at Arad." *Bulletin of the American Schools of Oriental Research* 254 (1984), 1–34.
Hestrin, Ruth. "The Lachish Ewer and the 'Asherah." *Israel Exploration Journal* 37 (1987), 212-23.
Hoffmann, G. "Aramäische Inschriften aus Nêrab bei Aleppo, neue und alte Götter." *Zeitschrift für Assyriologie und verwandte Gebiete* 11 (1896), 207–92.
Hoftijzer, Jacob. "The Prophet Balaam in a 6th Century Aramaic Inscription." *Biblical Archaeologist* 39 (1976), 11–17.
Honeyman, A. M. "Epigraphic Discoveries at Karatepe." *Palestine Exploration Quarterly* 81 (1949), 21–39.
_____. "The Phoenician Inscriptions of the Cyprus Museum." *Iraq* 6 (1939), 104–08.
Horn, Siegfried H. "The Amman Citadel Inscription." *Annual of the Department of Antiquities Jordan* 12, 13 (1967, 1968), 81–83.
_____. "The Ammān Citadel Inscription." *Bulletin of the American Schools of Oriental Research* 193 (1969), 2–13.
Jack, J. W. "The Lachish Letters Their Date and Import: An Examination of Professor Torczyner's View." *Palestine Exploration Quarterly* 70 (1938), 165–87.
Jaroš, Karl. "Zur Inschrift Nr. 3 von Ḥirbet el-Qôm." *Biblische Notizen* 19 (1982), 31–40.
Joüon, P. "Sur les Ostraca hébraïques de Lachish." *Revue des études semitiques* (1938), 84–88.
Kaiser, Otto. "Zum Formular der in Ugarit gefundenen Briefe." *Zeitschrift des Deutschen Palästinavereins* 86 (1970), 10–23.
Kaufman, Ivan T. "The Samaria Ostraca: An Early Witness to Hebrew Writing."

Biblical Archaeologist 45 (1982), 229–39.

Kaufman, Stephen A. "Reflections on the Assyrian-Aramaic Bilingual from Tell Fakhariyeh." *Maarav* 3 (1982), 137–75.

―――――. "Review Article: The Aramaic Texts from Deir 'Alla." *Bulletin of the American Schools of Oriental Research* 239 (1980), 71–74.

―――――. "Si'-Gabbari, Priest of Sahr in Nerab." *Journal of the American Oriental Society* 90 (1970), 270–71.

King, Philip J. "The Contribution of Archaeology to Biblical Studies." *Catholic Biblical Quarterly* 45 (1983), 1–16.

Knauf, Ernest A. "Supplementa Ismaelitica." *Biblische Notizen* 45 (1988), 62-81.

Knudsen, Ebbe E. "Fragments of Historical Texts from Nimrud—II." *Iraq* 29 (1967), 49–69.

Krahmalkov, Charles. "An Ammonite Lyric Poem." *Bulletin of the American Schools of Oriental Research* 223 (1976), 55–57.

Lagrange, M. J. "La nouvelle inscription de Sendjirly." *Revue biblique* 21 (new series no. 9) (1912), 253–59.

Lawton, Robert. "Israelite Personal Names on Pre-Exilic Hebrew Inscriptions." *Biblica* 65 (1984), 330–46.

Lemaire, André. "Les inscriptions de Khirbet el-Qôm et l'asherah de YHWH." *Revue biblique* 84 (1977), 595–608.

―――――. "A Note on Inscription XXX from Lachish." *Tel Aviv* 7 (1980), 92–94.

―――――. "Prières en temps de crise: les inscriptions de Khirbet Beit Lei." *Revue biblique* 83 (1976), 558–68.

―――――. "Who or What Was Yahweh's Asherah." *Biblical Archaeology Review* 10/6 (1984), 42–51.

Levine, Baruch A. "The Deir 'Alla Plaster Inscriptions." *Journal of the American Oriental Society* 101 (1981), 195–205.

Lidzbarski, Mark. "An Old Hebrew Calendar-Inscription from Gezer." *Palestine Exploration Fund Quarterly Statement* 41 (1909), 26–29.

Lipiński, Eduard. "From Karatepe to Pyrgi, Middle Phoenician Miscellanea." *Rivista di studi fenici* 2 (1974), 45–61.

Macalister, R. A. Stewart. "Twenty-first Quarterly Report on the Excavation of Gezer." *Palestine Exploration Fund Quarterly Statement* 41 (1909), 87–105.

Margalit, Baruch. "Some Observations on the Inscription and Drawing from Khirbet el-Qôm," *Vetus Testamentum* 39 (1989), 371-78.

Martin, M. "A Preliminary Report After Re-examinatin of the Byblian Inscriptions." *Orientalia* 30 (1961), 46–78.

McCarter, P. Kyle, Jr. "The Balaam Texts from Deir 'Allā: The First Combination." *Bulletin of the American Schools of Oriental Research* 239 (1980), 49–60.

Meshel, Ze'ev. "Kuntillet 'Ajrud, An Israelite Religious Center in Northern Sinai." *Expedition* 20 (1978), 50–54.

――――― and Carol Meyers. "Did Yahweh Have a Consort? The New Religious Inscriptions from the Sinai." *Biblical Archaeology Review* 5/2 (1979), 24–34.

Michaud, Henri. "Les ostraca de Lakiš conservés a Londres." *Syria* 34 (1957),

39–60.
Millard, Alan R. "Alphabetic Inscriptions on Ivories from Nimrud." *Iraq* 24 (1962), 41–51.
_____. "The homeland of Zakkur." *Semitica* 39 (1990), 47-52.
Mittmann, Siegfried. "Die Grabinschrift des Sängers Uriahu." *Zeitschrift des Deutschen Palästinavereins* 97 (1981), 139–52.
Moran, William L. "A Note on the Treaty Terminology of the Sefîre Stelas." *Journal for Near Eastern Studies* 22 (1963), 172–76.
Mosca, P. G. and J. Russell. "A Phoenician Inscription from Cebel Ires Daği in Rough Cilicia," *Epigraphica Anatolia* 9 (1987), 1–27.
Müller, D. H. "Altsemitischen Inschriften von Sendschirli." *Wiener Zeitschrift für die Kunde des Morgandlandes* 7 (1893), 33–70.
Naveh, Joseph. "Graffiti and Dedications." *Bulletin of the American Schools of Oriental Research* 235 (1979), 27–30.
_____. "Old Hebrew Inscriptions in a Burial Cave." *Israel Exploration Journal* 13 (1963), 74–92.
Nöldeke, Theodor. "Bemerkungen zu dem aramäischen Inschriften von Senschirli." *Zeitschrift der Deutschen Morganländischen Gesellschaft* 47 (1893), 96–105.
Noth, Martin. "Der historische Hintergrund der Inschriften von Sefîre." *Zeitschrift des Deutschen Palästinavereins* 77 (1961), 118–72.
O'Connor, Michael. "The Poetic Inscription from Khirbet el-Qôm." *Vetus Testamentum* 37 (1987), 224–29.
_____. "The Rhetoric of the Kilamuwa Inscription." *Bulletin of the American Schools of Oriental Research* 226 (1977), 15–30.
Puech, Émile and Alexander Rofé. "L'inscription de la Citadelle d'Amman." *Revue biblique* 80 (1973), 531–46.
Reifenberg, A. "A Newly Discovered Hebrew Inscription of the Pre-exilic Period." *Journal of the Palestine Oriental Society* 21 (1948), 134–37.
Ronzevalle, Sébastien. "The Gezer Hebrew Inscription." *Palestine Exploration Fund Quarterly Statement* 41 (1909), 107–12.
Sasson, Victor. "The 'Amman Citadel Inscription as an Oracle promising Divine Protection: Philological and Literary Comments." *Palestine Exploration Quarterly* 111 (1979), 117–25.
_____. "The Aramaic Text of the Tell Fakhriyah Assyrian-Aramaic Bilingual Inscription." *Zeitschrift für Alttestamentliche Wissenschaft* 97 (1985), 86–103.
Savignac, R. "Inscription phénicienne d'Ur." *Revue biblique* 37 (1928), 257–59.
Shea, William H. "The Amman Citadel Inscription Again." *Palestine Exploration Quarterly* 113 (1981), 105–10.
_____. "The Khirbet el-Qôm Tomb Inscription Again," *Vetus Testamentum* 40 (1990), 110-16.
_____. "Milkom as the Architect of Rabbath-Ammon's Natural Defences in the Amman Citadel Inscription." *Palestine Exploration Quarterly* 111 (1979), 17–25.
_____. "The Siran Inscription: Amminadab's Drinking Song." *Palestine Exploration Quarterly* 110 (1978), 107–12.

Singer, Suzanne. "Cache of Hebrew and Phoenician Inscriptions Found in the Desert." *Biblical Archaeology Review* 2/1 (1976), 33–34.
Smit, E. J. "The Tell Siran Inscription: Linguistic and Historical Implications." *Journal for Semitics* 1 (1989), 108–17.
Sollberger, Edmond. "Samsu-Iluna's Biblingual Inscriptions C and D." *Revue d'assyriologie et d'archaéologie orientale* 63 (1969), 29–43.
Speiser, E. A. "Nuzi Marginalia." *Orientalia* 25 (1956), 1–23.
Sukenik, E. L. "Inscribed Hebrew and Aramaic Potsherds from Samaria." *Palestine Exploration Quarterly* 65 (1933), 152–56.
_____. "Inscribed Potsherds with Biblical Names from Samaria." *Palestine Exploration Quarterly* 65 (1933), 200–04.
Teixidor, Javier. "Book Review Article: J. C. L. Gibson's *Textbook of Syrian Semitic Inscriptions*, vol. 3." *JBL* 103 (1984), 453-55.
_____. "L'inscription d'Ahiram à nouveau." *Syria* 64 (1987), 137–40.
_____. "Les tablettes d'Arslan Tash au Musée d'Alep." *Aula Orientalis* 1 (1983), 105-08.
Thistleton, Anthony C. "The Supposed Power of Words in the Biblical Writings." *Journal of Theological Studies* 25 (1974), 283–99.
Thompson, Henry O. and Fawzi Zayadine. "The Tell Siran Inscription." *Bulletin of the American Schools of Oriental Research* 212 (1973), 5–11.
Thureau-Dangin, F. "Rituel et amulettes contre Labartu." *Revue d'assyriologie et d'archaéologie orientale* [no vol.] (1921), 195–98.
Tigay, Jeffrey H. "A Second Temple Parallel to the Blessings from Kuntillet 'Ajrud." *Israel Exploration Journal* 40 (1990), 218.
Torczyner, Harry. "A Hebrew Incantation Against Night Demons from Biblical Times." *Journal of Near Eastern Studies* 6 (1947), 18–29.
Torrey, Charles C. "The Ahiram Inscription of Byblos." *Journal of the American Oriental Society* 45 (1925), 269–79.
_____. "New Notes on Some Old Inscriptions." *Zeitschrift für Assyriologie und verwandte Gebiete* 26 (1912), 77–92.
_____. "The Zakar and Kalamu Inscriptions." *Journal of the American Oriental Society* 35 (1915), 353–69.
Ussishkin, David. "Excavations at Tel Lachish 1978–1983: Second Preliminary Report." *Tel Aviv* 10 (1983), 92–175.
van Rooy, H. F. "The Structure of the Aramaic Treaties of Sefire." *Journal for Semitics* 1 (1989), 133–39.
van Selms, A. "Some Remarks on the 'Amman Citadel Inscriptions." *Bibliotheca Orientalis* 32 (1975), 5–8.
Vincent, L. H. "Les Fouilles de Byblos." *Revue biblique* 34 (1925), 161–93.
Virolleaud, Charles. "Fragments mythologiques de Ras-Shamra." *Syria* 24 (1944–45), 1–23.
Wehmeier, Gerhard. "Deliverance and Blessing in the Old and New Testaments." *The Indian Journal of Theology* 20 (1971), 30–42.
Weidner, Ernest. "Der Staatsvertrag Aššurniraris VI. [=V] von Assyrien mit Mati'ilu von Bit-Agusi." *Archiv für Orientforschung* 8 (1932), 17–34.

Weinfeld, Moshe. "Kuntillet 'Ajrud Inscriptions and Their Significance." *Studi epigraphici e linquistici* 1 (1984), 121–30.
Wiseman, D. J. "'Is it Peace?'—Covenant and Diplomacy," *Vetus Testamentum* 32 (1982), 311-26.
Woolley, C. Leonard. "The Excavations at Ur, 1926–7." *The Antiquaries Journal* 7 (1927), 410.
Yardeni, Ada. "Remarks on the Priestly Blessing on Two Ancient Amulets from Jerusalem." *Vetus Testamentum* 41 (1991), 176-85.
Youngblood, Ronald. "Amorite Influence in a Canaanite Amarna Letter (EA 96)." *Bulletin of the American Schools of Oriental Research* 168 (1962), 24–27.
Younger, K. Lawson, Jr. "Panammuwa and Bar-Rakib Two Structural Analyses." *Journal of the Ancient Near Eastern Society of Columbia University* 18 (1986), 91–103.
Zayadine, Fawzi and Henry O. Thompson. "The Ammonite Inscription from Tell Siran." *Berytus* 22 (1973), 115–40.
Zevit, Ziony. "The Khirbet el-Qôm Inscription Mentioning a Goddess." *Bulletin of the American Schools of Oriental Research* 255 (1984), 39–47.
_____. "A Phoenician Inscription and Biblical Covenant Theology." *Israel Exploration Journal* 27 (1977), 110–18.
Zwickel, Wolfgang. "Das 'edomitische' Ostrakon aus Ḥirbet Gazza (Ḥorvat 'Uza)." *Biblische Notizen* 41 (1988), 36–39.

Unpublished Material

Block-Smith, Elizabeth M. "The Cult of the Dead in Judah." A paper delivered at the 1990 Annual Meeting of the Society of Biblical Literature, New Orleans, Louisianna, November 18, 1990.
Drinkard, Joel F., Jr. "Vowel Letters in Pre-exilic Palestinian Inscriptions." Ph.D. dissertation, The Southern Baptist Theological Seminary, 1980.
Lawton, Robert B., Jr. "Israelite Personal Names on Hebrew Inscriptions Antedating 500 B. C. E." Ph.D. dissertation, Harvard University, 1977.
Levine, Baruch. "Survivals of Ancient Canaanite in the Mishnah." Ph.D. dissertation, Brandeis University, 1962.
Pitard, Wayne. "The Tombs of Ugarit and the Care of the Dead." A paper delivered at the 1990 Annual Meeting of the Society of Biblical Literature, New Orleans, Louisianna, November 18, 1990.

DATE DUE	
FEB 11 1997	
DEC 02 1996	
MAY 27 1997	
JUN 17 1997	
FEB 19 1998	
DEC 1 1997	
JUL 5 2007	
GAYLORD	PRINTED IN U.S.A.

GTU Library
2400 Ridge Road
Berkeley, CA 94709
For renewals call (510) 649-2500
All items are subject to recall.